To the Polson Institute
in the hope of fruitful
collaboration.
13 May 2002
Daniel J. Hogan

Human Dimensions of Global Environmental Change

Brazilian Perspectives

Edited by:

Daniel Joseph Hogan

Maurício Tiomno Tolmasquim

Human Dimensions of Global Environmental Change

Brazilian Perspectives

ACADEMIA BRASILEIRA DE CIÊNCIAS

RIO DE JANEIRO

2001

Ficha catalográfica elaborada pela Biblioteca Aristides Pacheco Leão/ABC

Human Dimensions of Global Environmental Change: Brazilian Perspectives/
edited by Daniel Joseph Hogan, Maurício Tiomno Tolmasquim. - Rio de Janeiro:
Academia Brasileira de Ciências, 2001.
392p.

ISBN: 85-85761-20-2

1.Meio-ambiente 2. Desenvolvimento sustentável. I. Hogan, Daniel Joseph.
II. Tolmasquim, Maurício Tiomno. III. Título.

CDD-363.7

CONTENTS

7 | Sustainability: a Key Challenge to Science & Technology
EDUARDO MOACYR KRIEGER

9 | **Preface**

13 | **Population**

15 | Demographic Aspects of Global Environmental Change: What is Brazil's Contribution?
DANIEL JOSEPH HOGAN

43 | Global Environmental Change and Health in Brazil: Review of the Present Situation and Proposal for Indicators for Monitoring these Effects
ULISSES E.C.CONFALONIERI

79 | Global Environmental Changes and Traditional Populations
MAURO DE ALMEIDA AND MANUELA CARNEIRO DA CUNHA

99 | **Institutions**

101 | Local Policies and Global Environmental Change
LEILA DA COSTA FERREIRA

121 | Brazil and Global Governance: the Case of Climate Change
EDUARDRO VIOLA AND HÉCTOR LEIS

157 | Technological Change and Development

159 | Energy and Development Strategies within the Context of Global Environmental Changes
MAURICIO TIOMNO TOLMASQUIM AND CLAUDE COHEN

193 | Science and Technology on the Threshold of the New Century
HENRIQUE RATTNER

215 | Behavioral Change

217 | How Green are Brazilians? Environmental Values, Attitudes and Behavior in Brazil
SOLANGE SIMÕES

249 | Global Food Risks: Enviromental and Heath Concerns in Brazil
JULIA SILVIA GUIVANT

279 | Consumption Patterns and Environmental Impacts in a Global Socioecosystemic Perspective
DONALD SAWYER

297 | Territory

299 | Amazonian Frontiers at the Beginning of the 21st Century
BERTHA K. BECKER

325 | Recent Changes in Land Use and Land Cover in Brazil
CLAUDIO A. G. EGLER

353 | Urbanization and Global Environmental Changes
HENRI ACSELRAD

385 | Authors

SUSTAINABILITY: A KEY CHALLENGE TO SCIENCE & TECHNOLOGY

There is increasing scientific evidence that the major problems presently faced by humanity are caused and nourished by human action. Disarrayed populational growth in various parts of the planet; degradation of agricultural and forest land; augment in the number of the endangered species; natural resource depletion; destruction of the ozone layer caused by greenhouse gas emissions; acid rain; global warming; alteration in the patterns of precipitation; sea level rise; environmental degradation caused by pollution; famine in ample parcels of the world population; extreme inequity in income and wealth distribution; social, cultural and ethnic conflicts... Hence, the roster is long and, more serious, alarming.

What the future holds in store for individuals and to humankind itself depends, to an extent, on the wisdom and use, destined by humanity, to science and technology. The development of science can provide humans with the knowledge of the biophysical environment and of social behavior needed to develop effective solutions to its global problems; without that knowledge, progress toward a safe world will be unnecessarily handicapped.

It is a major challenge to the international scientific community to contribute to the overcoming of the distresses presently afflicting societies, especially in developing countries. Without the continuous development and creative use of new technologies, society may limit its capacity for survival and for working toward a world in which the human species is at peace with itself and its environment.

As a result of planning and organizing the Open Meeting of the Human Dimensions of Global Environmental Change Research Community, held in Rio de Janeiro on October of 2001, the Board of the Brazilian Academy of Sciences decided to invite Brazilian scientists to present position papers on subjects concerned with transition to sustainability in the new millennium. This book represents a contribution of Brazilian science to the debates of the international scientific community on the quest for sustainability. I hope you enjoy your reading.

Krieger

Eduardo Moacyr Krieger
President of the Brazilian Academy of Sciences

PREFACE

The initiative of the Brazilian Academy of Sciences, in creating the National Committee on the Human Dimensions of Global Environmental Change, exemplifies the importance of leadership in furthering certain academic themes. Environmental studies have developed slowly within Brazilian social sciences and do not yet constitute a major area of concern. *Global* environmental change, in this context, has not been a priority, not even for researchers in the environmental area. These researchers were much more concerned – and continue to be so – with determinants and consequences of environmental change at the local (Brazilian) level. The gravity of the degradation of urban environments, pressure on natural resources and a limited tradition of public policies in the environmental area justify this emphasis.

Initial concern was directed to the "brown agenda," seeking to identify the environmental components of an unjust society. The differential effects of environmental change on different social groups revealed the additional onus borne by the poor, in terms of health and quality of life. Unlike the experience of social science in rich countries, the "green agenda" and concern for the integrity of natural systems appeared later. The attention given to ecologically important regions emerged from a concern for the effects of both environmental degradation and of environmental preservation in the lives of traditional populations. Today, Brazilian social sciences contemplate the full range of environmental problems.

In the last decade (with a lag in comparison to rich countries), the several disciplines of the social sciences have considerably expanded their research and teaching in the environmental area. This can be seen, for example, in the establishment of working groups on the environment within professional associations and the number of seminars dedicated to this topic. It is also seen in the creation of areas of concentration within traditional graduate programs and the creation of several specific programs on the environment. Both the creation of new interdisciplinary journals and the increasing attention of traditional disciplinary journals to environmental affairs, are notable.

All of these activities, however, have been overwhelmingly concerned with environmental problems at the local, regional and national levels. This is understandable, considering that it is at these levels that environmental changes are experienced by the population. The urgency of these problems has rarely left time for research on global changes. There is a second reason, however, for this emphasis. The temporal scales which direct work in the social sciences are distinctly different than those which prevail in the "hard" sciences. The long term, for social sciences, is much more limited than the time scales at which global environmental changes can be observed. The long-term view for social sciences is backward: we are continually seeking to understand historical change, the better to understand contemporary societies. The overarching concept of *sustainable development* has introduced a new temporality in traditional paradigms, and the earlier perspective may not be so dominant in the future.

But in this context, how was one to link contemporary social, economic, political and cultural realities with something so abstract as sea levels at the end of the twenty-first century?

It was in this academic vacuum that the initiative of the Brazilian Academy of Sciences interfered. The initial work was slow to get off the ground and the most significant result was the integration of a small segment of the socio-environmental community to the activities of the International Human Dimensions Program. This circumstance, in turn, was then complemented by the choice of Brazil as the host country for the IV Open Meeting of the Human Dimensions of Global Environmental Change Community. The preparatory discussions of the National Committee led us to propose a state-of-the-art evaluation of the Brazilian contribution to global environmental change.

The intention was twofold. In the first place, we were not convinced that the categories used by rich country social scientists were totally appropriate for coming to terms with global environmental change from the perspective of underdeveloped countries. We sought to organize an analysis of the question which would reflect Brazilian realities. In the second place, it was necessary to overcome the isolation of Brazilian environmental researchers with regard to the human dimensions community.

The present book is the result of this effort. Leading social scientists – with considerable experience in environmental studies – were invited to prepare papers on a set of themes defined by the National Committee. They were asked to reflect on the global environmental change agenda, from the perspective of their own research on these themes. These papers were then discussed at a seminar at the State University of Campinas, in June 2001, with the support of the Academy of Sciences and the São Paulo Research Foundation – Fapesp. Our hope is that the result will encourage a greater participation of Brazilian social scientists in this field.

A parallel product of this effort is a view of Brazilian thinking, which may be of interest to the international community. It is for this reason that the Brazilian Academy of Sciences resolved to publish this book in English, in time for presentation to the IV Open Meeting in October of 2001, in Rio de Janeiro.

The chapters in this book, then, mirror fundamental concerns of the social science/environmental community. The section on **population** rejects a neo-malthusian focus, concentrating on population distribution processes and health consequences of global environmental change. Special attention is given to so-called "traditional populations" and transformations in their use of resources.

A section on **institutions** focuses on the evolution of mechanisms and political instances important for understanding global change. Brazil's experience, both at the local and regional levels, and at international diplomatic levels, has been significant.

A section on **technological change and development** is concerned with the perspectives of decoupling energy and economic growth, with the evolution of science and technology in the context of a world divided into rich and poor, as well with sustainable alternatives for resource use in poor countries.

Behavioral changes are examined in three distinct contexts. In the first place, attitudes, values and opinions relevant to the environment are discussed. Secondly, the concept of a "risk society" is seen from the angle of food risks, and Brazil's recent experience permits an in-depth discussion of this issue. Finally, the difficult issue of consumption patterns is addressed. The stereotyped view that the North contributes to global change via consumption and the South via population, has handicapped an analysis of the environmental consequences of development/consumption in Brazil.

The last section discusses the basic issue of the use of **territory** for global environmental change. Natural resources are spatially specific, as are the economic activities associated with their use. It is the cumulative effects of all land use changes which produce global environmental change, but it is the localized presence of population and economic activities which are the driving forces. Brazil's national treasure – the Amazon forest – has been the center of attention of global change researchers when looking at Brazil. A chapter on the Amazon region explores this complex question. A second chapter reviews land use and tendencies for land cover change in Brazil, surely one of the most significant issues for this continental size country. Finally, determinants and consequences of global environmental change are examined at the level of urban spaces – home to four out of five Brazilians.

This panorama is a first step in the consideration of the human dimensions of global environmental change, from a Brazilian perspective. It is the hope of the National Committee that this picture will soon be enriched by the contributions of other environmental researchers to this new, and crucial, field.

Rio de Janeiro, October 2001

Daniel Joseph Hogan and
Maurício Tiomno Tolmasquim

POPULATION

Human Dimensions of Global Environmental Change
Brazilian Perspectives

Demographic Aspects of Global Environmental Change: What is Brazil's Contribution?

Daniel Joseph Hogan
State University of Campinas

INTRODUCTION

As one of the few continental-sized countries in the world, where both population density and population/resource ratios are favorable, Brazil poses a unique set of threats and opportunities regarding global environmental change. On the one hand, the burning of the Amazon forest – the issue which has mobilized international concern – both threatens biological diversity and is Brazil's largest single source of greenhouse gasses. Other threats to biodiversity and global warming are also significant, though have not yet generated the same media attention. In the case of the Amazon – the Earth's last major tropical forest resource – this concern is heightened.

At the same time, Brazil's economic growth, though stalled since the "lost decade" of the eighties, has solid foundations which will mean faster-paced growth in coming decades, a growth likely to be modeled on the same unsustainable lifestyles of developed countries. According to the *Human Development Report*, Brazil's annual per capita income is US$6 500. When we see that this value is $14 900 for Brasília and $10 500 for São Paulo, we get an idea of prospects for consumption patterns which will accompany development. The pressure which this largely urban growth will place on such natural resources as water, soils and forest cover and the increasing energy needs of a developed society will push Brazil further up the list of countries responsible for global environmental change.

On the other hand, one must consider Brazil's reduced dependence on fossil fuels for generating electric power; its major investment in renewable sources of fuel for automotive transport; a well-trained scientific establishment, very much in tune with international trends; a professional foreign service attentive to environmental concerns and disposed to assume a leadership role in North/South fora; and vast regions still relatively preserved in their (close to) original state. Furthermore, both demographic equilibrium and economic growth will take place in a national and international climate of environmental concern. These factors necessarily temper alarmist concern for the destructive potential of Brazilian development.

Where does population fit into this complex picture? Contrary to much of the popular literature on population and global environmental change, the present analysis does not emphasize population size and growth rates. The first section of the text, which provides an overview of current population trends, does not support the popular "demographic terrorism" frequently found in discussions of the environment. Rather than stopping at this conclusion, the text seeks to examine how current Brazilian tendencies are related to global environmental change. In this regard, attention is given to those issues which the growing consensus of the international community of researchers on the human dimensions of global environmental change has identified as useful starting points to understanding the driving forces of such change.

The second section, then, draws out the demographic content of land use and land cover change, industrial transformation and human security. In some cases, it is possible to use Brazilian research to examine issues; to a great extent, however, this analysis suggests a modest framework to guide Brazilian demographers' efforts to relate their studies to global environmental change.

AN OVERVIEW OF BRAZILIAN POPULATION TRENDS

Brazil enters the 21st century at a dramatic new moment in its population history. With its Demographic Transition nearly completed – low fertility and mortality leading to slower growth and an older population – and high urbanization indices, Brazil is poised for an era of demographic stability. For a country whose recent history was conditioned by extremely rapid city growth, large families and a youthful population, these are remarkable changes.

FIGURE 1: BRAZILIAN DEMOGRAPHIC TRANSITION, 1940 – 1996

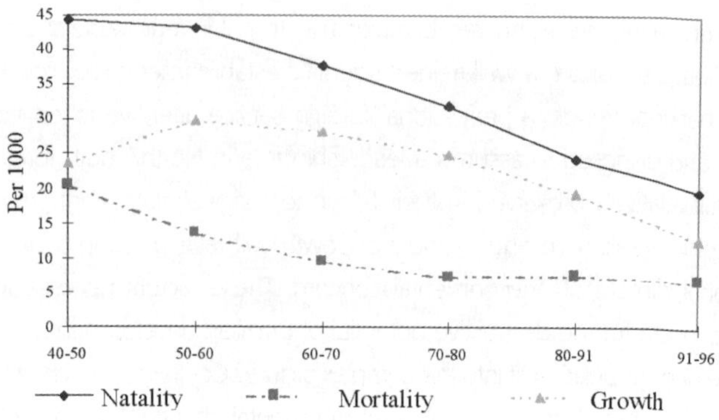

Source: apud Berquó and Baeninger, 2000

Having increased more than sixteen-fold since its first census in 1872, Brazil ranks in fifth place among world nations. Population is unevenly distributed over 8 547 403.5 km^2, reflecting the history of economic cycles which developed and populated successive regions. At 19.84 persons per square kilometer in 2000, Brazil has a low population density by world standards, although this index varies from 3.34 in the Amazon region in the North to 77.92 in the industrial heartland of the Southeast. Since 1970 – when Brazil was still in its rapid growth phase – population density has increased nearly 50% at the national level, while in the North and Center-West, density has more than doubled. The Amazon, the Pantanal and the savanna lands of the *cerrado*, objects of national and international concern for the integrity of their natural resources, continue to be sparsely populated, however.

TABLE 1. POPULATION OF BRAZIL

Year	Population
1872	9 930 478
1890	14 333 915
1900	17 438 434
1920	30 635 605
1940	41 236 315
1950	51 944 397
1960	70 070 457
1970	93 139 037
1980	119 002 706
1991	146 825 475
1996	157 070 163
2000	169 544 443

Source: IBGE.

TABLE 2. SELECTED DEMOGRAPHIC CHARACTERISTICS, BRAZIL

Annual Growth 1991-2000	Per cent Urban 2000	Sex Ratio 2000	Dependency Ratio 1996
1872	81.2%	96.9	58.7

Source: IBGE. The sex ratio is the number of males per 100 females; the dependency ratio is the population considered inactive (0 to 14 years and over 65 years) over the population potentially active (15 to 64 years).

Population densities, however, do not reveal another notable characteristic of Brazil's population distribution: its advanced degree of urbanization. In 2000, 81.2% of Brazil's population lived in cities. While this value includes the State of São Paulo (93.4% urban) and the industrialized Southeast region as a whole (90.5%), even Amazônia (North region) had nearly 70% of its population living in cities. In the Center-West, also an object of concern about loss of native vegetation cover, 86.7% of the population is urban.

TABLE 3. PERCENT URBAN IN BRAZIL, BY REGION, 1970-2000

Year	Brazil	North	Northeast	Southeast	South	Center-West
		41.7	41.8	72.7	44.3	**50.7**
		50.3	50.5	82.8	62.4	**70.9**
1991	75.6	**59.1**	60.7	88.0	74.1	**81.3**
		62.4	65.2	89.3	77.2	**84.4**
2000	81.2	**69.7**	69.0	90.5	80.9	**86.7**

Source: 1970-1996, apud Patarra, Baeninger and Cunha, 2000; 2000, IBGE.

The end of the rapid growth which marked the second half of the 20th century is within view. During the 1990s, Brazil grew at an average rate of 1.6% per year, down from 3.5% in the sixties. The early decades of the 21st century will see further declines in this rate, as the large cohorts of women born during the rapid growth phase complete their childbearing years. In the more-developed regions of the South and Southeast, fertility levels are already at or below replacement rates and other regions continue to converge to the national average. By the time population growth bottoms out in mid-21st century, Brazil will have dropped to seventh place among world nations.

TABLE 4. POPULATION DENSITY IN BRAZIL – PERSONS/KM²

Year	Brazil	North	Northeast	Southeast	South	Center-We
1970	11.10	**1.09**	18.54	43.64	29.00	**2.88**
1980	14.23	**1.76**	22.90	56.89	33.69	**4.36**
1991	17.36	**2.68**	27.52	68.16	39.32	**5.86**
1996	18.38	**2.92**	28.68	72.26	40.74	**6.51**
2000	19.84	**3.34**	30.60	77.92	43.43	**7.20**

Source: IBGE.

TABLE 5. TOTAL FERTILITY RATE OF BRAZIL, BY REGION, 1996

Brazil	North (urban)	Northeast	Southeast	South	Center-West
2.28	2.77	2.89	1.99	2.10	2.29

Source: IBGE. The total fertility rate is the average number of children a woman would have during her childbearing years (15-49), if subject to prevailing age-specific fertility.

Compared to European or North American countries, Brazil's transition from high to low rates of fertility and mortality was swift. It also proceeded at different rates in different parts of the country. Not only fertility, but especially mortality still varies among regions. Life expectancy at birth, the most universal index of mortality levels, varies from 70.2 years in the South to 64.5 in the poverty-ridden Northeast. This pattern is more evident when comparing *infant* mortality. While even rates in the industrial South and Southeast continue high by international standards, there is nearly a three-fold difference between the South and the Northeast.

TABLE 6. LIFE EXPECTANCY AT BIRTH IN BRAZIL, BY REGION, 1996

Brazil	North (urban)	Northeast	Southeast	South	Center-West
67.6	67.4	64.5	68.8	70.2	68.5

Source: IBGE. Life expectancy at birth represents the average number of years a newborn child would live if subject to prevailing mortality conditions.

TABLE 7. INFANT MORTALITY RATE IN BRAZIL, BY REGION, 1996

Brazil	North (urban)	Northeast	Southeast	South	Center-West
37.5	36.1	60.4	25.8	22.8	25.8

Source: IBGE. The infant mortality rate is the number of deaths to children under one year of age per 1000 live births, in a given period.

Such changes occur only once in a nation's history and change all aspects of life. As the accompanying population pyramids show, slow or zero growth produces an age structure in which numbers at different ages are similar. Among the changes Brazil will face in the 21st century is a larger elderly population; this may compensate for smaller younger cohorts in terms of the dependency ratio. There will also be a gap in the transition to this new structure, caused by the especially rapid fertility decline. This will raise the possibility of directing social investment to reducing historical inequalities and catching up with environmental infrastructure, rather than struggling to keep up with the needs of a growing population.

FIGURE 2: AGE STRUCTURE, BRAZIL 1950

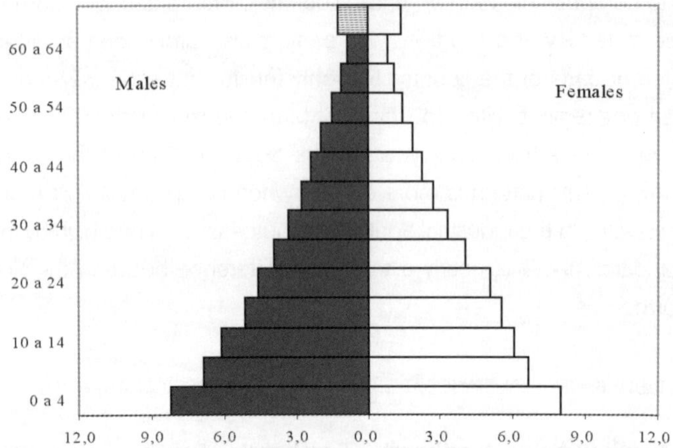

Source: apud Berquó and Baeninger, 2000

FIGURE 3: AGE STRUCTURE, BRAZIL 1996

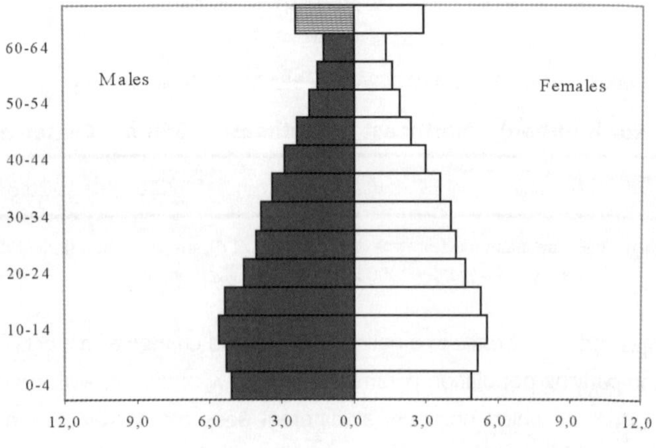

Source: apud Berquó and Baeninger, 2000.

FIGURE 4: PROJECTED AGE STRUCTURE, BRAZIL, 2000

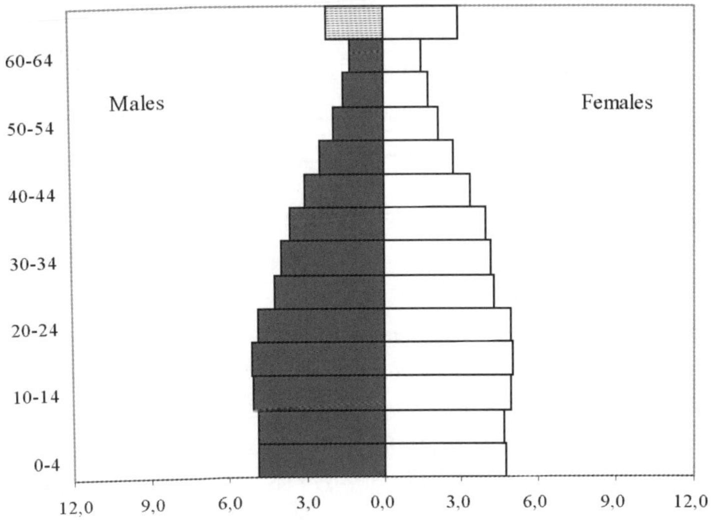

Source: apud Berquó and Baeninger, 2000.

FIGURE 5: PROJECTED AGE STRUCTURE, BRAZIL, 2020

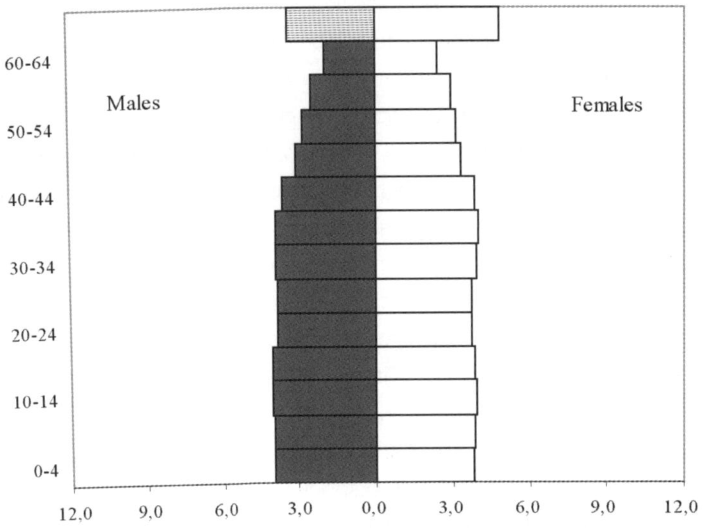

Source: apud Berquó and Baeninger, 2000.

The major causes of death in contemporary Brazil are circulatory illnesses (160.7 deaths per 100 000 persons), cancer (61.27 per 100 000 persons) and external causes – accidents and violent deaths in general (68.4 per 100 000 persons)[1]. This pattern, typical of developed countries, coexists with older patterns of morbidity and mortality; infectious and parasitic diseases still present mortality rates of 27.6 per 100 000 persons. Infant mortality rates, especially in some regions, reflect this situation. In the eighties and nineties the HIV-Aids pandemic expanded, reaching an incidence of 7.87 cases per 100,000 persons in 1995. It is estimated that 330,000 to 480,000 persons between 15 and 49 years of age are HIV-infected.

Young people are especially vulnerable to HIV/Aids and to mortality from external causes. Among men 20-29 years of age, mortality rates from external causes increased from 27.8% to 37.9% between 1991 and 1994. By 1995, Aids was the second cause of death for both sexes in the age group 20-49 years of age. Maternal mortality rates also continue high. Diseases such as dengue fever, malaria and tuberculosis – related to deteriorated domestic and ambient environments – have made an alarming comeback in recent decades.

Today an urban country, with four-fifths of its population living in cities, Brazil is also entering a new phase in terms of distribution trends. Considering the already high level of urbanization, rural-urban migration will necessarily be less important in the future than in the recent past. Inter-regional migration has also declined in intensity and metropolitan areas are growing more slowly, while short-distance movements and return migration to regions of origin have increased. An unknown factor in the near future is international migration: in the nineteen eighties, one million Brazilians emigrated to the United States, Europe and Japan (and in smaller numbers to other places). Historically a receiving country for immigrants, the globalization of the world economy suggests more fluid movements in the future.

Brazil, then, enters the 21st century with slower population growth and slower urbanization rates. It is faced with features common to both developed and underdeveloped countries; mortality patterns, for example, affected by both old and new diseases, represent major challenges. In terms of sustainability, the basic parameters are favorable. Much will depend on Brazil's ability to overcome obstacles to universal health care and to social justice in general, as well as to the development of institutional mechanisms to promote patterns of population distribution compatible with sustainable use of natural resources.

DEMOGRAPHIC ASPECTS OF GLOBAL ENVIRONMENTAL CHANGE

In examining global environmental change, then, the major demographic issues for Brazil are to be found in features of population change other than size or growth rates. The present text will not reproduce a general review of the demographic literature, with proposals for a framework for population-environment studies (for such a review, see Hogan, 2000a, 2000b) nor a general review of work in Brazil (see Hogan 2001). The principal point, which is certainly true in the case of Brazil – as the discussion of Brazilian population perspectives has shown – is that discussing population or demographic dynamics does not begin and end in discussion of population **size and growth rate**. Indeed, as earlier reviews have argued (Hogan, 1992, 1999), these are poor starting points for an analysis concerned with policy options. Given the *momentum* of population growth, only draconian, authoritarian measures could reduce Brazil's population size and growth rate at rhythms more pronounced than those observed today. Indeed, researchers have already gone beyond the neomalthusian-anti-neomalthusian polemics to investigate nexus which offer opportunities for both theoretical development and policy interventions. There are, however, several basic points to be retained from the earlier debate for the present discussion:

1. the initial handicap of breaking through the neomalthusian cloud of confusion slowed Demography's work in this field. This approach is still to be found, as Martine (2001) has noted, although it does not prevail in scientific/academic research in the social sciences.

2. the literature of the nineties reveals a range of approaches to the study of the population-environment relationship, examining demographic processes (fertility, mortality, migration); demographic structure (age, sex); demographic distribution (density, urbanization patterns); topical studies (water, CO2, land); integrated studies of river basins, forest regions, environmentally protected areas.

3. the two major issues to be faced are health/morbidity/mortality (E Ü P) and distributional questions – migration/urbanization/density (P Ü E).

4. in this volume, since health consequences of global environmental

changes are dealt with separately, the main focus here will be on population mobility and environment.

We begin the new decade with a clearer view of how to unpackage the society/nature relationship. The social sciences generally have been slow to develop the theoretical and methodological frameworks necessary for examining social aspects of environmental change. While the demands from the physical and biological sciences (whose work led them earlier to environmental questions) were important in spurring social scientists onward, the major impulse came from the environmental movement. By the seventies, it simply was no longer possible to ignore the mutual impacts of natural and social processes. But, most especially, it was impossible not to recognize the importance of the environmental movement. Social scientists did not come to the study of the environment as a logical derivation of their general theories: **society placed the environment on our agenda**.

In the case of global environmental change[2], however, social movements are still fragile. Most of the laws, confrontations, organizations, and institutionalization of the environmental question relate to **local** factors: this is where the population directly experiences the impacts of environmental change. The source of most of the pressure for research and intervention concerning **global** environmental change is the scientific community: atmospheric scientists, oceanographers, biologists and others. These fields organized earlier and identified earlier the human origin of global environmental changes. Although a primitive neomalthusianism is still very much present, it was these fields which first sought out the social science contribution. They knocked on our doors looking for "damage control." Sociologists tend to regard physical and natural sciences as "technical;" in this case, however, it was **our** "technical" capacities in conflict control, social change, environmental education, policy formulation and implementation which motivated harder sciences to look for help from us.

But if we were slow to come to grips with environmental change generally, global environmental changes were definitely outside our scope. The issue of time scale was a major inhibitor. For social scientists, the long-range view is backward: we continually seek to explain socio-historical change, the better to understand current events. Anything beyond a decade into the future, however, is *futurology*, the very term denoting a condescension for such efforts. The rise in sea level, the consequences of the growing ozone hole, species extinctions: these changes will have impacts on future generations but we have not yet developed appropriate tools for incorporating them into social theory.

The 1990s, however, witnessed intense efforts first to engage social scientists in this problematic and, second, to identify approaches capable of incorporating global environmental changes within our purview. The result of these efforts is a relatively short list of priority research areas which now mobilize international efforts among social scientists. These areas include:

- land use and land cover change;
- global environmental change and human security;
- institutional dimensions of global environmental change;
- industrial transformation;
- land-ocean interactions in the coastal zone;
- health consequences of global environmental change;
- public perceptions, values and behaviors regarding global environmental change.

The first four of these topics have been selected for the concentrated efforts of the International Human Dimensions Program; of these, demographic aspects can be identified for land use and cover change; human security; and industrial transformation. These themes will be discussed below. Furthermore, land-ocean interactions, as to their demographic content, can be analyzed in terms of land use and cover change, while health consequences and public perceptions of global environmental change are subjects of separate chapters in this book.

The realization that a social science perspective on global environmental changes had to deal with understanding the "driving forces" of these changes led to emphasis on studies at the local level. The challenge of this effort is to draw conclusions concerning the aggregate consequences and determinants of human actions. The task now is not to discover new realms of human action (which was the challenge for early environmental sociologists) but, first, to identify those social processes with impacts on **global** (as well as local) changes and, second, to link changes at the local level with global changes.

What does all this mean for a review of population and global environmental change? It means that for Demography, as for social sciences in general, much relevant research is being done under the rubric of environmental sociology or environmental demography without feeding into the global environmental change research community. The effort to engage demographers (and other social scientists) in the study of global environmental change means convincing them of the importance of the issue, not such a difficult task. It also means, however, **demonstrating the relevance of their current research to global environmental change studies**. Global environmental change studies will prosper and be legitimized within social science disciplines to the extent that the field is not an exotic niche for globetrotting vocations, but very much related to a wide range of ongoing research.

These considerations have structured the present text. While the first section mentioned some specific aspects of Brazilian population dynamics which are relevant to global environmental change studies, this section discusses Brazilian demographic research on the topics mentioned above.

To evaluate Brazilian Demography's contribution to these issues, we first carried out an exercise to map this contribution. Three sources were examined: the printed proceedings of the *Associação Brasileira de Estudos Populacionais*, whose biannual meetings began in 1978; the Association's journal, the *Revista Brasileira de Estudos de População*, which began publication in 1983; and two books which the Association published on population-environment issues (see Martine, 1992 and Torres and Costa, 2000). As a first step, all articles with any relation to the human dimensions of global environmental change were identified. In terms of the *Proceedings* – the single most significant library of Brazilian Demography – the overwhelming majority of these articles (80%) treated aspects related to land use and land cover change. Only 17 articles dealt with other environmental issues. LUCC here includes studies of frontier expansion and colonization; internal migration; and changing agricultural practices as related to processes of occupation of new lands. Of these articles, 10 dealt with environmental aspects of land use change in the Amazon; 14 with land use change in the Amazon without an environmental emphasis; 27 with environmental aspects of land use change in other regions; and 23 with land use change in other regions, with no environmental emphasis. In addition to health impacts of global environmental change, then, and as earlier analyses of demographic dynamics and environmental change had concluded, the principal issue examined by demographers is the distribution of population in space: the size, direction, characteristics, determinants and consequences of migratory movements. As discussed below (see, especially, Table 8), the period analyzed by these studies was the period of the most intense migratory movements, including frontier migration.

TABLE 8. CLASSIFICATION OF ARTICLES ON ENVIRONMENT IN PUBLICATIONS OF THE BRAZILIAN ASSOCIATION OF POPULATION STUDIES (ABEP).

Classification of articles	ABEP Proceedings (1978-2000)	ABEP Proceedings (1978-2000)	ABEP Proceedings (1978-2000)
Lucc	23	0	0
Lucc/Amazônia	14	0	0
Lucc/environment	27	1	8
Lucc/Amazônia/environment	10	1	1
Environment	17	2	10
Total	91	0	19

LAND-USE AND LAND COVER CHANGE

From both national and international perspectives, this may be the single most important Brazilian issue in considering global environmental change (although MacKellar et al., 1998, place great emphasis on energy; see the following section on industrial transformation for some considerations on this topic). Brazil's vast territory – and its economic and demographic occupation have supported the idea of endless frontiers, always an important dimension of Brazilian ideology. When coffee denuded the slopes and impoverished the soils of the Paraíba Valley in the 19th century, cattle-grazing took over and coffee production moved on to newly opened, fertile lands, first in São Paulo, then in Paraná. When soybean production replaced coffee in Paraná and required extensive new lands for its expansion, Mato Grosso and other Center-West states – and today the Amazon region – were transformed into large-scale monocultures.

Environmental concern with such movements, considering the unquestioned myth of the endless frontier, only arose when the "last frontier" of the Amazon was reached. By this time, a policy which considered forest loss as a rational economic investment for extending agriculture was a policy challenged by environmentalists. One of the early villains in the process was identified as the poor settler who, from ignorance, lack of credit or avarice, destroyed successive areas of forest. This settler was introduced to the Amazon through organized colonization schemes.

Demographers soon challenged this view and identified a more complex reality. Sawyer (1984) showed that official colonization schemes accounted for only a small portion of migratory movements to the Amazon; spontaneous movements were much more important. Even then, though, it was not the poor settler who was the most destructive. Better-financed second owners (who succeeded poorer pioneers, unable to overcome the lack of credit and technical assistance to establish successful farms) cleared more forest than the poorer pioneer settlers. While lack of opportunities in their regions of origin was indeed a major impetus for migration to the Amazon, it was not these poor original settlers who were responsible for the greater part of forest burning. Financial incentives led large investors to clear land for cattle raising, in a well-known process of soil loss and degradation. Martine called attention to the lack of absorption of migrants to Amazônia as early as the mid-80s: if in the past

> *migration to frontier regions was a significant counterpoint to rural-urban migration," recently "the frontier has been absorbing fewer and fewer migrants with a rural destination; in the last few decades, even in frontier areas, rural-urban migration is more important than migration to rural areas. (cited in Mueller, 1992, 25)*

Measurement and analysis of migratory movements to the Amazon (Oliveira, 1996; Moreira and Moura, 2000) have chronicled this process since the seventies. Sydenstricker (1992), for example, has examined family strategies in colonization schemes, finding that women remained in cities, helping to finance the establishment of farms in Rondônia, before joining their husbands. Population growth peaked in the 1970s, when the average annual growth rate was an historically high 4.7% per year. After pointing out the historical importance of migration in the demographic evolution of the Amazon region, Moreira and Moura suggest that:

> *There are strong indications, however, that lead us to suspect that there has been, in recent years, a change in this picture, with de-intensification of inter-regional migratory flows and greater intra-regional population movements, above all that of a rural-urban nature. (p. 63)*

In evaluating the consequences of migratory movements for land-use change, we must first appreciate how concentrated these movements were over time. Limiting analysis here to the Amazon region (North) and the cerrado (Center-West), it is clear that migration has peaked. In the North region, net migration added more than a half million new residents in the 1970s, but rapidly tapered off in the eighties and nineties. On the simplistic assumption of doubling the moves from the 1991-96 period to estimate the decade's total results, net migration for the nineties would be less than was found for the 1986-1991 period.

TABLE 9. NET MIGRATION, BY REGION, 1960-1996.[3]
Source: apud Patarra, Baeninger and Cunha, 2000; Cunha, 2000, 61.

Period	North	Northeast	Southeast	South	Center-West
1960-70	**-51,063**	-1,754,761	815,284	371.175	**746,611**
1970-80	**585,397**	-2,402,244	2,262,364	-1,613,377	**638,281**
1986-91	**131,218**	-876,534	-185,391	290,569	**290,559**
1991-96	**60,844**	-852,732	597,890	-30,510	**224,508**

In the Center-West, the same process is clear, although with some time lag. Migration was already impressive in the sixties, nearly reaching the levels found for the Southeast. While migration to the Southeast "exploded" in the seventies, movement to the Center-West continued at high levels until the early nineties, when migration flows declined.

Comparing this process with urbanization trends (Table 3), the picture that emerges clearly suggests that migration to the agricultural frontier was followed by a rapid reflux to cities, in both North and Central-West regions.

The driving force of land-cover change in these regions, then, cannot be identified as migration to rural, frontier areas. If this were the case, concern would diminish in the same rhythm as migration to these regions. The rapid urbanization of the Amazon reflects both the rapidly diminishing potential for agriculture to absorb migrants, confirming the more important role of large-scale capitalism as the major driver of deforestation (Bilsborrow and Hogan, 1999), and signals a new concern for consequences of urban growth. These cities respond to the commercial and service demands of primary sector producers but are also home to large numbers of unemployed and under-employed workers – similar to the situation prevailing in the other cities of Brazil. Two questions arise: (1) What will be the ecological footprint of these cities in their immediate hinterlands? (2) What employment opportunities will be developed as alternatives?

Large-scale economic interests which neither preserve native vegetation cover nor create jobs for unskilled agriculturalists, bear the greater responsibility. Furthermore, the demands for communication, transportation and service infrastructure generated by agribusiness will have a major "ecological footprint." Needs of the urban populations themselves, as well as the land-use profile of these cities, will add to these pressures. The so-called *Avanço Brasil*, a federal plan to integrate Brazilian territory through an ambitious public works program, represents the major pressures to be seen in the first decades of the 21st century. One attempt to evaluate this impact (Laurance et al., 2001) predicts serious environmental damage to the Amazon Forest.

In examining the literature in this field, including the Brazilian studies, one is struck by how rare it is to find studies which examine complex, industrializing countries where modern agribusiness is much more important than small family farms in producing land cover change. The major land-use changes underway in Brazil are in the Amazon and in the *cerrado* of the Center-West. For both regions, it is urgent to expand studies from the family farm, village level to the evolution of monoculture grain crops whose command and control centers are in São Paulo and other capitals. Micro-level studies can tell us how the small farmer is losing in this struggle; in terms of major driving forces, more attention must be paid to the factors which determine investment decisions, market opportunities, life-cycles of large-scale farming activities, tax structures (which lead to environmentally damaging incentives in fragile areas), international trade advantages, the sensitivity of national economic and political elites to land-cover issues. "Socializing the pixel" may help to record the persistent loss of forest cover and to identify the immediate agents of this change, but the mediating factors mentioned above are surely more crucial determinants. This challenge of linking the macro and micro levels is a perennial difficulty for the social sciences; in the absence of a grand unifying theoretical model, it will be necessary to complement the local studies now underway with macro approaches.[4]

INDUSTRIAL TRANSFORMATION

Energy and materials use are related to three important demographic trends. In the first place, there is an increase of **single-person households**. Such domestic arrangements as changing cohabitation patterns, population aging, and divorce and separation all mean that the rate of growth of households is now greater than the rate of population growth. Furthermore, this trend seems likely to continue. In addition to the social and cultural dimensions of these changes, it is important to note, in this discussion, that the economy of scale found in larger households is giving way to situations where much domestic infrastructure, with its energy and materials requirements, is multiplied. However, other than IIASA's work in this field (MacKellar et al., 1995), this has been a little explored issue; certainly in the Brazilian case there are no published studies which relate these changes to increasing energy consumption.

Energy needs are also related to the land-use issue. The pattern of **urban sprawl** which characterizes Brazilian cities – vs. population concentration in city centers with their urban infrastructure already in place – is a result of cultural values (in favor of single-family houses) and a real estate market distorted by land speculation. Both vegetation loss (from swallowing up farms and from eliminating native vegetation cover) to housing and highways and carbon emissions from a fragmented, fossil-fuel based transportation system result from the horizontal expansion of cities. Recent proposals, in São Paulo for example, to "densify" the urban core, taking advantage of built-up areas and the transportation/environmental infrastructure which support them, may be a sign of counter-trends. Brazilian experience, in this regard, is more similar to the USA than to Europe. The creation of new metropolitan areas in the nineties, with some form of regional government to promote rational land use and transportation systems, is another positive sign. One measure being promoted is the expansion of electricity-based urban/suburban trains (bearing in mind the preponderance of hydro-electricity in Brazil's energy matrix). New data from the 2000 census on commuting (unavailable since 1980) will help in measuring the importance of inter-municipal movements and their contribution to energy use and carbon emissions.

On the other hand, industrial transformation may also have other, contradictory, consequences for urban systems. Will the telecommunications revolution create more possibilities for **deconcentration**, smaller cities not necessarily connurbated in traditional urban hierarchies? Will this mean lower transport costs, less production of greenhouse gasses and a generally more benign relationship between man's occupation of the landscape and the integrity of natural systems?

Finally, an issue which merits attention is a **generational (cohort) perspective**. In Brazil and other countries with rapid fertility decline, there will be a significant gap in age pyramids for much of the 21st century. The "youth wave" now characterizing Brazil's age structure will move up the pyramid, with specific consequences for consumer choice and the structure of consumption. Is the younger generation to be as consumerist as their parents or more disposed to "environmentally correct" behavior? Will a wave in family formation mean more or less use of energy and materials? To consider scenarios of evolving energy-use profiles, on the basis of these age changes would be a useful exercise.

GLOBAL ENVIRONMENTAL CHANGE AND HUMAN SECURITY

The major issue discussed internationally under this heading is food security. While this is not irrelevant for Brazil, it is not agricultural potential but income concentration – and its consequences for nutritional levels – which is critical. The problems are not related to pressures on the resource base to produce food. While agricultural activities are indeed pressuring land cover and biological diversity, the Brazilian situation is unlike that which inspires concern in other places, such as sub-Saharan Africa. It is export-oriented agriculture (grains, coffee, oranges, sugar) which is responsible for land-clearing. The image of the landless poor forced to cut the last tree to be able to grow food is not a Brazilian one. Nutritional deficiencies exist and are well-documented by research in public health. But in terms of global environmental change, this is not a major issue.

Besides the issue of food, the importance of population displacement from environmental degradation is a research problem examined under the rubric of "human security." "Environmental refugees" have been defined as

> *...those people who have been forced to leave their traditional habitat, temporarily or permanently, because of a marked environmental disruption (natural and/or triggered by people) that jeopardized their existence and/or seriously affected the quality of their life."* (El-Hinnawi, 1985)

MacKellar et al. (1998) prefer the less sensational expression, *environmental stress migration*, to describe "migration in which the deterioration of the natural resource base is a significant push factor." (MacKellar et al., 1998, 162) Drought in Brazil's Northeast, construction of hydroelectric plants in all regions (Vainer, 1993), flooding or landslides provoked by deforestation or bad urban planning (Hogan, 1992, 1993; Torres and Cunha, 1992; Torres, 1997; Taschner, 2000; Hogan et al., 1999) and which result in population displacement are examples which have been studied in Brazil, but neither quantified nor conceptualized as a distinct kind of population movement. The IPCC (Intergovernmental Panel on Climate Change) has suggested that the greatest impact of climate change on populations will be migration away from low-lying coastal regions or islands. (IPPC, 1990) While the concept of environmental refugees has been criticized as too imprecise to be useful, there has been no systematic assessment in Brazil of how much population mobility is due to changes in environmental quality, much less any link to the concept of refugees.

In a recent review for the *Environmental Change & Security Project Report*, former US ambassador Richard E. Benedick presents an informed overview of population growth and perspective trends but, as is typical in this literature, does not go beyond the issue of population growth. (Benedick, 2000) The present discussion of the Brazilian case has concentrated on a *demographic* perspective.[5] As Homer-Dixon (1995) has concluded, the Malthusian-neomalthusian debate has become sterile. It is time to get on with just how demographic dynamics affect environmental change.

CONCLUSIONS

This discussion has emphasized that Brazil's contribution to global environmental change, in terms of its current demographic trends, is less related to size and growth rate than to population distribution tendencies. Considering the point Brazil has reached in its demographic transition (and the speed with which it has done so), policy interventions for slowing growth are neither appropriate nor efficacious.

On the other hand, Brazil's rich biodiversity and extensive forests and savannas offer opportunities for either environmental destruction – with global consequences – or environmental preservation – with beneficial effects for global environmental change. It is in this area – land use and land cover change – that Brazilian demography can contribute to an understanding of the driving forces of global change. To a lesser extent, but also important, are considerations on industrial transformation and human security.

While Brazilian demography has consistently dealt with land use and land cover change, the relations between industrial transformation or human security with demographic processes has been little examined. Furthermore, the existing literature has not been specifically directed to considerations of global environmental change. Both existing studies and the potential contribution of new research will be important avenues to connect the Brazilian research community to international human dimensions research.

NOTES

1. *Much of the demographic information in the following paragraphs is taken from CNPD, 1999.*

2. *Following Benedick, we may identify global environmental changes as climate change; depletion of the stratospheric ozone layer; loss of biological diversity; spread of arid lands, desertification and soil erosion; pollution of marine and freshwaters; destruction of forests; and the worldwide diffusion of hazardous substances.*

3. *One must bear in mind that the last two rows refer to 5-year periods (due to more detailed information for these years), which, while indicating the trend, cannot be simply doubled to reach a total for the decade. See Patarra et al. (2000, 28) for the caveats on the methodological differences among the different periods.*

4. *One of the difficulties here is that it has generally been the economists who have been able to deal adequately with such factors. As economists (with the honorable exception of some marginal fringe groups) are not in the forefront of the environmental question, this approach has been inadequately developed.*

5. *Criticism of this limited approach can be found in Hogan, 2000a, 2000b.*

6. *The other major factor to be considered in a demographic analysis of global environmental change is health and mortality, not examined here because it is the object of special attention in the chapter by Ulisses Confalonieri.*

BIBLIOGRAPHY

BECKER, B.K. Reflexões sobre Políticas de Integração Nacional e Desenvolvimento Regional. Ministério de Integração Nacional, September 2000.

BENEDICK, R.E. Human Population and Environmental Stresses in the Twenty-First Century. *Environmental Change & Security Project Report*, v.6, p.5-18, 2000.

BERQUÓ, E., BAENINGER, R. *Os Idosos no Brasil:* considerações demográficas. *Textos NEPO*, Campinas, v.37, 2000.

BILSBORROW, R.E., HOGAN, D.J. (Eds.). *Population and Deforestation in the Humid Tropics.* Liège, IUSSP, 1999. 291p.

CNPD. National Population and Development Commission. *Cairo + 5*: report on Brazil. CNPD, Brasília, 1999.

CUNHA, J.M.P. Migração no Centro-Oeste Brasileiro: as tendências e características do período 1986/96, p.51-106. In ANAIS DO II ENCONTRO DE DEMOGRAFIA DA REGIÃO CENTRO-OESTE E TOCANTINS, Brasília, SEDUH, 2000.

EL-HINNAWI, E. *Environmental Refugees.* Nairobi: UNEP, 1985.

FEARNSIDE, P. Human carrying capacity estimation in Brazil's Amazonian settlements as a guide to development policy. p.122-137. In: BILSBORROW, R.E., HOGAN, D.J. (eds.). Population and Deforestation in the Humid Tropics. Liège: IUSSP, 1999. 291p.

HOGAN, D.J. The Impact of Population Growth on the Physical Environment. European Journal of Population, v. 8, p.109-123, 1992.

HOGAN, D.J. Migração, ambiente e saúde nas cidades brasileiras. p.149-170 In: HOGAN, D.J., and VIEIRA, P.F. (eds.), *Dilemas Socioambientais do Desenvolvimento Sustentável.* Campinas: Editora da Unicamp, 1992.

HOGAN, D.J. População, Pobreza e Poluição em Cubatão. p.101-131 In: MARTINE, G. (ed.) *População, Meio Ambiente e Desenvolvimento*: verdades e contradições. Campinas: Editora da Unicamp, 1993. 207 p.

HOGAN, D.J. A relação entre população e meio ambiente: desafios para a demografia. In TORRES, H., COSTA, H. (eds.) *População e Meio ambiente:* debates e desafios. São Paulo: Editora do Senac, p.21-52, 2000.

HOGAN, D.J. *Socio-Demographic Dimensions of Sustainability.* Paper presented to the World Congress of Rural Sociology/ISA RC Environmental Sociology. Rio de Janeiro, 2000.

HOGAN, D.J. *Demographic Dynamics and Environmental Change in Brazil.* Paper prepared for presentation at the International Conference on Population, IUSSP, Salvador, 2001.

HOGAN, D.J., CUNHA, J.M.P., BAENINGER, R., CARMO, R.L. (eds.). *Migração e Ambiente em São Paulo:* aspectos relevantes da dinâmica recente. Campinas: Núcleo de Estudos de População, 2000. 518p.

HOGAN, D.J., CUNHA, J.M.P., BAENINGER, R., CARMO, R.L. (eds.) *Migração e Ambiente no Centro-Oeste.* Campinas: Nepo, forthcoming.

HOGAN, D.J., CUNHA, J.M.P., CARMO, R.L., OLIVEIRA, A.A.B. Urbanization Process and Socio-economic Vulnerability: the case of Campinas, Brazil. p. 108 In: *Abstracts of the 1999 Open Meeting of the Human Dimensions of Global Environmental Change Research Community.* Japan: Shonan Village, 1999.

HOGAN, D.J., VIERA, P.F. (eds.). *Dilemas Socioambientais e Desenvolvimento Sustentável.* 2.ed. Campinas, Editora da Unicamp, 1995. 234p.

HOMER-DIXON, T. The Ingenuity Gap: can poor countries adapt to resource scarcity? *Population and Development Review*, v.21, n.3, p.587-612, Sept. 1995.

JOLLY, C.L., TORREY, B.B. *Population and Land Use in Developing Countries.* Washington: National Academy Press, 1993.

KAHL, C.H. Population Growth, Environmental Degradation and State-Sponsored Violence: the case of Kenya, 1991-93. *International Security*, v.23, n.2, p.80-119, 1998.

LAURANCE, W.F. et al. The Future of the Brazilian Amazon. Science, v.291, p.438-439, Jan. 2001.

LONERGAN, S. Environmental Degradation and Population Displacement. *Global Environmental Change and Human Security Project*, Research Report # 1, Victoria, BC, GECHS Project, 1998.

LONERGAN, S., SWAIN, A. Environmental Degradation and Population Displacement. Aviso, v.2, May 1999.

MACKELLAR, F. L., LUTZ, W., PRINZ, C., GOUJON, A. Population, households and CO_2 emisions. *Population and Development Review*, v.21, n.4, p.849-865, 1995.

MACKELLAR, F. L., LUTZ, W., MCMICHAEL, A. J., SUHRKE, A. Population and Climate Change. P.89-193 in RAYNER, S. MALONE, E.L. (eds.). *Human Choice and Climate Change.* Columbus: Battelle Press, 1998, v. 1.

MARTINE, G. (ed.) *População, Meio Ambiente e Desenvolvimento*: verdades e contradições. Campinas, Editora da Unicamp, 1993. 207 p.

MARTINE, G. The Sustainable Use of Space: Advancing the Population-Environment Agenda. Population Environment Research Network cyber-seminar, p.22-26, Jan. 2001.

MOREIRA, M.M., MOURA, H.A. Análise Demográfica da Região Norte. p.61-151. In: PACHECO, C.A., PATARRA, N. (eds.) *Dinâmica Demográfica Regional e as Novas Questões Populacionais no Brasil.* Campinas, Instituto de Economia/Unicamp, 2000. 568 p.

MUELLER, C.C. Dinâmica, condicionantes e impactos socio-ambientais da evolução da fronteira agrícola no Brasil. *Documento de Trabalho*, Brasília, n.7,: Instituto Sociedade, População e Natureza,. 1992.

OLIVEIRA, L.A.P. Dinâmica populacional e social na região amazônica. p.74-102. In: MOURA, H.A., *A Pesquisa social na Amazônia*: avanços, lacunas e prioridades. F. Joaquim Nabuco. Recife. 1996. 274 p.

PACHECO, C.A., PATARRA, N. (eds.) *Dinâmica Demográfica Regional e as Novas Questões Populacionais no Brasil.* Campinas: Instituto de Economia/Unicamp, 2000. 568 p.

PATTARA, N., BAENINGER, R., CUNHA, J.M.P. Dinâmica Demográfica Recente e a Configuração de Novas Questões Populacionais. p.1-44 In: PACHECO, C.A., PATARRA, N. (eds.) *Dinâmica Demográfica Regional e as Novas Questões Populacionais no Brasil.* Campinas: Instituto de Economia/Unicamp, 2000. 568 p.

SAWYER, D.R. Fluxo e Refluxo da fronteira agrícola no Brasil: ensaio de interpretação estrutural e espacial *Revista Brasileira de Estudos de População*, v.1, n.1/2, p.3-34, 1984.

SAWYER, D.R. Campesinato e ecologia na Amazônia. P.211-234. In: HOGAN, D.J., VIERA, P.F. (eds.). *Dilemas Socioambientais e Desenvolvimento Sustentável.* 2.ed. Campinas: Editora da Unicamp, 1995. 234p.

SAWYER, D.R. População e meio ambiente na Amazônia brasileira. P.149-170 in MARTINE, G. (ed.) *População, Meio Ambiente e Desenvolvimento*: verdades e contradições. Campinas: Editora sda Unicamp, 1993. 207 p.

SAWYER, D.R. População, meio ambiente e desenvolvimento sustentável no cerrado. In HOGAN, D.J., CUNHA, J.M.P., BAENINGER, R., CARMO, R.L. (eds.) *Migração e Ambiente no Centro-Oeste.* Campinas: Nepo, forthcoming.

SYDENSTRICKER NETO, J.M. *Parceleiros de Machadinho: história migratória e as interrelações entre a dinâmica demográfica e o ciclo agrícola em Rondônia.* Dissertação (Mestrado) Unicamp, 1992.

SYDENSTRICKER NETO, J.M., TORRES, H.G. Mobilidade de migrantes: autonomia ou subordinação na Amazônia Legal? *Revista Brasileira de Estudos de População*, v. 8, p.1-2, 1991.

TASCHNER, S.P. Degradação ambiental in favelas de São Paulo. p.271-291. In: TORRES, H., COSTA, H. (eds.), *População e Meio ambiente*: debates e desafios. São Paulo: Editora do Senac, 2000. 351 p.

TORRES, H.G. *Desigualdade Ambiental na Cidade de São Paulo.* Tese (Ph.D.) Universidade Estadual de Campinas, 1997.

TORRES, H. G., CUNHA, J.M.P. População sujeita a riscos ambientais: o caso de Campinas. In: IX ENCONTRO NACIONAL DE ESTUDOS POPULACIONAIS. *Anais...* São Paulo: ABEP, 1992.

TORRES, H., COSTA, H. (eds.), *População e Meio ambiente*: debates e desafios. São Paulo: Editora do Senac, 2000. 351 p.

VAINER, C.B. População, meio ambiente e conflito social na construção de hidrelétricas. p.183-207. In: MARTINE, G. (ed.) *População, Meio Ambiente e Desenvolvimento*: verdades e contradições. Campinas, Editora da Unicamp, 1993. 207 p.

Global Environmental Change and Health in Brazil: Review of the Present Situation and Proposal for Indicators for Monitoring these Effects

ULISSES E.C.CONFALONIERI
Program on Global Environmental Changes and Human Health - PMAGS
FIOCRUZ, Rio de Janeiro

INTRODUCTION

The need for the establishment of national monitoring programs for the most important environmental parameters associated with human diseases is one of the goals suggested by the Pan American Conference on Health and the Environment in Sustainable Human Development (PAHO, 1995a; 1995b). However, it has been recognized that there is a paucity of indicators to demonstrate the relationships between environment and health. Accordingly, the need for the development of specific, reliable and practical environmental health indicators has been stressed since the many existing environmental indicators are only partially health-related and, therefore, of scarce utility as indicators of human exposure (KJELLSTROM & CORVALAN, 1995; WILLS & BRIGGS, 1995). One such strategy could be the identification of health events considered as "sentinels" for the development of a selective health/environmental surveillance system. This system could serve both for the analysis of the temporal trends in the occurrence of environmentally induced health effects and also for the indirect monitoring of key environmental variables in different ecosystems of the country. The basic underlying assumption for this proposal is that biological responses involved in human disease phenomena form the most important set of parameters for the assessment of environmental quality in general and, therefore, all measures in support of environmental protection are justified on the basis of their impact on human health (MOGHISSI, 1994). The increasing severity of the health impacts of multiple and widespread environmental exposures is stimulating the development of appropriate measures of these impacts, such as aggregate public health indicators (Hollander, et al., 1999).

In this paper we propose a well defined preliminary set of sentinel health events with linkages to a broad environmental context as an instrument for a global and regional ecological monitoring adapted to the Brazilian epidemiological and environmental profiles.

SENTINEL EVENTS

Although originally proposed from single preventable cases of diseases for the assessment of health services and for occupational health surveillance (RUTSTEIN, et al. 1983; RUTSTEIN, 1987) the concept of sentinel health events was subsequently widened in its scope to cover other situations. Thus statistical events not necessarily preventable such as rare clusters of diseases; conditions occurring in unusual populations; increases in strategic manifestations and even the use of in vivo test systems for bioessays in sentinel animal species which manifest diseases more quickly than humans were included (ALDRICH, 1991). A recent panel on the identification of sentinel health events as indicators of environmental contamination stressed the alarm function of the sentinel health events by accepting its definition as "A preventable disease, disability or untimely death whose occurrence serves as a warning sign" (ROTHWELL et al., 1991). However, this alarm function usually refers to the need for the improvement in the quality of preventive or medical care, that is, sentinel events are undesirable occurrences if we have adequate health services (KJELLSTROM & CORVALAN, loc.cit.). In this paper we have adopted the concept of sentinel events with a slightly different but analogous meaning as will be discussed later.

CRITERIA FOR THE SELECTION OF THE EVENTS

The following criteria were chosen for the selection of the indicators dealt with in this paper:

1. The variable severity of the biological outcomes, ranging from symptomless biochemical disturbances (e.g. body burden of pollutants) to explicit ilnesses and death.

2. A well established epidemiological relationship between the health outcome selected and a discrete environmental risk.

3. The non-occupational nature of the risks involved. Occupational risks were included only if the activities generating the risks also result in an important impact for the general environment such as widespread chemical contamination.

4. Health effects largely independent of risks conditioned by life styles and by voluntary exposures.

5. The practicability for obtaining the health data in terms of recognition and notification.

6. The linkage of the sentinel health events to anthropic modifications of the environment.

7. Events such as infections acquired from undisturbed foci in natural ecosystems were excluded.

8. Only the immediate biological, chemical and physical factors that affect health were considered as environmental hazards and the underlying social and economic conditions were not included.

9. Representativeness of the environmental situations, meaning that the widest possible range of ecosystems (urban, rural, wilderness) as well as of the nature of the exposure factors (physical, chemical, biological) were chosen.

10. The global character of the environmental modifications related to the health indicators, that is, environmental factors that, due to their widespread occurrence, can affect most of the population.

DESCRIPTION OF THE SELECTED SENTINEL HEALTH EVENTS

A brief description of the ten sentinel events used in this paper and the justification for their selection follows:

Mortality due to traffic accidents

It is estimated that traffic accidents are responsible for about 700.000 deaths worldwide each year (NAKAJIMA, 1994) and the Brazilian figure is about 28.000 deaths a year, which meant a mortality rate of 20/100,000 in 1987 (MELLO JORGE & LATORRE, 1994). A great part of the victims reported are pedestrians (two-thirds of all traffic-related deaths in one metropitan area studied by KLEIN, 1994), especially adolescents and young adults (TAKET, 1986; GRAHAM, 1993) Traffic accidents and deaths have been associated with traffic volume (ROBERTS et al.,1992; ROBERTS & CROMBIE, 1995), with vehicle speed

(MUELLER et al., 1990; ROCK, 1995) and, in general terms, with the motorization index of the countries, that is, to the number of vehicles relative to the population, which is rapidly increasing in many Latin American countries (GRAHAM, loc.cit; BANGDIWALA & ANZOLA-PEREZ,1987; SÖDERLUND & ZWI, 1995). Although behavioral factors strongly influence the occurrence of these accidents, the end result of risky driving, whether it is alcohol-induced or not, is the generation of threats for other drivers, commuters, pedestrians and cyclists. Thus a high rate of traffic accidents, measured by the resulting number of deaths, is an indicator of a degraded and dangerous environment, mainly in highly urbanized areas but also on highways and roadsides. Even knowing that most accidents do not result in fatalities but rather in injuries of variable severity, the use of mortality rates is preferred due to the ease of obtaining them.

Increases in emergency room visits/hospital admissions due to asthma in urban residents

There are evidences that the global prevalence of asthma, especially in children, is increasing (FULWOOD, et al., 1900; WHITE, et al., 1994; KOREN, et al., 1995). Although asthma attacks are not triggered only by extrinsic factors such as air pollutants it is now well established that increases in atmospheric pollution levels are frequently correlated with an increased number of exacerbations of chronic respiratory diseases and the corresponding rates of morbidity (SUNYER, et al., 1991; BATES, et al., 1995; SCHWARTZ, 1996) and mortality (IMAI, et al., 1986; SCHWARTZ, 1991; SALDIVA, et al., 1995; GUIDOTTI, 1996). In regard specifically to asthma attacks several authors observed correlations between higher levels of air pollution and higher frequencies of asthma episodes as measured by the number of visits to emergency rooms. PÖNKÄ (1991) observed that hospital admissions due to the aggravation of asthma in Helsinki were significantly correlated with ambient air concentrations of several pollutants and SCHWARTZ, et al. (1993), reported an increase in emergency room visits in Seattle, USA, to be correlated to the levels of respirable particulate matter. Significant correlations were also found between ozone (WHITE, et al, loc.cit.), nitrogen oxides (KEIDING & RINDEL, 1995) and sulphur dioxide levels (XU, et al., 1995; ROMIEU, et al., 1995) and increased number of contacts with the health services due to asthma attacks.

Due to the high urbanization rate in most countries, especially the developing ones (up to 80% of the population) and therefore the increasing number of big cities (eleven with more than one millon inhabitants in Brazil), a great number of people are being exposed to these air pollutants. In Latin America it was estimated that 81 million people (26.5% of the total urban population) are exposed to pollutant levels that exceed the WHO guidelines for adequate health protection (ROMIEU, et al., 1991) and that the total suspended particles alone were responsible for an annual excess morbidity of three million cases of chronic cough in children (WEITZENFELD, 1992). Thus this change in the chemical composition of the troposphere - due mainly to the burning of fossil fuels from vehicles - because of its widespread distribution and the size of the population affected, and also the fact that air quality is susceptible to the processes of climate oscillations, can be considered a global environmental problem.

Acute non intentional pesticide poisonings in rural areas

Since the Second World War there has been a rapid increase in the global production and use of pesticides in agriculture (WHO, 1990) and by 1990 some 2.5 million tons of pesticides were used worldwide (MILLER, 1991). Since 1945 some 15,000 individual compounds and more than 35,000 formulations have come into use as pesticides (DAVIES, 1987). Brazil is the 5th largest consumer of these chemicals in the world, amounting to 300,000 tons annually, and the national demand is increasing about 6.5 % each year (SCHWENK, 1996).

Global estimates of acute poisoning due to agricultural pesticides are inexact and have varied from one to two million cases per year, with fatalities being estimated from 20,000 to 40,000 (WHO, loc.cit.; FORGET, 1991; LEVINE & DOULL, 1992), the cases being reported both in developed (LABONTE, 1989; MEHLER et al., 1992) and developing countries (JEYARATNAM, 1990; IGBEDIOH, 1991). In Brazil most pesticide poisonings among farmworkers go unreported and official figures for the year 1993 were 2446 cases with 25 deaths (SINITOX, 1993). However a gross undereporting can be inferred taking the year 1985 as an example: the federal surveillance system reported 1749 cases all over the country but the surveillance sector of one State alone reported 1119 cases with 106 fatalities (BORTOLLETTO, 1995).

Although most of the cases of accidental intoxications by pesticides in Brazil occur in occupational settings (farmworkers and their families), usually due to the lack of protective clothing as well as the mishandling of the products, equipment and recipients (13,150 tons of empty recipients are thrown away in fields and rivers after each harvest - SCHWENK, loc. cit.) these cases can also be taken as a measure of the contamination of the environment. It is known that many of these products have a long persistence in the environment and drift and run-off processes are responsible for their dispersion for long distances (PLIMMER, 1990; OEHME, 1991; IGANIWURA, 1991; HARGRAVE, et al., 1992).

Although a chronic exposure of a much greater part of the population probably occurs as a result of the ingestion of pesticide residues in contaminated food, the resulting morbidity and mortality is difficult to estimate due to the long induction period of their health effects.

Thus the use of the rate of non intentional poisoning in agricultural settings due to its much easier identification seems to be a better indicator of the environmental contamination by these substances.

The incidence of skin cancer

There is a well established relationship between exposure to sunlight UV-B radiation and the development of skin cancer, especially of the non-melanocytic types (BENTHAM, 1993; ARMSTRONG, 1994), while this association is less certain regarding melanoma (LONGSTRETH, et al., 1995). The depletion of the stratospheric ozone layer and the resultant greater incident UV-B radiation at ground level is one of the better known global environmental changes with direct impacts on the health of human populations (MOAN & DAHLBACK, 1992; LLOYD, 1993; LAST, 1993). Reports in the last decade have shown an increasing incidence of all types of skin cancer in many parts of the world (GLASS & HOOVER, 1989; STENBECK, et al., 1990; BURTON, et al., 1993; KRICKER, et al., 1993; KO, et al, 1994) and one estimate indicated that a depletion of 1% in the ozone layer would result in a 2% increase in the incidence of non-melanoma skin cancer (LONGSTRETH, et al., loc cit.).

The specificity of the association UV-B/skin cancer makes this one of the most useful health indicators for the monitoring of the global environmental changes. Some authors have argued that public awareness about health effects of ozone depletion would encourage people to adopt behavioral changes and thus reduce exposure of body sites to increasing ultraviolet levels (DIFFEY, 1992).Although it seems to be plausible in the developed countries it would be much more difficult in tropical countries like Brazil, where exposure to sunlight for tanning is highly valued culturally.

Despite some difficulties for the adequate reporting of the cases of non-melanocytic skin cancer to surveillance systems due to its benign nature

(KRICKER, et al., loc. cit.), its high prevalence (about 25% of all cancers in Brazil) and easy recognition make it a useful environmental health indicator.

Body burden of mercury in inhabitants of areas of gold mining activities in freshwater sites

Recent reports indicate that the global concentrations of mercury are increasing in the atmosphere (SLEMR & LANGR, 1992), hydrosphere and soils (NRIAGU, 1990) and as a consequence human exposure is also increasing. There are several sources of mercury both natural (geologic deposits) and man-made such as industrial dumping into waterways and the atmosphere (in the latter case solid waste incineration and coal combustion account for 75% of the mercury entering the atmosphere - SLEMR & LANGER, loc.cit.). Projections were made that humans are likely to double their mercury releases during the next 50 years (WATSON Jr., 1979). Another important human source is the elemental mercury added to the gold ore collected by informal mining operations in many rivers and waterways in developing countries (JERNELOV & RAMEL, 1994). In Brazil this occurs especially in the Amazon region (NRIAGU, 1992). Estimates of mercury lost to the environment in the Amazon range from 100 to 168 tons annually (FERREIRA & APPEL, 1990; VEIGA & MEECH, 1995), which means about 80% of all non-recovered mercury used in Brazil. Reports from surveys in Brazil have shown high mercury concentrations both in abiotic components of the environment (soil, river water and sediments) and in animals, especially freshwater fish species in the Amazon (OLIVEIRA, et al., 1990; MALM, et al., 1990; PALHETA & TAYLOR, 1995) as well as in other important aquatic ecosystems (PFEIFFER, et al., 1990; HYLANDER, et al., 1994). Methylmercury is much more toxic than elemental mercury and methylation occurs when the element reaches freshwater bodies; bioaccumulation follows in the aquatic food chain, especially in muscle tissues of high trophic level fishes (FITZGERALD & CLARKSON, 1991). In terms of intoxication levels, children were found to be 5-10 times more sensitive to mercury than adults (CLARKSON, 1990). Recent surveys have pointed out that humans exposed to mercury-contaminated environments in Brazil have high burdens of this metal in their bodies (OLIVEIRA, et al., loc.cit.; MALM, et al., loc. cit.; HOMEWOOD, 1991; BRANCHES, et al., 1993; AKS, et al., 1995). Although in most investigations in the Amazon the many individuals shown to have important body burdens of mercury were gold miners occupationally exposed due to the inhalation of vapors from the burned amalgam, the best indicator for environmental contamination is the methylmercury body burden in the general population from these regions which become contaminated by the ingestion of freshwater fish. Despite the large area of the Amazon (about 4 million square kilometers) and the global importance of its waterways (it was estimated that informal gold mining in Brazil accounts for 2.5 to 11.0 % of the total world contamination by mercury - PFEIFFER et al., loc. cit.), we must keep in mind that this is just a special case of global mercury pollution.

Body burden of lead in children

It is known that due to their ever-increasing production and use, the contamination of the environment with toxic metals has become a global phenomenon. Human-induced changes have overwhelmed the natural biogeochemical cycle in many ecosystems. Anthropogenic emissions of lead exceed the fluxes from natural sources by 28-fold (NRIAGU, 1990). It was estimated that over the recorded production history of lead, more than 300 million tons of this metal have been discharged to the environment as a contaminant (SMITH & FLEGAL, 1995); world lead consumption has steadily increased over the period 1965-1990 and was about 5.6 x 10 tons in 1990 (IPCS, 1995).Lead is considered as a uniquely persistent and ubiquitous pollutant, and airborne lead can be deposited in soil and water thus reaching humans through the food chain and in drinking-water; atmospheric lead is also a major source of lead in household dust (IPCS, loc. cit.). Human exposure comes from many environmental sources such as air, dust, food and water (AMODIO-COCCHIERI, et al.1996) and the major point sources identified were industrial lead emissions; exhaust from engines burning gasoline; peeling lead paints; lead-soldered cans; food and drinking water; soil and lead-glazed earthenware (ARNETZ & NICOLICH, 1990). Significant associations between ambient air lead measurements and blood lead concentrations have been found and hematological, neurobehavioral, cardiovascular and renal changes have been attributed to lead at rather low levels (ARNETZ & NICOLICH, loc. cit.). Children are recognized as particularly susceptible to lead's toxic effects and although clinically observed lead poisoning is rarely reported, many children have blood lead levels of clinical and public health concern (LEVALLOIS & RHAINDS, 1992); impaired neuropsychological development and classroom performance is the most important effect. Lead-based paint, found mainly on houses built before 1980, seems to be the most common source of high-dose lead poisoning among children (STAES et al., 1994) but other sources such as food from soldered cans (PIRKLE, et al., 1995) and lead-contaminated soil (WEITZMAN, et al., 1993) have also been incriminated. In Brazil a few studies have identified excessive body burdens of lead mostly in children from districts highly affected by industrial emissions such as those from smelter plants (SANTOS FILHO et al.,1993a, b; SILVANY-NETO, et al, 1995). CROCETTI, et al;. (1990) estimated that for the year 1984 the total number of metropolitan children exposed in the USA above the criterion level to be about 6.4 million and those living in poor inner city and old dwellings were at greater risk. Although in some countries the regulation of some lead-containing products (gasoline additives; solder; paint) has reduced some emissions of this metal to the environment, with the consequent decreases in blood lead levels and cognitive improvements in moderately lead-poisoned children (RUFF, et al., 1993) recent data indicate that current regulations of lead uses and emissions may be inadequate to alleviate the environmental lead problem (SMITH & FLEGAL, loc. cit.). Although comprehensive surveys on Brazilian children not especially exposed to industrial emissions were not performed, high lead body burdens are nevertheless expected to be present due to the living conditions of most of the infants in the country and also due to the high national production and use of lead in the past and presently.

Noise-induced hearing loss before the age of 18 years

It is generally recognized that noise pollution is becoming an increasingly serious environmental problem (CONSENSUS CONFERENCE, 1990) and, as a consequence, acoustic overstimulation an important public health problem. It is well acknowledged that exposure to continuous or intermittent noise of sufficient level and duration can produce permanent auditory damage in the form of sensorineural hearing loss. This seems to occur due to the degeneration of cells in the peripheral cochlear organs and also in the neuronal central auditory pathway (CONS. CONF., loc. cit.; BROOKHOUSER, 1994). Although occupational noise remains the most commonly identified cause of noise-induced hearing loss (BROOKHOUSER, 1992), the proliferation of potentially harmful noise sources in the general environment, at home, in recreational environments and on farms has placed increasing numbers of unsuspecting people at risk of developing NIHL (BROOKHOUSER, 1994, loc. cit.). Thus non-occupational NIHL can also be an important problem since individuals in the general population, including young children, are exposed to many sources of noise such as traffic, jet engines, subways, farm machinery, rock concerts, headphones, discos, firecrackers, domestic appliances, toys and hospital incubators (MITCHELL, 1984; BROSTE, et al., 1989; BABISCH & ISING, 1989; GUPTA & VISWAKARMA, 1989; CONS. CONFER., loc. cit; CLARK, 1991; SMOORENBURG, 1993; PLATH, 1994). It has been noted that high tone hearing loss in teenagers has become evident in recent years (AXELSSON, et al., 1981) and reports based on investigations of personal histories and audiometric configurations detected high prevalences of NIHL in adolescents and children, either from multiple noise sources (WOODFORD & O'FARREL, 1983; COSTA, et al., 1988; BROOKHOUSER, 1992, loc. cit.; LAVERY, et al., 1993) or related to specific sources such as aircraft noise (GREEN, et al., 1982; CHEN & CHEN, 1993).

Hearing disability prevalence estimates in most countries is about 10% of the population (PLATH, loc.cit.) and estimates from the National Institutes of Health (USA) indicated that more than 20 million Americans are exposed on a regular basis to industrial or recreational noise that could result in hearing loss. Incidence figures also show that an estimated 15 of every 1000 persons under age 18 have some type of hearing impairment (LANG, 1994). It is also known that hearing loss affects approximately 28 million people in the USA and approximately 10 million of these impairments are at least partially attributable to damage from exposure to loud sounds (CONS. CONFER., loc cit.).

The use of NIHL in youngsters as an indicator of environmental degradation is useful since it can avoid confounding with occupational NIHL and with age-related changes in hearing. Furthermore it has been experimentally demostrated that young animals suffer more damage to the inner ear than adult animals exposed to identical noise levels (FALK, et al., 1974; PRICE, 1976). It has also been suggested that children may be hypersusceptible to the effects of noise and that exposures deemed safe for adults may adversely affect the hearing of young children (MILLS, 1975; BROOKHOUSER, 1994, loc. cit.). Although the younger population, especially adolescents, can voluntarily expose themselves to damaging sounds through personal equipment and in enclosed spaces (e.g. music) which do not characterize noise as an environmental problem, this should be part of the issue since they are also likely to be exposed to other general noise sources. Another factor to be considered is that although noise is still the most common exogenous cause of hearing loss, recent investigations pointed to the existence of a potentiation between noise and other environmental factors such as volatile organic solvent exposures in the development of NIHL (MORATA, 1990; RYBAK, 1992). The use of this indicator seems therefore appropriate due to the present widespread sources of acoustical overexposure in a mechanized society heavily dependent on noise producing devices.

Mortality due to diarrhea before the age of 5 years

The association of childhood diarrheas with water contaminated by intestinal pathogens in developing countries is well established. These pathogens usually reach the water from improperly disposed human feces due to the absence of disposal facilities and to the non-availability of piped water in low income communities. Several authors have demonstrated an association between the improvement of sanitary control measures and the reduction in the incidence and mortality due to childhood diarrhea (ESREY & HABICHT, 1986; YOUNG & BRISCOE, 1987; VICTORA, et al., 1988; GROSS, et al., 1989; MERTENS, et al., 1992; MANUN'EBO, et al., 1994; VANDERSLICE, et al., 1994; VANDERSLICE & BRISCOE, 1995). The lack of basic sanitation infrastructure creates environmental contamination problems of great magnitude, especially in urban areas where population densities are higher but can also be an important problem in rural areas. It is estimated that about 75 million people in Brazil are at a special risk of infectious diarrhea for not having access to safe water supplies or sewage systems in their houses (BRAZIL, 1995). Basic environmental sanitation (water/sewage) is considered a high priority area for investment in the main international documents dealing with environmental problems in developing countries (WORLD BANK, 1993; UNITED NATIONS, 1992; SOUZA & VIANNA, 1993).Recent mortality estimates for diarrhea under 5 years were around 3.3 million cases worldwide for 1992 (BERN, et al., 1992). Diarrheal diseases are still an important cause of morbidity and mortality in Brazil. The last official national mortality statistics published (for the year 1989) indicated that intestinal infectious diseases were responsible for 11,85 % of all deaths up to four years of age (BRAZIL, 1993). However the actual incidence rates for diarrhea morbidity cannot be known since its reporting is not compulsory and most episodes can be controlled with appropriate support measures. Since most fatal outcomes from diarrhea are probably adequately reported in death certificates, their incidence rate can be estimated and used as an important environmental health indicator.

Outbreaks of non-occupational leptospirosis following rainfall in urban areas

In urban as well as in rural flood-prone areas of different countries, outbreaks of leptospirosis following heavy rainfall episodes and flooding have been reported (DAMUDE & JONES, 1979; ANDRADE & BRANDÃO, 1987; PARK, et al., 1989;VENKATARAMAN, et al., 1991; MORSHED, et al.,1994; EPSTEIN, et al., 1995). The main reservoir of the **Leptospira** species involved in these outbreaks in urban settings is the brown rat **(Rattus norvegicus)**, an ubiquitous species which colonizes underground habitats (sewage and water collection systems) in cities all over the world. The populations of this synantropic animal are larger in areas where basic sanitation measures such as regular garbage collection and appropriate disposal is not common, as often happens in poor neighborhoods. These reservoirs contaminate the surroundings by shedding the leptospires in urine and human infection occurs after the direct contact with contaminated soil or water. Chances are greatly enhanced when water accumulated in floodings invade precariously built habitations and their surroundings. The resulting outbreaks of the disease indicate that there is an exposed population at greater risk due to living in inadequate environments, especially regarding poor drainage of contaminated water. In the past decades several outbreaks of this infection were reported during the summertime rainy season in major cities in Brazil (OLIVEIRA, et al., 1977; CALDAS, et al., 1979; VERAS, et al., 1985; SOUZA, 1986; COSTA, 1988; ANDRADE & BRANDÃO, loc.cit.). We can therefore consider the seasonal outbreaks of this transmissible disease an indicator of an inadquate man-made environment as shown by uncontrolled urbanization and deficient sanitation of human settlements, resulting in a high contamination of the environment by a biological agent of disease. Furthermore since these outbreaks result from the interaction of a certain type of urban environment and the natural phenomenon of heavy rainfall, if it is proven that an increasing number of storms is now occurring as a result of greater instability in climate (HAINES, 1993), these epidemics could also be considered as part of the health consequences of global climate changes.

Outbreaks of vampire bat bites in humans in natural ecosystems

In the past decades several reports have been published of outbreaks of vampire bat **(Desmodus rotundus)** attacks on humans in Latin American countries, both with and without rabies transmission (NEHAUL, 1965; VERLINDE, 1975; THOMAS & HARAN, 1981; SCHNEIDER & SANTOS-BURGOA, 1995). Data recently compiled showed that for the year 1993 these hematophagous animals were the second most frequent species transmitting rabies to humans in Brazil (25 % of all cases), second only to dogs (SCHNEIDER & SANTOS-BURGOA, loc. cit.); this was also true in Mexico (ANONYMOUS, 1991). There are generally two different ecological situations where bat attacks to humans have been reported: in rural areas and in natural ecosystems; in the latter case, in Brazil, mostly in the Amazon region. In the majority of these cases the underlying motive for the attacks seemed to be the same: bat populations deprived of their abundant and readily obtained primary food sources (animals) sought for alternative hosts (humans) to feed. In the case of rural areas it has happened when animals such as cattle or pigs were rapidly eliminated from an area (McCARTHY, 1989; SCHNEIDER, 1991; LOPEZ, 1992; BATISTA-DA-COSTA, 1993; SCHNEIDER, 1995). In the case of the wild vertebrates they are reduced when humans establish permanent or temporary settlements in remote areas (ALMANSA & GARCIA, 1980; SCHNEIDER, 1991; LOPEZ, 1991). A commonly observed situation when humans were massively attacked in the Amazon are the gold mining camps. The animal species which serve as food for the bats are drastically reduced due to overhunting or are chased away by deforestation or noise produced by machinery and airplanes (CONFALONIERI, 1990, personal observation; LOPEZ, 1991; UIEDA & COELHO, 1991; SCHNEIDER & SANTOS-BURGOA, loc. cit.; COELHO, 1995).

Vampire bat attacks on humans can therefore, in many cases, be taken as a measure of man-induced environmental modifications represented by the elimination of native species of animals in a given area of the natural ecosystems. Certainly it means only a small biodiversity loss in terms of the affected ecosystems as a whole, but these same activities that cause faunal depletion usually interfere with several other components of the same environment, which can result in the loss of habitats. Furthermore it provides us with one of the few direct associations between a biodiversity depletion factor and a human health event as represented by the bat attacks and their consequences.

DISCUSSION

Global environmental changes have been defined as those which "...alter the envelopes of the Earth System and hence are experienced globally...and those that occur at discrete sites but are so widespread as to constitute a global change" (VITOUSEK, 1992); or "...occur at a global level and affect the total global system and those that occur at local or regional levels but have consequences for the total global system"(UK/GER, 1993). Other concepts were modifications "... that may alter the capacity of the Earth to sustain life" (USA, 1990) and the changes which result "..when large global and regional systems are altered at a rate that is much faster than those systems are accustomed to experiencing" (CGCP, 1993). Within these broad definitions we can include phenomena such as global warming, the thinning of the ozone layer, changes in atmospheric chemistry, loss of biodiversity, land degradation, chemical pollution, groundwater depletion and urbanization (CPHA, 1992; CGPP, 1993; LAST, 1993; PHILLIPS, 1993; USA, 1990; 1996).

All these biospheric changes affect human health but some do so typically in an indirect way (e.g. biodiversity loss) and others are hazards that are either not tangible (e.g. depletion of the ozone layer) or have delayed health effects (e.g.global warming).This means some methodological difficulties for the estimation of their health impacts (McMICHAEL, 1992; 1993; HAINES, et al., 1993; McMICHAEL & MARTENS, 1995) and, accordingly, a limited range of choices to select from a group of human disease parameters for the assesment of global environmental quality. Although the parameters discussed in this paper could be applied to several countries (except the vampire bat bites), especially those in Latin America, they were chosen to address not only aspects of the disruption of the global ecosystem but also the diverse and complex social/environmental and epidemiological situations of Brazil. This is characterized by a wide range of ecosystem types, from vast extensions of poorly disturbed natural ecosystems (e.g. the Amazonian tropical rain forest) to the largest conurbations of the world (18 million inhabitants) and by a high incidence and impact of both the degenerative diseases linked to modern industrial life styles and of those transmissible diseases associated to poverty. Environmental destruction in the country has been the result of either large scale industrial activities utilizing advanced technology or by predatory subsistence extractivism practiced by large segments of the socially excluded population.

The characteristics of the health events proposed here (Table 1) cover a broad area in regard to the nature of the hazards, to the environmental settings considered as well as the biological effects on humans. In regard to the practicability for their use, some are already routinely reported to the national epidemiological surveillance system such as the pesticide poisonings, diarrhea, leptospirosis and traffic deaths. Others are less well reported (skin cancer, asthma and bat bites) and the enhancement of their reporting capacity through new approaches is needed. Body burden of lead and mercury and NIHL are not available from ordinary health data due to some technical difficulties for their measurement and special surveys are needed for obtaining them. These events should be considered as environmental health indicators according to the definition "summarized measures of a hazardous exposure or health outcome" which should have the attributes of representing an underlying causal relationship between an environmental exposure and a health consequence and also must be a reasonably valid estimate of the underlying causal factor (PASTIDES, 1995). These can be constructed by linking aggregate data (linkage-based)or by identifying environmental indicators with a health linkage (exposure-based) or health indicators with an environmental linkage (outcome-based) (KJELLSTROM & CORVALAN, loc. cit). In proposing this framework we have prioritized, without neglecting the strength of the epidemiological associations, the widest possible representative coverage of the population and territory, as well as its significance in the overall national picture in terms of global environmental exposures and also some of the current environmental health policy issues we think should be regarded as priorities.

There are special complexities in the identification of environmental health indicators since the incidence of many environmentally related diseases cannot be easily traced back to specific environmental exposures (KJELLSTROM & CORVALAN, loc.cit.). Indicators that are linkage-based, that is those obtained from linkages of health and environmental monitoring data, should be the best since they consider parameters from both the human population and the causal environmental factors to which they are exposed. However they are rarely available, especially in developing countries, due to the costs involved in keeping specific surveillance systems for that purpose; therefore, simpler and more cost-effective strategies must be sought. The events discussed in this paper can be regarded as "health outcome-based" in the sense that they occur as biological responses to varying exposure levels and, consequently, can show variable rates of occurrence and/or degrees of severity.

RUTSTEIN (1983; 1987) emphasized sentinel health events as individual cases with specific etiologies, the singularity of their occurrence, the ability for prevention and also their utility for the assessment of the quality of medical services. In the present framework proposal, dealing mostly with broad environmental exposures, we have included mainly population-based indicators, instead of single occurrences. However these indicators retain the preventability aspect and the main objective of their use should be, besides assessing the quality of the environment, the avoidance of unnecessary disease, disability and premature death consequent to environmental exposures. However, since they are not rare events, their warning sign character is given by occurrence rates higher than previously defined background levels. Another aspect to be considered in this set of events is the fact that several of them (NIHL, traffic deaths, lead burden, diarrhea, asthma) were chosen also for their special importance in children and adolescents. This is important not only because of the psychosocial aspects involved (premature deaths, potential years of economically productive life lost, increase in long-term costs of medical care, etc.) but also from the biological point of view, due to their frequently greater susceptibility to environmental determinants of disease and to the long induction period of the illnesses resulting from those exposures. This latter aspect makes prevention more possible, especially if cumulative exposures are involved, and can also give a general idea of exposure levels in older populations.

Concerning the utility of this strategy for the assessment of health services, this can be equally applied in this particular context but rather to the monitoring of the quality of environmental protection programs and services. In defining a set of environmentally related sentinel health events, authorities make the objectives of these activities explicit to the population, shifting from an approach based on environmental parameters per se to others based on the health and welfare of the human communities (PENA, 1990).

We believe that this proposal further expands the concept of sentinel health event strategy as applied to environmental issues, with the following innovative applications:

a. Its use as a specific monitoring technique of the health consequences of global changes, including some environmental degradation aspects with less direct links to health.

b. Its warning sign character stems mainly from the impacts of environmental exposures at young ages.

c. The promotion of a clearer definition of the objectives of environmental protection services as far as the health of the population is concerned.

It should also be stressed that, following the better known aspects of the use of sentinel health events for the evaluation of the quality of medical care (PENA, loc.cit.), one technical advantage of this strategy for the assessment of environmental quality is that data collection is selective when compared to a more comprehensive and detailed analysis of the situation of the environment as a whole and, therefore, operational costs are reduced.

It seems that this framework should be considered a cost-effective and epidemiologically reliable starting point for national environmental health monitoring. As future epidemiological and environmental studies show stronger links between other environmental variables and health outcomes, new events could be added to this set. Among these we could anticipate the effects of chemicals on birth defects and on reproductive problems due to hormone disruptions or excess mortality from extreme regional weather conditions when related to global climate changes. The following step would be the assignment, from the currently known baseline values, of criterion levels and incidence rates, scores for the range of epidemiological situations to be found in order to obtain a quantitative composite index with a more general application.

TABLE 1 - CHARACTERISTICS OF SENTINEL HEALTH EVENTS FOR GLOBAL ENVIRONMENTAL MONITORING

Health Event	Type of Hazard	Environmental Setting	G.E.C. Addressed	Epidemiological Association*	Surveillance**
Skin cancer	Physical	Atmosphere	Ozone Depletion	1	2
Asthma	Chemical	Urban	Atmospheric Pollution	2	2
Traffic Mortality	Physical	Urban	Urbanization	1	1
Body Lead	Chemical	Urban	Chemical Pollution	2	3
Body Mercury	Chemical	Rural/Wild	Chemical Pollution	2	3
Diarrhea	Biological	Urban/Rural	Water	1	1
N.I.H.L.	Physical	Urban	Urbanization	2	3
Pesticide Poisoning	Chemical	Rural	Chemical Pollution	1	1
Leptospirosis	Biological	Urban	Urbanization	1	1
Bat Bites	Biological	Wild	Biodiversity Loss	2	3

*

1 - very strong/specific;

2 - moderately strong/specific

* *

1.- routinely reported to surveillance systems;

2 - reporting existent but needs improvement.

3 - non-existent reporting system

BIBLIOGRAPHY

ANONYMOUS (1991). Reunion de Consulta sobre la Atencion a personas expuestas a la rabia transmitida por vampiros. *P.A.H.O.*, Washington, D.C.

AKS, S.E. et al.. (1995). Blood-mercury concentrations and Renal biomarkers in Amazonian Villagers. *Ambio*, v.24, p.103-104.

ALDRICH, T.E. (1991). Sentinel Event Strategies in Environmental Health. *Ann. Rev. Publ. Health.*, v.14, p.205-217.

ALMANSA, I. C. & GARCIA, R.C. (1980). Incidencia del murciélago hematófago Desmodus rotundus sobre los indígenas Yanomami de Venezuela. *Donana, Acta Vertebrata.*, v.7, p.1113-1117.

AMODIO-COCCHIERI, R. et al. (1996). Lead in human blood from children living in Campania, Italy. *J. Toxicol. Environ. Health.*, v.47, p.311-320.

ANDRADE, J. & BRANDÃO, A.P. (1987). Contribuição ao conhecimento da epidemiologia da leptospirose humana, com especial referência ao Grande Rio, Brasil, de 1970 a 1982. *Mem. Inst. Oswaldo Cruz*, v.82, p.91-100.

ARMSTRONG, B.K. (1994). Stratospheric Ozone and Health. Int. J. Epidemiol., v.23, p.873-885.

ARNETZ, B.B. & NICOLICH, M.J. (1990). Modelling of Environmental lead contributors to blood lead in humans. *Intern. Arch.. Occup.. Envir. Health.*, v.62, p.397-402.

AXELSSON, A. et al. (1981). Noisy leisure time activities in teenage boys. *Am.. Ind.. Hyg. Assoc. J.*, v.42, p.229.

BABISCH, W., & ISING, H. (1989). [The effect of music in discotheques on hearing ability.] Soz. *Praventivmed.*, v.34, p.239-242.

BANGDIWALA,S.I. & ANZOLA-PEREZ, E. (1987). Accidentes de Transito. Problema de salud en paises en desarrollo de las Americas. *Bol. Ofic.Sanit. Panam.*, v.103, p.130-139.

BATES, D.V. et al. (1990). Asthma attack periodicity : a study of hospital emergency visits in Vancouver. *Environ. Res.*, v.51, p.51-70.

BATISTA-DA-COSTA, M. et al. (1993). An outbreak of Vampire bat bite in a Brazilian Village. *Trop. Med. Parasitol.*, v.44, p.219-220.

BEARER, C.F. (1995). How are children different from adults? *Environ. Health., Perspect., v.103(sup.l 6): 7-12*

BENTHAM, G. (1993). Depletion of the Ozone Layer: Consequences for non-infectious human diseases. *Parasitol.*, v.106, p.539-546.

BORTOLETTO, M.E. (1993). "Tóxicos, Civilização e Saúde", FIOCRUZ, CICT/NECT, Serie Políticas de Saúde nº 12, Rio de Janeiro, 136 p.

BRANCHES, F.J.P. et al. (1993). The price of gold: Mercury exposure in the Amazonian rain forest. *J. Toxicol. Clin. Toxicol.*, v.31, p.295-306.

BRASIL.(1996). "Estatísticas de Mortalidade-Brasil, 1991". Ministério Saúde, Fundação Nacional de Saúde, Brasília, XII + 523 p..

BRASIL, Ministério da Saúde. (1995). "Plano Nacional de Saúde e Ambiente no Desenvolvimento Sustentavel. Diretrizes para Implementação". Brasília, DF. junho/1995, 104 p..

BROOKHOUSER, P.E. et al. (1992). Noise-induced hearing loss in children, *Laryngoscope*, v.102, p.645-655.

BROOKHOUSER, P.E.(1994). Prevention of noise-induced hearing loss. *Prev. Med.*, v.23, p.665-669.

BROSTE, H. S. et al. (1989). Hearing loss among high school farm students. *Am. J. Publ. Health.*, v.79, p.619-622.

BURTON, R.C. et al. (1993). An Analysis of a Melanoma Epidemic. *Int. J. Cancer.*, v.55, p.765-770.

CALDAS, E.M. (1979). Estudo epidemiológico do surto de Leptospirose ocorrido na cidade de Salvador, Bahia, em maio e junho de 1978. *Rev. Inst. Adolfo Lutz*, v. 39, p.85-94.

CANADIAN GLOBAL CHANGE PROGRAM. (1993). "Global Change and Canadians." ROYAL SOC. CANADA., Ottawa.

CHEN, T.J. & CHEN, S.S. (1993). Effects of aircraft noise on hearing and auditory pathway function of scholl-age children. *Int. Arch. Occup. Envir. Health.*, v.65, p.107-111.

CLARK, W.W. (1991). Noise exposure from leisure activities: a review *J. Acoust.. Soc. Amer.,* v. 90, p.175-181.

CLARKSON, T.W. (1991). Human Health Risks from Mercury and Monomethymercury in fish. *Environm. Toxicol. Chem.,* v.9, p.957-961.

CLEARY, D. (1996) Mercury contamination and health risk in the Brazilian Amazon.An Ethical Dilemma (Editorial) *Rev. Inst. Med. Trop. S. Paulo*, v.38, p.247-248.

COELHO, G.E. (1995). "Aspectos relacionados às agressões humanas por morcegos hematófagos potenciais transmissores da Raiva, no Estado de Roraima, Brasil." Faculdade. Cienc. Saúde, Univ. de Brasília, Brasília, DF, mimeo. 29 p.

CONFALONIERI, U. E. C. (1990). Personal observation.

CONFALONIERI,U.E.C., (1993).Transportation and Health in Brazil I. Premature deaths due to traffic accidents in the State of Rio de Janeiro, 1983-1987. *Bull. Intern. Soc. Doctors. Environm.*, v. 3, p.4-9.

CONSENSUS CONFERENCE (1990). Noise and Hearing Loss. *JAMA*, v.263, p.3185-3190.

CORDEIRO, R. et al., (1996a) Distúrbios neurológicos em trabalhadores com baixos níveis de chumbo no sangue I.: neuropatia periférica *Rev. Saúde Publ.,* v.30, p.248-255.

CORDEIRO, R. et al. (1996b) Distúrbios neurológicos em trabalhadores com baixos níveis de chumbo no sangue II. Disfunções neurocomportamentais *Rev. Saúde Publ.,* v.30, p.358-363.

COSTA, O.A. et al. (1988). Hearing loss at age 7, 10 and 13 - an audiometric follow up study. *Scand. Audiol. Sup.l.,* v.30, p.25-32.

COSTA, V.R. (1988). Vêm aí as águas de março. *Ciência Hoje*, v. 9, p.76-77.

CPHA. (1992). Human and Ecosystem Health. Canadian Perspectives, Canadian Action. *Canadian Publ. Health., Assoc.* Ottawa, 21 p.

CROCETTI, A.F., MUSHAK, P. & SCHWARTZ, J. (1990). Determination of numbers of lead-exposed US children by areas of the United States: an integrated summary of a report to the US Congress on childhood lead poisoning. *Environ. Health Perspect.,* v.89, p.109-120.

DAMUDE, D.F. & JONES, C. J. (1979). The Problem of Human Leptospirosis in Barbados. *Trans. Roy. Soc. Trop. Med. Hyg.*, v.73, p.169-177.

DAVIES, J.E. (1987). Changing Profile of Pesticide Poisoning. *N. Engl. J. Med.*, v. 316, p.807-808.

DIFFEY, B.L. (1992). Ozone depletion and skin cancer. *Brit. Med. J.,* v.304, p.1176-1177.

EPSTEIN, P.R. et al. (1995). Climate and Disease in Colombia. *Lancet.*, v.346, p.1243-1244.

ESREY, S.A & HABICHT, J.P. (1986). Epidemiologic Evidence for Health benefits from Improved water and sanitation in developing countries. *Epidem*. Rev., v.8, p.117-128.

FALK, S. et al. (1974). Noise-induced inner ear damage in newborn and adult guinea pigs. *Laryngoscope*, v. 84, p.444-453.

FERREIRA, R.C.H., & AP.EL. L.E. (1990). Fontes e usos de Mercúrio no Brasil. *Technical Contributions to Forest 90* (First Intern. Symp. on Environm. Stud. Tropical Rain Forest), Manaus, Brazil, October, 1990. 15 p.

FINKELMAN, J. (1996). Phasing out leaded gasoline will not end lead poisoning in Developing Countries. *Envir. Health., Perspect.,* v.104, p.10-11.

FITZGERALD. W.F. & CLARKSON, T.W. (1991). Mercury and Monomethylmercury: Present and Future Concerns. *Envir. Health., Perspect.,* v.96, p.159-166.

FORGET, G. (1991). Pesticides and the third world. *J. Toxicol. Environ. Health.,* v.32, p.11-31.

FULWOOD, R. et al. (1990). Asthma -United States, 1980-1987. *MMWR.,* v.39, p.493-497.

GALAL-GORCHEV, H. (1991). Global overview of dietary lead exposure. *Chem. Speciat. Bioavailab*. v.3, p.5-11.

GLASS, A. & HOOVER, R.N. (1989). The Emerging Epidemic of melanoma and squamous cell skin cancer. *JAMA*. v.262, p.2097-2100.

GOLDMAN, L.R. (1995) Children - Unique and Vulnerable. Environmental risks facing children and recommendations for response. *Environ Health., Perspect.,* v.103, p.13-18.(sup.l 6)

GRAHAM, J.D. (1993). Injuries from Traffic Crashes: Meeting the Challenge. *Ann. Rev. Publ. Health.,* v.14, p.515- 543.

GREEN, K.B. et al. (1982). Effects of Aircraft noise on Hearing ability of school age children. *Arch. Envir. Health.,* 37, 284-289.

GROSS, R. et al. (1989). The Impact of Improvement of water sup.ly and sanitation facilities on diarrhea and intestinal parasites: a Brazilian experience with children in two low-income urban commities. *Rev. S. Publ.* 23, 214-220.

GUIDOTTI, T.L. (1996). Ambient air quality and human Health: Current concepts. Part 2. *Can. Respir.* J. 3, 29-39.

GUPTA, D. & VISHWAKARMA, S.K. (1989). Toy weapons and firecrackers: a source of hearing loss. *Laryncoscope*. 99, 330-334.

HAINES, A. (1993).chpter 6. The Possible effects of climate change on Health, *In:* Critical Condition. Human Health and the Environment (E. Chivian et al., Eds), p.. 151-170, MIT Press.

HAINES, A. et al. (1993). Global health watch: monitoring impacts of environmental change. *Lancet*. 342, 1464-1469.

HARGRAVE, B.T. et al. (1992). Organochlorine pesticides and polychlorinated bifenyls in the Arctic ocean food web. *Arch. Environ. Contam. Toxicol.* 22, 41-54.

HOLLANDER, A.E.M. et al., 1999. Na Aggregate public Health Indicator to Represent the Impact of Multiple Environmental Exposures. *Epidemiology,* 10(5):606-617

HOMEWWOD, B. (1991). Mercury Poisoning cnfirmed among Amazon Villagers. *New. Sci.* 9, 18.

HYLANDER, L.D. et al. (1994). Mercury levels in Alto Pantanal: A Screening Study. *Ambio*. 23, 478-484.

IGANIWURA, T.T. (1991). Health and Environmental Hazards of Pesticides. *Rev. Environ. Health.* 91, 47-52.

IGBEDIOH, S. (1991). Effects of Agricultural Pesticides on Humans, Animals and higher Plants in Developing Countries. *Arch. Envir. Health.*, 46, 218-224.

IMAI, M. et al. (1986). Mortality from Asthma and chronic bronchitis associated with changes in sulphur oxides air pollution. *Arch. Environ. Health.*, v.41, p.29-35.

IPCS (1995). Inorganic Lead. *Environ. Health Criteria,* Geneva, n.165.

JERNELÖV, A & RAMEL, C. (1994). Mercury in the Environment. *Ambio.*, v.23, n.166.

JEYARATNAM, J.(1990). Acute Pesticide Poisoning: A Major Global Health Problem. *Wld. Health., Stat.* Q. v.43, p.139-144.

JONES, R.R. (1987). Ozone depletion and skin cancer. *Lancet II,* p.443-446.

KEIDING, L.M. & RINDEL, A.K. (1995). Respiratory Illnesses in children and air pollution in Copenhagen. *Arch. Environ. Health.,* v.50, p.200-206.

KJELLSTRÖM, T. & CORVALÁN, C. (1995). Framework for the Development of Environmental Health Indicators. *Wld. Health., Stat. Quart.,* v.48, p.144-154.

KLEIN, C.H. (1994). Mortes no trânsito no Rio de Janeiro, Brasil. *Cad. Saúde Publ.,* Rio de Janeiro, v.10, p.168-176. (Sup.l 1)

KO, C.B. et al. (1994). The Emerging Epidemic of Skin Cancer. *Brit. J. Dermatol.,* v.130, p.269-272.

KOREN, H. (1995). Associations betweeen criteria air pollutants and asthma. *Environ. Health Perspect.*, v.103), 235-242. (sup.l. 6)

KRICKER, A. et al. (1993). Health, Solar UV Radiation and Environmental change. *IARC Technical Report,* Lyon, n.13, 151p. Apendix.

LABONTE, R. N. (1989). Pesticides and Health Public Policy. *Can. J. Publ. Health.,* v.80, p.238-242.

LANG, L. (1994). Environmental Impact on Hearing. Is Anyone Listening? *Environ. Health., Perspect.* v.102, p.924.

LAST, J.M. (1993). Global Change: Ozone Depletion, Greenhouse Warming and Public Health. *Ann. Rev. Publ. Health.*, v.14, p.115-136.

LAVERY, B. et al. (1993). Adolescent risk-taking: An analysis of problems beaviours in problem children. *Exp. Child. Psychol.*, v.55, p.277-294.

LEVALLOIS, P. & RHAINDS, M. (1992). L'intoxicacion au plomb d'origine environnementale: un probleme de santé publique persistant. *Bull. Inform. Santé Environm.*, v.3, p.1-3.

LEVINE, R.S. & DOULL, J. (1992). Global estimates of acute pesticide morbidity and mortality. *Rev. Environ. Contam. Toxicol.*, v.129, p.29-50.

LOGANATHAN, B.G. & KANNAN, K. (1994) Global Organochlorine contamination trends: an overview *Ambio*, v.23, p.187-191.

LLOYD, S.A. (1993). Stratospheric Ozone Depletion. *Lancet*, v. 342, p.1156-1158.

LONGSTRETH, J. D. et al. (1995). Effects of Increased Solar Ultraviolet Radiation on Human Health. *Ambio.*, v.24, p.153-165.

LOPEZ, A. (1991). "Presentacion del Brote epidemico de Choque. Madre de Dios (Peru)."*In:* Reunion Consulta Atencion.Personas Expuestas Rabia Transmit.. Vampiros. OPS, Washington, v.1-5, Abril, 1991. 9 p.

LOPEZ, R. et al. (1992). Outbreak of human rabies in the Peruvian jungle. *Lancet.*, v. 339, p.408-412.

MAIZLISH, N. et al., (1995). Surveillance of Work-Related Pesticide Illness: An Ap.lication of the Sentinel Event Notification System for Occupational Risks (SENSOR). *Am. J. Publ. Health.*, v.85, p.806-811.

MALM, O. et. al. (1995). Mercury Pollution due to Gold mining in the Madeira River Basin., Brazil. *Ambio.*, v.19, p.11-15.

MANUN'EBO, M.N. et al. (1994). Influence of demographic, socioeconomic and environmental variables on childhood diarrhoea in a rural area of Zaire. *J. Trop. Med. Hyg.*, v.97, p.31-38.

McCARTHY, T.J. (1989). Human depredation by vampire bats (**Desmodus rotundus**) following a hog cholera campaign. *Am. J. Trop. Med. Hyg.*, v.40, p.320-322.

McMICHAEL,A.J. (1993). Global Environmental change and Human Population Health: A conceptual and Scientific challenge for Epidemiology. *Int. J. Epidemiol.*, v.22, p.1-8.

McMICHAEL, A.J. & MARTENS, W.J.M. (1995). The Health Impacts of global climate change: grop.ling with scenarios, predictive models and multiple uncertainties. *Ecosyst. Health.*, v.1, p.23-33.

MEHLER, L. et al. (1992). Acute Pesticide Morbidity and Mortality: California. *Rev. Environ. Contam. Toxicol.*, v.129, p.51-66.

MELLO JORGE, M.H.P.; CATORRE, M.R.D.O. (1994). *Acidentes de trânsito no Rio de Janeiro*, Brasil: Dados e Tendências.

MERTENS, T.E. et al. (1992). Excreta disposal behaviour' our and latrine ownership in relation to the risk of childhood diarroea in Sri Lanka. *Int. J. Epidemiol.*, v.21, p.1157-1164.

MILLER. Jr, G.T.(1991). " Environmental Science." 3rd Ed., Wadsworth, Belmont.

MILLS, J. (1975). Noise and children: a review of the literature. *J. Acoust. Soc. Amer.*, v.58, p.767-779.

MITCHELL, S.A.(1984). Noise pollution in the neonatal intensive care nursery, p. 17-25 *In:* Early Identification of Hearing Loss in Infants. Seminars in Hearing no 5, (T.Mahoney, Ed), Thieme Stratton, Inc.

MOAN, J & DAHLBACK, A. (1992). The relationship between skin cancers, solar radiation and ozone depletion. *Brit. J. Cancer.*, v.65, p.916-921.

MOGHISSI, A.A. (1994). Life Expectancy as a measure of effectiveness of Environmental Protection. *Environm. Internat.*, v.20, p.691-692.

MORATA, T. C. (1990). *An Epidemiologic Study of the Effects of exposure to Noise and Organic Selvents on worker's Hearing and Balance.* Dissertation (Ph. D.) - Univ. of Cincinnati.

MORIOKA, I. et al, (1996). Hearing Impairment among young Chinese in a Rural Area. *Publ. Health,* v.110, p.293-297.

MORSHED, M.G. et al. (1994). Seroprevalence of Leptospirosis in a rural flood prone district in Bangladesh. *Epidem. Infect.*, v.112, p.527-531.

MOURE, V.M.. et al. (1995). Vehicle speed and risk of a severe crash. *Epidemiol.*, v.69, p.258-262.

MUELLER, B.A. et al. (1990). Environmental factors and the risk of childhood pedestrian motor vehicle collision occurrence. *Am. J. Epidemiol.,* v.132, p.550-560.

MULLAN, R.J. & MURTHY, L.I., (1991) Occupational sentinel health events: an updated list for physician recognition and public health surveillance. *Am. J. Ind. Med.,* v.19, p.775-799

MURRAY, C.J.L. & LOPEZ, D.L. (Eds), (1996) The Global Burden of Disease, p. 22, WHO/Harvard School of Public Health/World Bank, Harvard U.Press

NAKAJIMA, H. (1994). La violence sur la route fait 700.000 morts chaque annee dans le monde. *J. Med. Liban.*, v.42, p.80-82.

NASCHENVENG, R.A.(1983).Consequencias sociais da utilização de defensivos agrícolas. *Rev. Brasil. Saúde Ocup.,* v.11, p.7-15.

NEHAUL, B. B. G.(1965). Rabies transmitted by bats in British Guiana. *Am. J. Trop. Med. Hyg.,* v.43, p.550-553.

NRIAGU, J. (1990). Global Metal Pollution. Poisoning the Biosphere? *Environment,* 32, 7-11; 28-33.

NRIAGU, J.(1992). Mercury Pollution in Brazil. *Nature,* v.356, p.389.

OEHME, M. (1991). Further Evidence for long-range air transport of polychlorinated aromates and pesticides: North America and Eurasia to the Arctic. *Ambio.,* v.20, p.293-297.

OLIVEIRA, E.F. et al. (1990). Mercúrio via cadeia trófica na Baixada Cuiabana: Ensaios Preliminares". p.. 202-210, *In.* Riscos e Consequências do Uso do Mercúrio, Seminário Nacional (Brasilia, 1989) Hacon, S. et al. (eds), FINEP, Rio de Janeiro.

OLIVEIRA, V.J.C. et al. (1977). Considerações sobre novo surto epidêmico de leptospirose humana na Grande Recife, Brasil, em 1975. *Rev. Inst. Adolfo Lutz.*, v.37, p.33-36.

PAHO (1995). Health and Environment in the Sustainable Human Development of the Americas *Division of Health and Environment,* PAHO, Washington, DC 20 p.

PAHO (1995). Regional Plan of Action to implement the Pan American Charter on Health and Environment in Sustainable Development. Washington, 11 p.

PALHETA, D. & TAYLOR, A. (1995). Mercury in environmental and biological samples from a gold mining area in the Amazon region of Brazil. *Sci. Total. Environ.*, v.168, p.63-69.

PARK, S.K. et al. (1989). Leptospirosis in Chonbuk Province of Korea in 1987: a study of 93 patients. *Am. J. Trop. Med. Hyg.*, v.41, p.345-351.

PASTIDES, H (1995) .An Epidemiological perspective on Environmental Health Indicators. *World Health.*, Stat. Q., v.48, p.139-143.

PENNA, M.L.F. (1990). Condição marcadora e Evento Sentinela na avaliação de serviços de saúde, p. 185-191, IN: Projeto GERUS, R. de Janeiro.

PFEIFFER, W.C. et al. (1990). A Ameaça do Mercúrio nos Garimpos. *Ciência Hoje,* v.11, p.10-12.

PFEIFFER, W.C. et al., (1993). Environmental fate of mercury from gold-mining in the Brazilian Amazon. *Environ. Rev.,* v.1, p.26-37.

PHILLIPS, D.R. (1993). Urbanization and Human Health. *Parasitology.*, v.106, p.S93-S107.

PLATH, P. (1994). [Hearing loss caused by leisure activities-induced noise]. *H.N.O.,* v.42, p.483-487.

PLIMMER, J.R. (1990). Pesticide loss to the atmosphere. *Am. J. Ind. Med.,* v.18, p.461-466.

PÖNKA, A. (1991). Asthma and low-level air pollution in Helsinki. *Arch. Environ. Health.,* v.46, p.262-269.

PRICE, G. (1976). Age as a factor in susceptiblity to hearing loss: young versus adults ears. *J. Acoust. Soc. Amer.,* v.60, p.886-892.

RATCLIFF, H.E. ; SWANSON, G.M. & FISCHER, L. J. (1996) Human exposure to mercury: a critical assessment of the evidence of adverse effects *J. Toxicol. Environ. Health,* v.49, p.221-270.

ROBERTS I. et al. (1992). Child pedestrian mortality and traffic volume in New Zeland. *Brit. Med. J.,* 305, 283.

ROBERTS, I. & CROMBIE, I. (1995). Child pedestrian deaths: sensitivity to traffic volume - evidence from the USA. *J. Epidemiol. Comm. Health.,* v.49, p.186-188.

ROGAN, W.J. (1995) Environmental poisoning of Children Lessons from the Past *Environ. Health., Perspect.*, v.103 (sup.l 6), p.19-23.

ROCK, S.M. (1995). Impact of the 65 mph speed on accidents, deaths and injuries in Illinois. *Accid. Anal. Prev.*, v.27, p.207-214.

ROMIEU, I.; WEITZENFELD, H. & FINKELMAN, J. (1991). Urban air pollution in Latin America and the Caribbean. *J. Air. Waste Mgmt. Assoc.*, v.41, p.1166-1171.

ROMIEU, I.; et al.(1995). Effects of urban air pollutants on emergency visits for childhood asthma in Mexico City. *Am. J. Epidemiol.,* v.141, p.546-553.

ROTHWELL, C.J et al. (1991). Identification of sentinel health events as indicators of environmental contamination. *Environm. Health., Perspect.,* v.94, p.261-263.

RUFF, H.A. et al. (1993). Declining blood lead levels and cognitive changes in moderately lead-poisoned children. *JAMA*, v.269, p.1641-1646.

RUTSTEIN, D.D.(1976) Measuring the Quality of Medical Care: a clinical method. *New Engl. J.Med.,* v.294, p.582-588.

RUTSTEIN, D.D. et al. (1983). Sentinel Health Events (occupational): a basis for physician recognition and public health surveillance. *Am. J. Publ. Health.,* v.731, p.1054-1062.

RUTSTEIN, D.(1987). "Monitoring Progress and failure: sentinel health events (unnecessary diseases, disabilities and untimely deaths)", p. 195-212 *In:* Abelin, T. et al., Measurement in Health Promotion and Protection. WHO Regional Publications, European Series no 22.

RYBAK, L.P. (1992). Hearing The Effect of Chemicals. *Otolaryng. Head. Neck Surg.,* v.106, p.677-686.

SALDIVA, P.H.N. et al. (1994). Association between air pollution and mortality due to respiratory diseases in children in São Paulo, Brazil: a preliminary report. *Environ. Res.*, v.65, p.218-225.

SALDIVA, P.H.N., et al. (1995). Air pollution and mortality in elderly people: a time-series study in São Paulo, Brazil. Arch. *Environ. Health.*, v.50, p.159-164.

SANTOS-FILHO, E. (1993a). Teores de chumbo e mercúrio em cabelo de crianças residentes em Cubatão, na região sudeste do Brasil. *Rev. Saúde Publ.,* v.27, p.81-86.

SANTOS-FILHO, E. (1993b). Concentrções sangúineas de metais pesados e praguicidas organoclorados em crianças de 1 a 10 anos. *Rev. Saúde Publ.,* v.27, p.59-67.

SCHNEIDER. M.C. & SANTOS-BURGOA, C. (1996). Algunas Consideraciones sobre la Rabia Humana Transmitida por Murciélago. *Rev. Salud. Publ.,* México, v.37, p.354-362.

SCHNEIDER, M.C. (1991). Presentacion del Brote Epidemico de Apicás, Mato Grosso (Brasil). *Reun. Consulta. Atenc. Pers. Expuestas a la Rabia Transmit. por Vampiros,* OPS, Washington, 1-5 April, 1991, 25 p.

SCHWARTZ, J. (1991). Particulate Air pollution and daily mortality in Detroit. *Environ. Res.,* v.56, p.204-213.

SCHWARTZ, J. (1996). Air Pollution and Hospital Admissions for Respiratory Disease. *Epidemiology*, v.7, p.20-28.

SCHWARTZ, J. et al. (1993). Particulate air pollution and hospital emergency visits for asthma in seattle. *Am. Rev. Resp. Dis.,* v.147, p.826-831.

SCHWENK, P, (1996). O vilão dos Agrotóxicos. *Parabólicas*, v.3, n.14, p.8.

SELIGMAN, P.J. & FRAZIER,T.M., (1992)."Surveillance: the sentinel event ap.roach,", *In:* Public Health Surveillance, (Halperin,W & Baker Jr.E.L., Eds), p. 16-25, Van Nostrand, N.York

SIMCOX, N.J. et al.(1995) Pesticides in household dust and soil: exposure pathways for children of agricultural families *Environ. Health., Perpect.,* v.103, p.1126-1134.

SINITOX. (1995). "Estatística Anual de casos de Intoxicação e envenenamento Brasil 1993." FIOCRUZ/CICT. 74 p.

SLEMR, F & LANGER, E. (1992). Increase in global atmospheric concentrations of mercury inferred from measurements over the Atlantic ocean. *Nature,* v.355, p.434-437.

SMITH, D.R. & FLEGAL, R.A. (1995). Lead in the Biosphere: recent trends. *Ambio.,* v. 24, p.21-23.

SMOORENBURG, G.F. (1993). Risk of noise-induced hearing loss following exposure to chinese firecrackers. *Audiology.,* v.32, p.333-343.

SÖDERLUNG, N & ZWI, A.B. (1995). Traffic-related mortality in industrialized and less developed countries. *Bull. W.H.O,* v.73, p.175-182.

SOUZA, D. (1986). Considerações sobre enchentes e leptospiroses humanas no município de São Paulo. *Rev. Esc. Enferm.* USP., v.20, p.243-250.

SOUZA, E.B. & VIANNA, S.M. (1993). "Políticas de Investimento em Saúde e Meio Ambiente no Brasil." PIAS, Serie Estudos nº 1, OPS/ OMS, Washington. 78 p.

STAES, C. et al. (1994). Retrospective Study of the impact of lead-based paint hazard remediation on children blood lead levels in Saint Louis. *Am. J. Epidemiol.,* v.139, p.1016-1026.

STENBECK, K.D. et al. (1990). Patterns of treated non-melanoma skin cancer in Queeensland- the region with the highest incidence rates in the world. *Med. J. Austral.,* v.153, p.511-515.

SILVANY-NETO, A.M., et al. (1996). Evolução da Intoxicação por chumbo em crianças de santo Amaro, Bahia -1980, 1985 e 1992. *Bol. Ofic. Sanit. Panam.,* v.120, p.11-20.

TAKET (1986). Accident Mortality in children, Adolescents and Young Adults. *Wld. Health., Stat. Q.,* v.39, p.232-256.

THOMAS, J.G,. & HARAN, H.J. (1981). Vampire bat bites seen in humans in Panamá: their characterization, recognition and management. *Milit. Med.,* v.146, p.410-412.

UIEDA,W. & COELHO, G. MENEGOLA, I. (1992). Morcegos Hematófagos IV. Agressões no Estado de Roraima e considerações sobre sua incidencia na Amazônia. In: SEMINÁRIO NAC. RAIVA, São Paulo. *Anais...*

UK/GER (1993). Global Environmental Change. The UK Research Framework. IACGEC /GER, Swindon, 55 p.

UNITED NATIONS (1992). Agenda 21, Section II, Chapter 21 - Environmentally sound mamagement of solid wastes and sewage-related issues. United Nations Conference on Environment and Development. Rio de Janeiro, 3-14 June, 1992.

USA (1990). U.S. Global Change Research Program Act of 1990. Public Law, 101-606 (11/16/990) 104 Stat.. 3096-3104.

USA (1996). Our Changing Planet. An Investment in Science for the Nation's Future. Nat. Sci. Technol. Council, Washington, 154 p.

VANDERSLICE, J. et al. (1994). Drinking-water quality, sanitation, and breast-feeding: Their interative efects on infant health. *Bull. W.H.O.,* v.72, p.589-601.

VANDERSLICE, J. & BRISCOE, J. (1995). Environmental interventions in Developing countries: Interactions and their implications. *Amer. J. Epidemiol.,* v.141, p.135-144.

VEIGA, M.M. & MEECH, J.A. (1995). Gold Mining Activities in the Amazon: Clean-up Techniques and Remedial Precedures for Mercury Pollution. *Ambio.,* v.24, p.371-375.

VENKATARAMAN, K.S. et al. (1991). Human leptospirosis: a recent study in Madras, India. *Trans. Roy Soc. Trop. Med. Hyg.*, v.85, p.304.

VERAS, F.M. et al. (1985). Estudo epidemiológico dos casos de leptospirose registrados em Fortaleza, na epidemia de 1985. *Rev. Med. Univ. Fed. Ceará,* v.25, n.1/2, p.55-62.

VERLINDE, J.D. et al. (1975). A local outbreak of paralytic rabies in Surinam children. *Trop. Geogr. Med.,* v.27, p.1137-1142.

VICTORA, C.G. et al. (1988). Water Sup.ly, Sanitation and Housing in Relation to the Risk of Infant Mortality from Diarrhea. *Int. J. Epidem.*, v.17, p.651-654.

VITOUSEK, P.M. (1992). Global environmental change: an introduction. *Ann. Rev. Ecol. Syst.,* v.23, p.1-14.

WATSON, Jr, W.D. (1979). Economic considerations in controlling mercury pollution, p. 41-77, *In:* The Bio-geochemistry of mercury in the environment.), Nriagu, J.O., Ed), p. 41-77, Elsevier /North Holland, N. York.

WEITZENFELD, H. (1992). Contaminacion atmosferica y salud en America Latina. *Bol. Of. Sanit. Panam.,* v.112, p.97-109.

WEITZMAN, M. et al. (1993). Lead-contaminated soil abatement and urban children's blood lead levels. *JAMA*, v.269, p.1647-1650.

WHITE, M.C., et al. (1994). Exacerbations of childhood asthma and Ozone pollution in Atlanta. *Environ. Res.,* v.65, p.56-68.

WHO (1990). Public Health Impact of Pesticides used in Agriculture. *WHO/ UNEP,* Geneva, 128 p.

WILLIS, J.T. & BRIGGS, D.J. (1995). Developing indicators for environment and health. *Wld. Health., Stat. Quart.,* v.48, p.15.

WOODFORD, C.M. & O'FARREL. M.L. 1983. High-frequency loss of hearing in secondary school students: an investigation od possible etiologic factors. *Lang. Speech. Hear Sucs. Schools,* v.14, p.22-28.

WORLD BANK (1993). World Development Report. 1993. Investing in Health. Oxford. U. Press

XU, R.P., et al. (1995). Association of air pollution with hospital outpatient visits in Beijing. *Arch. Environ. Health.,* v.50, p.214-220.

YOUNG, B. & BRISCOE, J. (1987). A case-control study of the effect of environmental sanitation on diarrhea morbidity in Malawi. *J. Epid. Comm. Health.,* v.42, p.83-88.

Global Environmental Changes and Traditional Populations

Mauro de Almeida and Manuela Carneiro da Cunha

TRADITIONAL PEOPLES: FROM CULPRITS TO VICTIMS, FROM VICTIMS TO AGENTS

Until the eighties, a combination of poverty, demographic growth and resource degradation was perceived as a syndrome of underdeveloped countries. Poverty would have a causal role leading, through overpopulation, to the erosion of land and water resources on a global scale. In other words, poor people in third-world countries were pointed out as a source of pressure over natural resources, and thus as culprits of an ever-increasing environmental problem (Martine, 1993). The argument took many forms, among which that of the widely-publicized "tragedy of the commons", which pressed the points of birth control and of privatization as a means to prevent overuse of resources by the poor (Hardin, 1968; cf. McCay and Acheson, 1987).

The looming 'tragedy' supposedly provoked by low-capital, large-family, nature-dependent farmers suggested three combined strategies as part of the proposed solution: technological development (exemplified by the green revolution, designed to increase the productivity with the introduction of new seeds and techniques), property rights over land and dissemination of birth control techniques. It was current practice, even in environmentally protected areas, to ignore any active or positive role for people living in those areas.

This paradigm changed gradually along the two last decades of the XXth century. First, in the wake of a few notorious cases, e.g. the paving of the BR-364 road that connects the Brazilian states of Rondonia and Acre (Barbira-Scazzocchio, 1980; Hall 1989; Schmink and Wood, 1992), it became partially accepted that poor people were not always the culprits of environmental disasters, but rather victims of tragedies associated with ill-conceived development plans that disrupted lifestyles as well as water and forest environments.

A different paradigm gained strength in the late eighties, linking *traditional and indigenous peoples* to environmental issues in a novel way. 'Traditional population' and 'indigenous people', according to this new approach, were neither the *poor* who unwittingly destroyed natural resources, nor passive victims of progress. Instead, they started to appear in public discourse on world environmental problems as legitimate stakeholders, as actors endowed with significant knowledge on the natural environment and experienced with customary institutions which in many cases had worked well in the past. The paradigm gained acceptance by global organisms (WCED, 1987: 12, 114-117; UNEP, 1995), multilateral banks (Munansinghe, 1993; Cernea, 1994) and conservation institutions (McNeely, 1995), and became part of projects such as the Pilot Program for the Conservation of Tropical Forests (PPG-7) (May, n.d.).

At the same time, the role of traditional and indigenous peoples as potential partners of measures for environmental protection became an expanding research front with the participation of biologists and social scientists (for an Amazonian sample, see Almeida, 1990; Anderson, 1990; Nepstad and Schwartzman, 1992; Allegretti, 1994; Lescure and Emperaire, 1997; Emperaire, 2000; see also for a broader coverage Redford and Padoch, 1992; Ghai and Vivian, 1992; Croll and Parkin, 1992; Diegues Jr., 199 ;Campbel, 1996; Ruiz-Pérez and Arnold, 1996; Ostrom, Gardner and Walker, 1997; Gibson, McKean and Ostrom, 2000). Several countries sponsored environmental programs involving traditional and indigenous populations: an example is Brazil, which created in 1992 a National Center for Traditional Populations, whose mission is the co-management of Extractive Reserves.

The gist of these contributions is summarized in the following statement from a United Nations document:

> *"Indigenous peoples with a historical continuity of resource use practices often possess valuable knowledge about the behavior of complex ecological systems in their own localities. (...). Where indigenous peoples have depended on local environments for the provision of resources over long periods of time, they often develop a stake in conserving biodiversity. Their practices for biodiversity conservation were built through lengthy trial and error processes. This implies an intimate relationship between the application of ecological knowledge with belief systems (...) It is vital ... that the value of knowledge-practice-belief complex of indigenous peoples as it relates to biodiversity management is fully recognized. Conserving this knowledge might best be accomplished through promoting the community-based resource management systems of indigenous peoples" (Miller, Allegretti, Johnson & Jonsson, in UNEP, 1995: 1017-1018).*

This implies that traditional populations are not *hindrances* to globally conceived conservation systems, but *agents* and *partners* in the management of action on the environment problems. Populations inhabiting environments rich in forests, beaches, rivers, lakes, mangroves and other natural landscapes or reserves of resources are not just *poor*; they are in fact badly paid for the conservation services they provide, and they are deprived of territorial rights and managerial authority. The solutions emerging from this assessment include new strategies to face environmental and demographic problems: to recognize the value of conservation and traditional knowledge services; to improve the quality of life of local populations and their reproductive health, in particular with a focus on rights of women and children (as opposed to merely controlling the birth rate); and to distribute property and income in a fair way, recognizing existing territorial rights.

A RETURN TO THE BLAME-THE-POOR PARADIGM?

There is no unanimity, however, on the link between sustainable development and environmental conservation. The limitations of 'extractive', low-impact activities economies facing changing markets have been stressed early on (Homma, 1994), and the issue of who qualifies as "traditional population" has been discussed (Ferreira, 1996; Brosius, Tsing and Zerner, 1998).

The "island" style of conservation systems has been accordingly reemphasized (Kramer, von Schaik and Johnson, 1997; Brandon, Redford and Sanderson, 1998). These critics recognize that biologically rich tropical landscapes were inhabited by indigenous populations such as the Shuar, Yanesha and Quéchua (Brandon, 1998: 408), some of them for millennia and with ecologically sound practices, as shown by Philippe Descola (1986). However, although recognizing that in the past biodiversity might have coexisted with human use, conservative conservationists argue that social change "could seriously affect [the indigenous peoples'] relationship with wildlands and wildlife and patterns of resource use" (Brandon, 1998: 414). And although the role of traditional "insiders" (such as the Shuar, or the riverine inhabitants of the Solimões) in protecting areas from "outsiders" is recognized, critics assume the ongoing "decline of traditional institutions that have regulated resource use" (Brandon, 1998: 428-30) rather than the evolution of institutions (cf. Ostrom, 1997 and others). They see as an "impossible mission" to address both conservation and human needs of poor dwellers.

Among recent arguments there is the idea that even if traditional peoples coexisted in the past with biodiversity, this was but the result of their poverty. Thus, increasing well-being in poor areas near conservation targets "may be a poor strategy, acting as magnets for migrants or creating growth poles", a view which is paralleled by Katrina Brandon's commendation of "golf courses, vacation homes and intensive agriculture (assuming that the use of agricultural chemicals is controlled)" as comparatively better strategies for insulating parks (Brandon, 1998: 436).

We cannot fail to notice that these arguments essentially rephrase Hardin's diagnoses from the 1960s with a twist: then, it was poverty that led poor to cause resource erosion and environmental "tragedies"; now it is the loss of poverty which puts their traditional environments 'at peril'.[1]

THE SOCIAL CONTEXT OF CHANGES

The changes in discourse and paradigms that were mentioned above did not emerge in a vacuum. Rather, they were, to a large extent, the result of political moves and social movements.

As 'development' programs gave way to global ecological concerns, Southern countries' governments, organized under the G-77, rallied in favor of the old model of development, and labeled environmental constraints a new form of (eco)-imperialism. With poor populations in need of land, it was argued, underdeveloped countries could not afford the luxury of abstaining to use their territory. It was also rightly felt that since the main polluters were in the then G7 group of northern countries, they were the ones who should pay for conservation in the South. 'Sustainable development' was the compromise notion that arose from this confrontation. Although G-77 governments presented a reasonably united front in favor of 'development', domestic opposition in many of G-77 countries stressed that 'developmental' policies were not actually meeting the needs of the landless nor of the poor in general, still less of the local and indigenous populations that were being overrun and dispossessed by 'progress'.

Considering that local governments were not adequately taking local people into consideration, a number of indigenous groups and organizations, later followed by other traditional people, took the initiative, through the mediation of national and international NGOs, of appealing to international public opinion and to international and multinational institutions. Brazil was one of the major countries involved. In the seventies, under military dictatorship and experiencing an economic boom, the so-called 'Brazilian miracle', it had embarked in a World Bank subsidized ambitious infra-structural program. The building of the ill-fated 'Transamazônica' road brought dire consequences to indigenous groups, aptly named "Victims of the Miracle" by Shelton Davis. NGOs concerned with indigenous rights popped up in Brazil by the end of the seventies, as others were created in such unlikely places as Scandinavia, to address the same issues.[2] By 1981, the World Bank issued a document on the impacts of 'development' on indigenous peoples (R. Goodland). Independent teams' appraisal of the impact and implementation of the projects were made mandatory for loan contracts with the World Bank. In the case of the major iron complex of Carajás, for example, Brazilian anthropologists led by Lux Vidal, of the University of São Paulo and hired as an independent team by the State-owned mining company Cia do Vale do Rio Doce, had sufficient leverage under the new World Bank rules to be able to suspend disbursements until indigenous land in the region had been adequately demarcated.

On another front, the initiative of indigenous coalitions and NGOs in the international sphere was to prove momentous. In 1982, the UN, following the recommendations of a UN/NGO conference in 1977, created the Working Group on Indigenous Populations (WGIP), as a low-ranking annual forum under the UN Sub-commission on the Prevention of Discrimination and the Protection of Minorities, itself a part of the UN Commission of Human Rights. Despite its relatively modest origins, this Forum has grown from convening some thirty participants, in 1982, to more than 900, by 1998. From a representation of indigenous peoples mainly from the Americas, it now congregates delegations of more than 160 indigenous groups, from the five continents (A.Muehlebach , in press). Similarly, the active participation of indigenous delegations and supporting NGOs in the meetings of the Parties to the Convention for Biological Diversity resulted in the growing importance of what was initially but a sub-item of an article in the Convention. Article 8j, dealing with indigenous and local knowledge, practices and innovations (on which more is said below) was highlighted to such extent that a special expert group was convened and a permanent Forum was created, in 2000, to address this specific issue.

Indigenous issues grew in visibility at the UN. In 1989, a new Convention (International Labor Organization Convention 169 Concerning Indigenous and Tribal Peoples in Independent Countries) was opened to signatures. It replaced Convention 107, drafted more than thirty years earlier, in 1957, in the then prevailing assimilationist discourse framework. By 1990, the UN General Assembly proclaimed 1993 to be the International Year for the World's Indigenous People. And in 1992, at the "Earth Summit" in Rio de Janeiro, indigenous and local people's issues were to gain new prominence in several declarations and legal instruments, as will be discussed below.

From 1977 and its very inception in UN fora, indigenous discourse stressed the link between their land claims and the environmental destruction going on in their territories (A. Muehlebach, in press). At the very local level, ecological discourse was interwoven with cosmology. Davi Yanomami, a Global 500 Prize winner, whose intercultural intellectual labor was admirably described by anthropologist Bruce Albert, was able to elaborate a political mythology that, while espousing ecological categories, would also represent a "shamanic critique" to the very western concept of nature (B. Albert 1993). Thus, the link between ecological concerns and indigenous people was forged at many different levels and was most certainly not just the mimetic outcome of first-world discourse. It is worth noting that Davi Yanomami found inspiration in Chico Mendes (B. Albert 1993: 367) and that, conversely, Chico Mendes and the rubber-tappers' movement found inspiration in indigenous land struggles in Brazil (Carneiro da Cunha and Almeida, 2000).

INTELLECTUAL RIGHTS AND INDIGENOUS PEOPLE

'Biodiversity' became the name of the game in the nineties, due to the promises it holds of extracting valuable commodities from wild or cultivated biological resources, from knowledge and even from the very genes of local people living in remote areas (K. ten Kate and S. Laird 1999). It introduces what Escobar has called a 'techno-nature' regime as opposed to the more traditional and much less sophisticated 'capitalist nature', premised on 'development' and delivering plantations and cattle pasture landscapes (Escobar, 1999). Moreover, capitalist and informational (or techno-) nature are incompatible in the long run, since the use of industry's improved seeds destroys the very wealth of genetic resources needed for combating diseases and pests, thus generating what James O'Connor has called the second contradiction of capitalism (J.O Connor 1988 and J. Martinez-Alier 1994). This move to include techno-nature in the picture is consistent with the shift in the role traditional people were deemed to play.

The major legal watershed in this process was the UN Convention for Biological Diversity, opened for signatures by countries in 1992 and widely ratified with the notable exception of the United States. At the same "Earth Summit" meeting held in Rio de Janeiro in June 1992, the Rio Declaration on Environment and Development stated in its Principle 22 that:

"Indigenous People and their communities and other local communities have a vital role in environmental management and development because of their knowledge and traditional practices. States should recognize and duly support their identity, culture and interests and enable their effective participation in the achievement of sustainable development."

Another document issued at the same 1992 Conference, Agenda 21, devoted a full chapter to "Recognizing and Strengthening the Role of Indigenous People and Their Communities" (chapter 26).

But the Convention on Biological Diversity dealt with a further issue, namely intellectual rights. Not only did it make explicit reference in its Preamble and in several articles to the role of "many indigenous and local communities embodying traditional lifestyles", it also brought in issues of respect, maintenance and dissemination of traditional knowledge within the article dealing with in situ conservation (art.8). Most notably, it introduced a provision for the equitable sharing of the benefits arising from the utilization of indigenous and local knowledge, innovation and practices.

However, the Convention for Biological Diversity (CBD) has to be analyzed in a more general perspective. Among other things, it settled the long-standing dispute on sovereignty over genetic resources. Were genetic resources a common 'world heritage', therefore opened to prospection and appropriation, or were they subject to countries' control? As is widely known, this dispute had to do with the perception that wealth in genes and wealth in technology were differentially located. Technology was in the North, while biological diversity was in the South. Since technology from the North could not be freely appropriated, the G-77 argued, why should the South allow for free appropriation of its genetic resources? The CBD addressed the issue through market mechanisms: trade was to be the solution, and access to technology would compensate for access to biological resources. Sovereignty over genetic resources in their territory was recognized to countries, which in turn were asked to put in place a clear legal procedure for accessing biological resources.

While G77 States were satisfied with this approach, some of them, including

Brazil, began expanding the notion of sovereignty to that of ownership. Genetic resources were to be the property of the State. Local people's social movements and advocates started questioning the separation that was established between the territory over which traditional people had control and the genetic resources over which their control was not being recognized, in stark contradiction with the acknowledgement of their stewardship.

The issue of intellectual rights is the one that has stirred the greatest debate. While the CBD recommends that people share in the benefits that their knowledge, practices and innovations can bring, several procedures could be envisaged. The less imaginative is to expand existing intellectual property rights mechanisms to collectivities. This seems to be the path followed by WIPO, the World Intellectual Property Organization, which launched a two-year (1998-1999) labor-consuming fact-finding set of missions to gauge if they could 'protect' traditional knowledge using the procrustean bed of copyrights, trademarks, origin certifications or even patents (WIPO 2000). There are several inherent difficulties for so doing, not the least of which is that all existing Intellectual Property Rights have a time limit and rely on exclusivity; whereas traditional people argue for rights in perpetuity and many will favor non-exclusivity. IPRs as we know them are themselves a roughly two-centuries old and evolving historical institution, that is undergoing unprecedented expansion in recent times. Traditional societies, for their part, have construed a multitude of knowledge rights regimes (S. Harrison, 2000). How is one to organize the interface not between just two very different systems, but rather one globalized IPR system and a multitude of different local regimes with specific colonial and pre-colonial histories?

Furthermore, what is the logic under which we will be operating? Is it just to use knowledge, practices and innovations for the market or is it to promote its continued existence? Is it present knowledge we are valuing or is it present and future knowledge? That is, are we focusing on available knowledge or rather on processes of producing knowledge? "What is traditional in traditional knowledge" as has been aptly conveyed in the final document of the Convention of the Parties to the CBD in Buenos Aires, in 1996, " is not its antiquity but the way it is acquired and used". In short, knowledge is dynamic, and so are institutions.

The issue therefore cannot be discussed in the abstract. Let us take as an example one proposal that is on the table and that has been gaining ground in many Latin-American and African Countries, although it originated in Southeast Asia. I am referring to the Community Intellectual Rights as the Third World

Network originally proposed them. Note that the term property is no longer there. The basic idea is that traditional knowledge should stay in the public domain for anyone's use, but that originators should share in the benefits when it is used for commercial purposes for an indefinite time. This is in stark contrast with contemporary practice of IPRs, which requires exclusivity and a limited time frame. In other words, expectations of both parties are opposed. Free access and public domain versus monopoly and secrecy. Unlimited time frame for intellectual rights versus prescription of intellectual rights after a certain time.

A source of such contrastive approaches is to be found in the very concept of traditional people. In one case, they are seen best as relics of the past whose knowledge should be valued but who have no further participation as such in the future. As against this view, traditional people have a claim to be part of our common future.

POSITIVE EXPERIENCES AND SOME LESSONS

The upper Jurua river is an important center of biodiversity. Indigenous and traditional people make it also an area of high cultural diversity (Ricardo, 1999; Carneiro da Cunha and Almeida, in press). As it stands today, in its Brazilian southernmost part, three different territorial regimes are at play in separate but contiguous parts of the area. A National Park was established in the eighties. Several Indian territories for the Ashaninka, Arara-Jaminawa and Kaxináwa were demarcated from the eighties up to the year 2000. In 1990, the Alto Juruá Extractive Reserve was established, for the sustainable use of rubber tappers. The case exemplifies the positive potential of combining several and distinct regimes of protection of nature, as an effective barrier against the pressure of loggers and ranchers.

This case also illustrates some of the factors that may be important for the success of protected areas. A prominent one is the integration of local action and nation-wide structures. Thus, local-level grass-roots organizations (e.g. the Associations of Rubber Tappers and Indigenous Associations) articulate with regional and national level organizations (the National Rubber Tappers' Council, Indigenous federations), and grass-roots organizations interact with state and federal agencies (the National Center for Traditional Population). Another important factor at work is the existence of well-established institutions through which local residents actively enforce their own use plans and development projects.

Finally, these grass-roots managed conservation systems benefit from ongoing cooperation with scientific and technical institutions, such as the University of Acre, the University of Campinas and the University of Sao Paulo. Other examples in the Amazon region of ongoing articulation between environmental policies and organized traditional populations are the Upper Rio Negro area (Ricardo, 2000) and the Solimões River lowland (Lima and Ayres).

To sum up, traditional peoples may be innovative. The social movement of rubber tappers in South America led to the innovative concept of Extractive Reserves, containing the important notion that forests, lakes and coastal areas may be used and managed by traditional people as a collective heritage to be shared by groups who commit themselves to the task of keeping them for permanent and sustainable use. The Southeast Asian lesson is that traditional knowledge be kept in the public domain for shared use and improvement by people. In both cases, traditional populations and their intellectual allies have taken innovative positions, although they may clash against current market dogmas.

Some current critics of the role of indigenous and traditional people in the management of nature assume that city dwellers evolve towards a better way of using the environment and develop conservation strategies, while indigenous and other local groups would allegedly evolve in the opposite direction. Indigenous peoples of the present are even at times described as if they were the living remnants of a paleolithic humankind that is supposed to have caused faunal extinctions in the past.

> *We have argued instead that 'traditional populations' is a future-oriented concept, not a label for a frozen past - as illustrated in the above discussion of intellectual property rights, and by new institutions such as the extractive reserves. The social groups who define themselves and are recognized as "traditional populations" for negotiation purposes are those who,* besides a record of sustainable use of nature, commit themselves to the goals of sustainable development and environmental conservation *(Carneiro da Cunha and Almeida, 2000).*

After less than a decade of experiences which followed the early nineties consensus on the importance of traditional groups as part of resource conservation, as illustrated in the cases of the upper Jurua River, of the Amazon floodplain, and of the upper Rio Negro, it is premature to dismiss as "mere slogans" the goals of articulating conservation interests and human goals. The problem of collective resource use by traditional populations, as well as the issues of how to distribute the benefits of traditional knowledge, remain difficult. It is also true that "*a struggle is underway to define ownership of the meaning of biodiversity*" (Brandon, Redford and Sanderson, 1988). Indeed, as has been pointed out by Arturo Escobar,

> "*... there is no single objective definition of biodiversity, but rather a multiplicity of views from the perspective of each social actor in his respective locality, be it the World Bank or the Pacific Communities" (Escobar and Pedrosa, 1996: 125).*

Biodiversity has different meanings and uses for plantation owners and loggers, for biodiversity prospectors, for professional conservationists and for local populations. All these agents interact on a common field where alliances are possible. One important set of alliances was forged among social movements and conservationists in recent years. This gain should not be jeopardized by authoritarian and socially unfair solutions such as militarization and privatization in favor of the rich. This shift would increase the cleavage between rich and poor and menace the important policy gains obtained with the affirmative struggle of such leaders as Chico Mendes.

NOTES

__1.__ The "peril" image comes from Brandon, Redford and Sanderson 1998.

__2.__ This was IWGIA, International Working Group on Indigenous Affairs in Copenhagen. Many already existing NGOs, such as Survival International or OXFAM in Britain, started to address the same issues.

BIBLIOGRAPHY

ALLEGRETTI, Mary H. (1994). Policies for the Use of Renewable Natural Resources: the Amazonian Region and Extractive Activites". In: Miguel CLUSENER-GODT, Miguel, SACHS, Ignacy, (eds.) *Extractivism in the Brazilian Amazon: Perspectives on Regional Development.*, MBA Digest 18, Paris, UNESCO, p. 14-33.

ALMEIDA, Alfredo W. Berno de. (1995). *Quebradeiras de Côco Babaçu: Identidade e Mobilização.* São Luís: Movimento Interestadual das Quebradeiras de Coco (São Luís) e Terre des Hommes (Suíça).

ALMEIDA, M. W. B. (1996). The Management of Conservation Areas by Traditional Populations: the Case of the Upper Juruá Extractive Reserve. In: REDFORD, MANSOUR(1996), p.137 - 158.

ALMEIDA, Mauro W. B.(1990). As Colocações como Forma Social, Sistema Tecno-lógico e Unidade de Recursos Naturais. *Terra Indígena*, Araraquara, Centro de Estudos Indígenas - UNESP, v. 7, n 54,p. 29-40, jan.-mar. 1990.

ALMEIDA, Mauro W. B. (1996). Household Extractive Economies. In: PEREZ, Ruiz, Arnold (1996), p. 119 -142

ALMEIDA, Mauro W. Barbosa de. (1995). Le Statut de la Terre et les réserves extractivistes". *Cahiers du Brésil Contemporain, Numéro Espécial 1995: Le Statut de la Terre*, p. 169-190, 1995.

ANDERSON, A. B. (organizador). (1990). *Alternatives to Deforestation: Steps Towads sustainable Use of the Amazon Rain Forest.* New York: Columbia University Press, 1990.

ANDERSON, Anthony B., MAY, Peter, BALICK, Michael J. (1991). *The Subsidy from nature: Palm Forests, Peasantry, and Development on an Amazon Frontier.* New York: Columbia University Press, 1991.

ANDERSON, Anthony et al. (1994). *O Destino da Floresta. Reservas Extrativistas e Desenvolvimento Sustentável na Amazônia.* Rio de Janeiro: Relume Dumará, Instituto de Estudos Amazônicos e Fundação Konrad Adenauer, 1994.

BANCO MUNDIAL. 1992. *Relatório sobre o Desenvolvimento Mundial 1992: Desenvolvimento e Meio Ambiente.* Rio de Janeiro: Fundação Getúlio Vargas e Banco Mundial, 1992.

BARBIER et al. 1995. *The Economic Value of Biodiversity*. UNEP 1995,: p. 822-914.

BARBIRA-Scazzocchio, Françoise (ed.) (1980). *Land, People and Planning in Contemporary Amazonia.* Cambridge, Centre of Latin American Studies Occasional Publication No. 3.

BRANDON, Karina, REDFORD, Kent H., SANDERSON, Steven E. (eds). (1998). *Parks in Peril: People, Politics and Protected Areas.* Washington, The Nature Conservancy and Island Press, 1998.

BRANDON, Karina. (1998). Comparing Cases: A Review of Findings and Perils to Parks: The Social Context of Threats. In: K. BRANDON, K. REDFORD and S. SANDERSON *Parks in Peril: People, Politics and Protected Areas.* p.375-414 and p. 415-439.

BROSIUS, J. Peter; TSING, Anna L. , ZERNER, Charles. (1998). Representing Communities: Histories and Politics of Community-Based Natural Resource Management. *Society and Natural Resources*, 11, p.157-168., 1998

CAMPBEL, Bruce (ed.). (1996). *The Miombo in Transition: Woodlands and Welfare in Africa.* Bogor (Indonesia), Center for International Forestry Research - CIFOR.

CARNEIRO CUNHA, M.M., ALMEIDA, Mauro W. B. de. (2000). "Indigenous People, Traditional People, and Conservation in the Amazon". *Daedalus/ Journal of the American Academy of Arts and Sciences.* Vol. 129, n.2, p. 315-338, 2000.

CERNEA, Michael. (1994). *Sociology, Anthropology and Development: An Annoted Bibliography of World Bank Publications* 1975-1993. Washington, The World Bank, 1994.

CROLL, Elisabeth, PARKIN, David (eds.) (1992). *Bush Base: Forest Farm. Culture, Environment and Development.* London, Routledge, 1992.

DAVIS, Shelton. *Victims of the Miracle.*

DESCOLA, Philippe.(1986) *La Nature Domestique. Symbolisme et Praxis dans l'Écologie des Achuar.* Paris: Maison des Sciences de l'Homme, 1986.

DROWER, Michael 1995. MCNEELY, Jeffrey A (org.). (1995). *Expanding Partneships in Conservation.* Washington, IUCN e Island Press, 1995.

ESCOBAR, Arturo and Alvaro Pedrosa (eds.). (1996). *Pacífico: Desarrollo O Diversidade? Estado, Capital y Movimientos Sociales en el Pacífico Colombiano.* Bogotá, CERECE e ECOFONDO.

ESCOBAR, Arturo. (1996). "Viejas y Nuevas Formas de Capital y los Dilemas de la Biodiversidade". In: Arturo ESCOBAR & Alvaro Pedrosa 1996, p. 109-131.

FERREIRA, Lúcia C. (1996). *A Floresta Intransitiva: Conflitos e Negociações na Mata Atlântica.* Ph.D. Thesis, Universidade Estadual de Campinas, 1996.

GHAI, Dharam, VIVIAN, Jéssica M. (1992). *Grassroots Environmental Action: People's Participation in Sustainable Development.* Londres e New York: Routledge, 1992.

GIBSON, Clark C., MCKEAN, Margaret A., OSTROM, Elinor (eds). (2000). *People and Forests. Communities, Institutions, and Governance*. London and Massachusets, MIT Press, 2000.

HOMMA, Alfredo K. O. (1994). Plant Extractivism in the Amazon: Limitations and Possibilities. In: l CLUSENER-GODT, Miguel, SACHS, Ignacy, (eds.) *Extractivism in the Brazilian Amazon: Perspectives on Regional Development*, MBA Digest 18, Paris, UNESCO, p. 14-33/34-57.

KENT, H., REDFORD, MAUSOUR, Jane A. (eds.) (1996). *Traditional Peoples and Biodiversity Conservation in Large Tropical Landscapes.* Arlington (Virgínia): America Verde Publications/The Nature Conservancy, 1996.

KRMER, R., SCHAIK, C. von, JOHNSON, J., eds. (1997). *Last Stand: Protected Areas and the Defense of Tropical Biodiversity.* Oxford, Oxford University Press, 1997.

LAURE, Emperaire (ed.).(2000) *O Extrativismo Vegetal na Amazônia.* São Paulo, Editora da UNESP, 2000.

LESCURE, Jean-Paul, PINTON, Florence and EMPERAIRE, Laure. (1994). People and forest Products in Central Amazonia: The Multidisciplinary Approach of Extractivism. In: *Extractivism in the Brazilian Amazon: Perspectives on Regional Development*, Miguel Clüsener-Godt and Ignacy Sachs, eds., MBA Digest 18, Paris, UNESCO, p 58-87.

LEVEGUE, C. (ed.). (1995). "Human Influences on Biodiversity: Information requirements for the sustainable use of biodiversity". In UNEP 1995: 783-791.

MARTINE, George (ed.). (1993). *População, Meio Ambiente e Desenvolvimento: Verdades e Contradições.* São Paulo, Editora da Universidade Estadual de Campinas.

MARTINE, George. (1993). "Introdução. A Demografia na Questão Ecológica: Falácias e Dilemas Reais". In: *População, Meio Ambiente e Desenvolvimento: Verdades e Contradições*, George Martini (ed.), São Paulo, Editora da Unicamp, 1993. p 21-42.

MAY, Peter H et al (eds.). n.d. *Políticas Públicas Coerentes para uma Amazônia Sustentável: O Desafio da Inovação e o Programa Piloto.* Brasília, Friends of the Earth and GTA (Grupo de Trabalho Amazônico).

MCAY, Bonnie J. , ACHESON, James M. (eds.) (1987). *The Question of the Commons. The Culture and Ecology of Communal Resources.* Tucson, The University of Arizona Press, 1987.

MCNEELY, J.A., GADGIL, M, LEVEQUE, C. Padoch and K. REDFORD, K (eds.). (1995). Human influences on biodiversity. In: UNEP 1995: 711-822.

MCNEELY, Jeffrey A (org.). (1995). *Expanding Partneships in Conservation.* Washington, IUCN e Island Press, 1995.

MILLER, K., M. H. ALLEGRETTI, N. Johnson, Jonsson,, B.(eds.). (1995). Measures for conservation of biodiversity and sustainable use of its components. In UNEP 1995 p.915-1062.

MUNANSINGHE, Mohan (ed.). (1993). *Environmental Economics and Natural Resource Management in Developing Countries.* Washington, CIDIE and World Bank, 1993.

NEPSTAD, Daniel C. and SCHWARTZMAN, Stephan (eds.) (1992). *Non-Timber Products from Tropical Forests: Evaluation of a Conservation and Development Strategy. Advances in Economy Botany, vol. 9.* New York, The New York Botanical Garden, 1992.

OSTROM, Elinor, GARDNER, Roy , WALKER, James (eds.) (1997). *Rules, Games, & Common-Pool Resources.* Ann Arbor, The University of Michigan Press, 1997.

REDFORD, Kent H. and Christine Padoch (eds.). 1992. *Conservation of Neotropical Forests: Working from Traditional Resource Use.* New York: Columbia University Press.

REDFORD, Kent H. e Christine Padoch (organizadores). (1992). *Conservation of Neotropical Forests. Working from Traditional Resource Use.* New York: Columbia University Press, 1992.

Ruiz-Pérez, Manuel and J.E.M. Arnold (eds.). (1996). Bogor and Oxford: Center for International Forestry Research and e ODA (Overseas Development Agency), 1996.

SCHMINK, Marianne, WOOD, Charles H. (1992). *Contested Frontiers in Amazonia.* New York: Columbia University Press, 1992.

UNEP (United Nations Environment Programme) (1995). *Global Biodiversity Assessment*, ed. V.H. Heywood and R. T. Watson, Cambridge, Cambridge University Press and UNEP, 1995

WCED (The World Commission on Environment and Development). (1987). *Our Common Future.* Oxford, Oxford University Press. WCED (The World Commission on Environment and Development). 1987. *Our Common Future.* Oxford, Oxford University Press, 1987

YOUNG, M. D. (1992). Sustainable Investment and Resource Use: Equity, Environmental Integrity and Economic Efficiency. Paris, UNESCO and The Parthenon Publishing Group, 1992

INSTITUTIONS

Local Policies and Global Environmental Change

Leila da Costa Ferreira
State University of Campinas (UNICAMP)

INTRODUCTION

Today there is nearly a scientific consensus with regard to the contribution of human activities to the rise in temperature on the planet. The consequences of this process are still unknown, but the first signs are perceptible. The theme is necessarily global since the atmosphere belongs to all, and without a concentrated effort of activities the answers will be punctual due to the seriousness of the problem.

In that respect, the present paper is intended to analyze and discuss formulation and implementation of environmental policies relevant to global changes that have some impact, direct or indirect, on public policies of sustainability at the local level, and verify whether there is any efficacy in these "institutional innovations".

For this purpose, the present paper has as its general objective the analysis of experiences at the level of local power through a study of the formulation and implementation of public policies with socioenvironmental features, adopted by several city administrations in the South and Southeast of Brazil after the promulgation of the 1988 Constitution, when the environment issue entered local agendas as a relevant indicator for the formulation and implementation of public policies (Ferreira,2000).

Past papers (Ferreira,1998 and 1999) have shown a total disconnection between environmental policies and other public policies at state and federal levels in Brazil. I aim here to analyze these issues at the local, regional level, examining the policies concerning environment during the last municipal administration (1989 up to now) in large cities in the states of São Paulo, Rio Grande do Sul, Minas Gerais, Paraná and Rio de Janeiro where new political actors have been elected. They are: São Paulo, Porto Alegre, Belo Horizonte, Curitiba and Rio de Janeiro.

The present paper is organized in the following way: first, an introduction with a brief account of the importance of discussion about sustainability indicators in contemporary cities; second, the analysis of the Brazilian situation and the results of a survey carried out in the South and Southeast of the country, and lastly, the conclusion which highlights the relation between sustainability and the Brazilian cases.

Mega cities will be changing their address this millennium. In the year 2015, four of the five largest urban agglomerations on the planet will be in developing countries. The developed metropolises, exceeded in their population by centers such as Lagos (Nigeria) and Dacca (Bangladesh) will maintain their economic power through a network of global cities, where the headquarters of multinational companies and most of the money that circulates in the world are found. The accelerated urbanization of poor countries means that the population in cities will xceed that of the countryside by 2006 for the first time in history.

At the global level, less developed metropolises will not only concentrate population, but poverty as well. Growing at a faster and faster rate, they will barely be able to provide so many people with adequate housing, transport and basic sanitation. However, they will not be the only ones to face these problems. Even metropolises at the top of the global hierarchy, such as New York, have already been affected by traffic jams, pollution and violence. In spite of their size or location, the cities will be facing at least two common challenges: environmental degradation and urban tension growth caused by an increasing inequality between their dwellers.

As the "The Situations of Cities in the World" report points out, the progress of a nation is closely correlated with the viability of its urban areas. Despite a small decline, the urbanization rate will continue to rise steadily at the beginning of the millennium. We know that the number of urban dwellers living in complete poverty grew rapidly in the 1980s, especially in Latin America, Africa, and in less-favored Asian economies (Global Report on Human Agglomerations, 1996).

The situation of the metropolises , such as that of São Paulo and Mexico City, where the population is projected to increase by about 8 million in the first decade of 2000, tends to repeat itself in other poor countries. Studies by UNCHS - United Nations Center for Human Settlements - estimate that as from 2000, nearly 50 per cent of the population in poor countries will be in cities, of which at least thirty are expected to have a population of over 5 million (WRI/UNCH,1996) The inequality between cities is one of the major costs of the urbanization. United Nations Center for Human Settlements shows that the average household income in cities of industrialized countries is 38 times higher than that in African cities. In Latin America, the difference is 17 times, but even in cities of industrialized countries, the 20 per cent richer have an income 10 times higher than that of the 20 per cent poorer (UNCH, 1996).

To a growing inequality a State crisis was added, that deprived governments of most of their investment power in infrastructure and social services. As a result, to an increasing portion of the population urban life has also become synonymous of unemployment, deprivation, violence, slums, traffic jams and pollution. In the average of 236 cities surveyed by United Nations Center for Human Settlements, the time to commute to work is 34 minutes, 30 per cent of the households are situated below the poverty line, 41 per cent of the employment is in the informal sector, 6.4 per cent of the children die before the age of five, and 70 murders per 100 thousand inhabitants/year are recorded. These indicators are likely to worsen as urbanization grows faster in poor countries: on average 5 per cent per annum as against 0.7 per cent in highly industrialized countries. As a consequence, from 1995 to 2015, the urban population in less industrialized countries is projected to grow 52 per cent, whereas in the industrialized the rate will be of 7 per cent (UNCH, 1996).

Simultaneously, an increase in the number of large cities in poor regions is expected. In 1950, there were only about 100 urban agglomerations with a population of over 1 million in the world, most of which in rich countries. In 2015, according to the UN (1996) there will be 527 large cities, and 3 out of 4 will be located in less industrialized countries. The fast urbanization and the concentration of problems in the poor countries make the improvement in local administration even more essential. On examining the 700 "best urban practices" in 100 countries, United Nations Center for Human Settlements arrived at some key words for this transformation: decentralization, State partnerships with other social actors, and democratic participation. As Castells (1999) emphasizes, the solutions must be per sector, and take into account the growing citizen participation in decision-making.

One of the fundamental problems in all this process, and which has drawn attention from city researchers is the urban traffic and the growing rate of motorized individuals among the population, both in highly industrialized countries and in less industrialized countries. Personal property rates, as well as those of car use have grown steadily since the 1950s, and in a more noticeable way as from the 1970s (Macedo, 2000).

The consequences of the problems derived from the relation between transport and environment occur at the local level with losses caused by traffic jams (pollution and its impact on health), at the regional level with economic conflicts resulting from disputes between states and municipalities over auto industries, and at the global level with pollutant emissions contributing to the aggravation of climate changes on the planet.

According to Macedo (2000), in the European Union for example, a scenario of public policies especially aimed at the control of emissions through economic instruments to reduce the contribution of greenhouse effect gases can be identified. The argument for the creation of "environmental-taxes" has become stronger and stronger, and the guidelines established at the Fifth Framework Program of the European Community were projected to go into effect in the first decade of 2000. According to the author, the documents of the European Community brought forward by the transport and environment commissions help outline the scenario for the regional approach, showing the need of providing mobility options in public transport.

There are other aspects of the analysis, which must be commented on, such as the relation between standards of consumption and their connection with the transport issue. Whitellegg (1997) broadens the discussion, introducing the concept of "time pollution", where waste of time caused by traffic jams is included as a negative impact.

The debate on the use of different instruments concerning transport, public policies and their environmental, economic and social impacts has been gaining considerable ground in highly industrialized countries and must be repercussions in Brazil (Pucher and Lefèvre, 1996; Proost and Braden, 1998, Vasconcelos, 1996 and Macedo, 2000, et al.).

As a matter of fact, in half a century, Brazil has undergone one of the fastest urbanization processes in the world: of 46 per cent in 1940, the cities sheltered 80 per cent of the population in 1996, and are projected to reach 88 per cent in 2025. Industrialization has made the urban centers responsible for 90 per cent of everything produced in the country. This process has led to a concentration of people in large metropolis, such as in São Paulo and in Rio de Janeiro, the third and fifteenth largest cities in the world, respectively. The present trend, which is likely to be maintained in the future, is the population growth inland. As a reflection of this, medium-sized cities, with a population between 100 thousand and 500 thousand inhabitants were the ones to record the highest absolute population growth in the 1990s. Together they sheltered 36.7 million Brazilian people in 1996 as against 31.9 million in 1991 (IBGE, 1996).

The trend toward an increase in agglomerations also implies changes in urban administration. To deal with the problems concerning several cities (litter, waterworks, safety, transport and pollution), new ways of action must be planned, with the internalization of the environmental problems in the process of formulation and implementation of different public policies being crucial for this process.

SUSTAINABILITY AND PUBLIC POLICIES IN CITIES IN THE SOUTH AND SOUTHEAST OF BRAZIL

The growth of the cities is closely related to the increase in the number of people that choose them to live in. The rising concentration of the population in the urban area is accompanied by a deterioration of the quality of life, especially in cities in Latin America, Africa and Asia (Lidstone, 1995), as seen in the introduction to this paper. On the other hand, a reframing of the urban dynamics is an aspect to be taken into account, which directs us to the question of city administration. In some cities in the world there are already initiatives toward the promotion of an integrated management of the urban activities, which enhances the life quality of the population and preserves the environment balance (Vig and Kraft, 1998).

Nevertheless, in the Brazilian case, due to historical features, the process of industrialization and urbanization has brought to metropolitan areas – and presently, because of the process of "countermetropolization", is bringing to regions within the states – the intensification of social and public service privations. It has also brought environmental deterioration (especially of the water resources) and the strangling in the infrastructure of the cities (mainly in the sanitation, housing and transport sectors) (Hogan et all, 2000). This situation causes the appearance of millions of small local dramas in the country as a whole, serious problems of pollution, housing, health, extra needs of schools, organization of provisions systems, especial programs for critical poverty, development of projects for basic sanitation, and so forth.

The cities start to face a serious situation, which requires swift interventions in matters outside traditional policies for the urban area. They involve large projects of infrastructure, social policies, programs of employment and environmental policies, and even local strategies for dynamizing the economic activities. Although the municipalities are situated in the front line of the problems, they are placed in the last rank in the public administration. There is a generalized relocation of the problems to the local sphere; whereas the administrative and political structures are kept centralized.This process has generated a kind of institutional impotence that makes any modernization of the local administration extremely difficult. With the process of urbanization, the problems have been relocated, but the corresponding decision system has not.

Nonetheless, it can be stated that, while the Brazilian environmental policy (Viola, 2000; Guimarães, 2000) has declined at the federal level with regard to political importance, experiences of sustainability within local public policies have occurred in the municipal sphere in several Brazilian states. Various cases have already been considered emblematic, such as that of the city of Curitiba (Menezes, 1996).

However, what seems truly innovative is that this process is reaching several other Brazilian cities, and the case of the State of São Paulo could be mentioned here as a significant example in this process (Ferreira, 1998). All the same, these experiences have occurred in various other states, as we will see now.

In fact, the establishment of public agencies aimed especially at the environment issue and the formulation of environmental policy is quite recent in Brazil (Guimarães, 2000; Viola, 1997). The State of São Paulo was the precursor of actions focused on the question of environment (Ferreira, 1992). At the municipal level, the significant growth in the specific action toward the environment occurs after 1988. The Organic Laws reflect the full development of this subject in the agendas of local governments. However, the creation of most governmental agencies in the three spheres of government and the fundamental legislation directed toward the environment issue date from the 1980s. Therefore, they were devised during the period of a State crisis, which meant restricted budget and difficulties in the coordination of the different public policies. In spite of that, at the municipal level (Ferreira, 1998; Jacobi, 1994) there are already several legal and institutional instruments to enable the local power to face the process of socioenvironmental degradation.

In a society of extreme social exclusion as in the Brazilian, the fact that environmental issues are always present on the agenda of the local governments in the cities studied, implies their political relevance. These cities all have Environment Agencies, which are quite well structured in the institutional and budgetary perspectives.

The Municipal Environment Agency of Porto Alegre, which dates from 1976, is in fact the oldest environment agency in the country; named as "the green city", Porto Alegre has a long tradition of concern with environment issues since AGAPAN (the local environment non governmental organization) is one of the oldest Brazilian environmental associations, with a significant impact of action even in its relation with the federal level.

According to the data of the survey, the city hall stated that the transport sector was the major cause of air pollution in the city. As a result, in June 1992, a network for air quality monitoring, carrying out daily readings, was installed. As from this diagnosis, whose data showed concentrations of sulfur dioxide above the limits established by the federal legislation, the municipal government, through an agreement with Petrobrás, decided on the compulsory use of "metropolitan diesel", which has a much lower content of sulfur than the common diesel used.The legislation went into effect in late 1992, and as from then a high drop in the levels of sulfur dioxide in the city has been achieved, thus positioning Porto Alegre below the maximum limit established by the legislation, even at the most critical places of traffic.

The need to extend the monitoring network was also noticed when other important standards used for air quality control in the city were analyzed, such as carbon monoxide and particulates. To carry out this task, the Municipal Environment Agency purchased in 1995, a laboratory for the analysis of these materials. In 1999 a covenant between the Environment Agency, Petrobrás and the Federal University of Rio Grande do Sul was being implemented for the study of air quality in the metropolitan area of Porto Alegre.

On the other hand, Belo Horizonte has a Planning and Environment Agency that dates from the early 1990s, but acts on a well-articulated basis with the Urban Activities Agency and the Health Agency.

According to technicians of the Municipal Environment Agency, through a covenant with the government of the State of Minas Gerais and the City Hall, since 1988 they have had a program, the "Operation Oxygen" for checking the emission levels by diesel powered vehicles circulating in the city. As stated by them, its objectives have been achieved up to that moment as the emission of dark smoke in the city has reached much lower levels since the program started. In order to extend the services concerning air monitoring, in 1996, the Municipal Agencies of Belo Horizonte, Betim and Contagem created the "Air Team", sponsored by the federation of industries of the State of Minas Gerais.

Nevertheless, the major programs implemented during the period analyzed in that city were: environmental recovery of the Lagoa Pampulha (haven of the highest-income classes in the city); works of sanitation in the slums, recovery of parks, squares and gardens, action of Community, Popular, Sector and Regional Boards for the implementation of a Participatory Budget and the attempt to integrate the various public policies, especially in the articulation between the city, health and housing spheres, as seen before.

During the period analyzed, our attention was drawn to the increasing public audiences, responsible for the discussion about noise and air pollutions and the implementation of "Operation Oxygen" and the "New Green Program".

As for Rio de Janeiro, before 1993, it had an Environment Superintendence linked to the Urban Agency. In 1992, mayor asked the Green Party to help him form the Municipal Environment Agency. Until 1994, this agency worked as an agency extraordinary, having a more symbolic role. That year it became effectively an Environment Agency, whose role was fundamental for the elaboration of the local Planning Directive.

Emphasized here must be the importance the relation between sustainability and transport had to local administrators, for the first action proposed by the agency was the implementation of bicycle lanes in the city. In addition, the Environment Agency launched a campaign against dark smoke emitted mainly by metropolitan buses, and was developing studies for the creation of an Ideal Network for Air Monitoring, in which varied information on pollution from the industries, major traffic corridors and climate data is crossed.

Moreover, differently from the federal environmental policy (Viola, 1997) in some of these cities there was a real concern about integrating the environment issue with the social issue, at least at the formulation level of their Planning Directive, or else in the formulation of their Organic Laws as the cases of Porto Alegre and Belo Horizonte illustrate.

These cities present technical and human resources (they all have interdisciplinary teams working in the agencies) more suitable to face so difficult a task, that is, linking the environment issue to the social problems originated from the Brazilian urbanization process. Stressed here must be the concern of the municipal agents with the problem of water and air pollution, the question of sewage treatment, and especially the question of a more equitable distribution of infrastructure in the urban space. Besides, the strategy for the participatory planning, which is an integrated part of these studies of the cases analyzed, suggests that the preoccupation of these administrators was about the real implementation of new actions, with the participation of several actors in this process, and not a merely technical and bureaucratic process.

Curitiba is definitely a significant example of the implementation of environmental policies, however it clearly differs from the case of Porto Alegre and Belo Horizonte in the way the population is encouraged to take part in the process.

Nevertheless, the case of Curitiba is regarded as an example of good practice in urban environmental administration, especially with respect to the integration of land use and transport (Rabinovitch and Leitman, 1996). The legislation for land use and occupation stimulates a high-density occupation along the axles that support the city. As the subway would be a very expensive option, the city designed the public transport services based on an adapted bus system.

It must yet be mentioned that what draws the attention is the relevance socioenvironmental diagnoses have gained in the process of formulation and implementation of public policies with socioenvironmental characteristics in all the cities studied.

Surprising is the case of the city of São Paulo. This city had an Environment Department from 1989 to 1992, creating the Municipal Green Agency in 1993. In the metropolis, there would be even conditions, taking technical and human resources into account, for an actual implementation of a municipal environmental policy. In a globalize city, with highly qualified universities, an intellectualized middle class that has a quite refined ecological awareness, the environment issue played a completely secondary role during the administrations analyzed in greater detail, as if the "prior" social problems, according to the local "policy makers", were not completely linked to the process of environment degradation in the metropolis.

Nonetheless, the project to limit car use ("rodizio") implemented in the metropolitan area of São Paulo by the State Environment Agency must be commented on, due to its relevance and also the fact that the local City Hall adopted the "rodizio" as well, though only during the hours of great traffic jam, up to now.

"Operation Rodizio" was implemented in the city of São Paulo from 1996 to 1998. According to Macedo (2000), the "rodizio operation" was an emblematic case of intersector action for the resolution of environmental problems since it had a technical and scientific basis of varied origins, and involved processes of public investigation and participation in the elaboration of a strategy to limit individual use of vehicles so as to avoid critical episodes of air pollution, in accordance with the "precaution principle".

According to the author, a positive attitude was verified by a high degree of adhesion (94 per cent in 1996 and 96 per cent in 1997 and 1998). The debate resulting from the 1996 campaign, with media publicity, made the question of air pollution and life quality in the city known to the public. It was expected that the fact would generate pressure on the local power so that measures would be taken concerning the public transport; however, that did not occur (Macedo, 2000).

In fact, the environmental problems in the city of São Paulo are particular since they are situated in a region of the Brazilian territory, with geographical features and a very specific historical and social inheritance; and are also general for they illustrate a process in course in different parts of the world, such as in the case of Rio de Janeiro itself, which is also nearing the condition of an "unsustainable city".

CITIES, SUSTAINABILITY AND GLOBAL CHANGES IN BRAZIL

The literature concerning debate on the question of State and environmental problems (Lester, 1989; Paehlke, 1989; Milbrath, 1989; Orr, 1992; Vig and Kraft, 1998; Cahn, 1995) demonstrates the importance of the role of various social actors in the process of formulation and implementation of environmental policies in an attempt to find solutions for the problems posed. In that regard, the administrations of Brazilian municipalities which started their activities after the promulgation of the 1988 Constitution should facilitate the job of these different actors since one of the most fundamental propositions of the Brazilian Constitution is institutionalization of the participatory process. Nevertheless, it must be added that this is a controversial debate, especially when it concerns global changes. Due to the complexity of the themes, it seems that only the possibility of action of different social actors alone would not solve the question (Olphus, 1997; Cahn, 1995).

In the last decade, the dynamics of world order has been deeply marked, from an ecological point of view, by either the emergence or intensification of global environmental problems. Several authors have analyzed the environmental policies as from this perspective (Paehlke, 1989; Lester; 1989; Vig and Kraft, 1998; Cahn, 1995; Viola, 1997; et al.).

However, the analysis of environmental policies at the local level is still quite recent (Ferreira, 1996; 1998). According to Amy (1994), these studies reflect the development of a "state of the art", that is, after a few decades of the implementation of environmental policies, a further analysis of this process is required, and studies at the local level are suggestive of such. On examining the development of the two preceding decades, it can be stated that, during the 1970s and 1980s, the United States and other highly industrialized countries adopted dozens of environmental policies and created new institutions to develop programs regarding the environment (Dunlap, 1997). However, for the 1990s, the environment agenda seems to have become more complex (Buttel, 1997).

Consideration of the continuity of environmental policies for the latter decade requires a careful evaluation. As a matter of fact, several governments are not yet sufficiently equipped to resolve global environmental problems. According to Beck (1992), institutional reforms and, especially, new methods in the decision processes are essential to the actual achievement of these policies. Incentives to behavioral changes should even complement governmental actions.

It is important, in the first place, to comment on the major axles of change that reach the complex society of this century, and influence the parameters of new ways of State administration (Beck, 1992). An immediate implication of this new reality is that there is no need for such a centralized State, since the population living in urban area can resolve most of their problems locally. This new reality is what led the highly industrialized countries to adopt a State frame profoundly different from that of the Brazilian, with ample participation of the local governments (Castells, 1999; Giddens,1991, Cahn, 1995). This implies, at another level, that the eternal dichotomy between decentralization or not, has lost its meaning as the community's public space gains fundamental weight, in terms of perspectives, thus reflecting the evolution of representative democracy for decentralized and participatory systems (Alier, 1997).

The designs of urban development have been the object of increasing attention of governmental organs, finance agencies, civil society entities and experts from different areas of knowledge. They have stimulated important articulations for a democratic administration of the cities, ethical urban planning, the right to citizenship, that is, dignified conditions of urban life for all citizens. What has presented itself, as innovative most recently, is the treatment of these specific questions in an integrated and systemic way, having the urban dynamics as its core. In this perspective, the approach of the urban system as part of the global ecosystem forwards the discussion of how to ensure a sustainable urban development taking the global environmental problems into account. In spite of the fact that this view has been gaining more ground, it is yet to go through an intense clash with a more traditional concept of urban development, which considers reality in a sectored and fragmented way.

In the entire world, urban areas are regarded as privileged places for providing jobs, introducing innovations, enhancing economic opportunities. Urban centers show great agility to build a network of relations concerning the economy, politics and culture linking rural areas, small, medium and large sized cities. At the milestone of these transformations, a significant portion of the world population started to have access to an unprecedented level of consumption and riches. Obviously, this part of the population that enjoys a high level of consumption, and the other one that enjoys a moderate one (fulfilling their basic needs) owe the living standard they have to the cities. On the other side of the urban space are the excluded, those who do not have their basic material needs fulfilled. Although it occurs in a differentiated way, cities all over the world face this picture of inclusion-exclusion (Orr, 1992; Beck, 1999).

Though on the one hand public spaces are created and urban life is socialized, on the other hand, the supply of public services and collective equipment increases, as well as the number of people who live on islands of degraded and peripheral areas, with no means to inform themselves and have access to this equipment.

On approaching these themes, the present paper tries to focus the analysis upon the legacy of modern rationality, where everyday life, reflection, science and technique mingle, many times confusing us to the point of our questioning the enlightenment heritage preached so much by us (Best and Keller, 1991).

Beck (1992) comments that it may be pointless to summon the promises of modernity to open the democratic and emancipator pathways of the contemporary crisis of the end of the century. As far as he is concerned, that is not "one more" crisis of the modern world. In fact, there is a simultaneous crisis of the processes of social regulation and their possible emancipating potential, here being included the consecrated ways of conceiving and thinking upon their tension within the categories of modernity. Concretely, it means that the commitment solutions already experienced historically between State, Market and Community - and their corresponding rationalities - have invariably led us to oceans of accumulated irrationalities, demonstrated by environment degradation, population growth, disparities and inequalities between center and periphery, deprivation and hunger that live together with abundance, ethnic and religious wars, dependence of the individual concerning consumer goods, the savage way of destination of the right to the labor market. The list is actually quite extensive.

According to Giddens (1991) modernity is a two-edged sword. The development of the modern social institutions and their proliferation at world level have generated greater opportunities for the human beings to enjoy a safer and more rewarding existence than in any other pre-modern system. However, modernity has also a dark side, which has become clearer in the present century.

Environmental sociology, both in the international and in the Brazilian spheres, (Buttel, 1996; Dunlap, 1996; Yearley, 1996, Ferreira, 1998, et al.) has attempted to ponder new ways of social action that may show the pathway to cognitive, political and subjective alternatives within the contemporary context.

With respect to political alternatives, the sole principle of the market confines the State and deprives the sociability patterns of legitimization. However, at the same time this movement hides other sociability's, practices and cultures that modernity has marginalized, while showing them as "political zed spaces". They are those of the households, that of the production, citizenship, and world, each of which forming a handful of social relations that on connecting themselves into their various interfaces by political action rupture with their amorphous and merely interactive place to which the institutional and political field had designated them. As a whole, they unfold an active multiculturalism that, in time becomes capable of identifying relations of power, and conceiving ways of turning them into relations of shared authority. To acknowledge that these areas of negotiation and conflict may constitute a "new political common sense" means to acknowledge that there is no answer in the strengthening of the sole principle of the State, but in its relative deregulation. Nor is this answer in the predominance either of the market or the community, incapable on their own of ensuring a social regulation that is not fragmented and dispersed.

In this perspective, it makes it impossible to consider politics exclusively at the national or local level. The world must be conceived as a "public space" (as Habermas suggests,1997), or else, as a field (as Bourdieu suggests, 1983), or even as an arena, as suggested by the North American sociology (Walace, 1995); a civil society in which projects and distinct, antagonistic or complementary views face each other. To recognize the various types of knowledge and practices which refuse the ruling of the fragmentation and dispersion is a sociological exercise present in several kinds of social actions, including those regarding institutional disputes, especially when related to the ecological problems, for that is of a symbolic nature that encompasses a wide range of values, including post-materialistic values, so they create and accommodate new and highly sophisticated public demands.

The urbanization process in course these last decades has taken place especially in Latin America, as seen before in the introduction to this paper. The rate of urban population growth in these countries is 3.7 per cent per annum, whereas it is 1.1 per cent per annum in most countries of the North (Lidstone, 1995). On the other hand, it is mainly in the large cities that this demographic process is occurring. Out of 82 cities with a population of over 4 million projected for the year 2000, 81 are situated in poor countries; out of 23 mega cities with a population of over 10 million expected for the same year, 17 are located in poor countries, and only 2 out of the 10 largest cities in the world are in highly industrialized countries (Tokyo and New York).

The analysis of the internalization process of institutional and political indicators of sustainability in public policies in cities of the south and southeast of Brazil is inspired upon three orders of concern already outlined above, and could be summarized in the following way.

Firstly, the need to follow up, systematize and learn more about innovative local environmental administration with increasing potential (in the perspective of examples); in addition, analyze the impact of the existence or not of innovative participatory practices in the formulation, follow up and control of local environmental policies, and finally, diagnose the interest in establishing networks between sectors openly engaged in the defense of life quality, highlighting the spaces of dialogue/interaction and the dissemination of the information about public policies.

The data obtained in the survey which directs the discussion of the present paper indicate, as seen before, that, despite the fact that the situation of Latin America , and particularly that of Brazil is quite worrisome with regard to their urbanization process, a few Brazilian cities are trying to consolidate institutional and legal instruments associated with sustainability, conferring more unity upon the matter so as to facilitate its learning and understanding by the local authorities, and ensure the efficacy of its practical application. Nonetheless, with respect to the question of transport and its relation to air pollution control, the actions implemented were irrelevant, due to the seriousness of the question regarding global changes.

Obviously, from the institutional perspective, the fact that we are "equipping" ourselves in order to advance toward internalization of sustainability indicators in their various dimensions is considered important; nonetheless, the actions implemented are far from the concerns with global issues.

In that respect, the discussion about institutional and political indicators of sustainability in the cities and global changes gain crucial relevance upon the political agenda at the beginning of this 21st century. That's the reason for the importance of discussing the environmental problems and strategies for their solution in the Brazilian cities.

BIBLIOGRAPHY

ALIER, J. M. (1997). Justiça Ambiental e Distribuição Ecológica de Conflitos. In: FERREIRA, Leila C. (org.). *A Sociologia no Horizonte do Século XXI.* São Paulo : Editora BomTempo.

AMY, D.(1994). Environmental Dispute Resolution: The Promise and the Pitfalls. In: VIG, J. and KRAFT,M.(Ed.). *Environmental Policy in The 1990s.* Washington, D.C: CQ PRESS.

BECK, U.(1992). *Risk Society: towards a new modernity.* London: Sage.

---------- (1999). *O que é Globalização. Equívocos do Globalismo e respostas à globalização.* São Paulo: Paz e Terra.

BEST, S.; KELLNER, D.(1991). *Postmodern Theory. Critical Interrogations.* The Guilford New York: Press.

BOURDIEU, P (1983). *Sociologia. Textos escolhidos.* São Paulo: Ed. Ática.

BUTTEL, F.; LARSON, O. (1996).Environmental and Resource Sociology: Theoretical Issues and Opportunities for Synthesis. *Rural Sociology*. v. 61, n.1.

---------- (1997). Social institutions and environmental change. In: Redclift, M and Woodgate, G. (ed). *The international handbook of environmental sociology.* Massachusetts: Edward Elgar Publishing Limited. Cheltenham.

CAHN, M.A.(1995). *Environmental Deceptions.The Tension Between Liberalism and Environmental Policymaking in the United States.* New York: State University of New York.

CASTELLS, M. (1999). *A Sociedade em Rede*. São Paulo: Ed. Paz e Terra.

DUNLAP, R (1993). From Environmental to Ecological problems. In: Calhoun, C. and Ritzer,G. (ed). *Social Problems.* New York: MacGraw-Hill.

---------- (1997). The evolution of environmental sociology: a brief history and assessment of the American experience. In: *The international handbook of environmental sociology.* Op. Cit

FERREIRA, Leila C.; Viola, E. (org.) (1996). *Incertezas de Sustentabilidade na Globalização.* Campinas: Ed. Unicamp.

FERREIRA, Leila C. (1998). *A Questão Ambiental: Sustentabilidade e Políticas Públicas no Brasil.* São Paulo: Bomtempo Editorial.

---------- (1999). La búsqueda de alternativas de sustentabilidad en el poder local. In: Ziccardi, A e Luján, S. (ed). *Ciudades Latinoamericanas. Modernización y Pobreza.* Cidade do México: Universidad Nacional Autónoma de México.

---------- (2000). Agenda de Desenvolvimento Sustentável. In: Projeto PNUD-BRA/98/017. *Agenda de Desenvolvimento humano e sustentável para o Brasil do século XXI.* Brasília: PNUD. Instituto de Política.

GHAI,D. (1995). Grassroots- *Environmental Action*. People's participation in *Sustainable Development.* London and New York: Routledge.

GIDDENS, A..(1991). *As Consequências da Modernidade.* São Paulo: Ed. UNESP, 1991.

GUIMARÃES, R.(2000). Brazil and Global Environmental Politics: Same Wine in New Bottles? In : ISA, Rio de Janeiro. Anais...
* Trabalho apresentado na. Rio de Janeiro.

HABERMAS, J.(1987). *The theory of Communicative Action.* Vol .2. Beacon Press.

HOGAN, D et al (2000). *Migração e Ambiente em São Paulo.* Campinas: NEPO/UNICAMP.

IBGE (1996). *Censo Demográfico.* Instituto Brasileiro de Geografia e Estatística. Rio de Janeiro.

JACOBI, P. (1995). Alcances e Limites de Governos Locais Progressistas no Brasil. *Cadernos CEDEC*, São Paulo, n.34.

LESTER, J.(1989). *Environment politics and policy.* Theories and Evidence. Durhan and London: Duke University Press..

LIDSTONE, J.(1995). *Global Issues of our Time.* Cambridge: Cambridge University Press.

YEARLEY, S. (1996). Sociology, *Environmentalism and Globalization*. London: Sage publications.

MACEDO, L. V. (2000). *Em busca do transporte coletivo sustentável para a região metropolitana de São Paulo: uma experiência de política ambiental nos anos noventa.* Dissertação (Mestrado) PROCAM. USP, São Paulo.

MACDONALD, G (1997). *Latin American Environmental Policy in International Perspective.* Boulder: Westview Press.

MENEZES, C. L. (1996). *Desenvolvimento Urbano e Meio Ambiente. A experiência de Curitiba.* Campinas: Papirus.

MORRISON, R.(1995). *Ecological Democracy.* Boston: South End Press.

ONU (1996). *As Situações das Cidades no Mundo.* Relatório Global sobre as Aglomerações Humanas. Nações Unidas. (mimeog.).

ORR, D.W. (1992). *Ecological Literacy. Education and Transition to Postmodern World.* New York: New York Press.

PAEHLKE, R.(1989). *Environmental and the Future of Progressive Politics.* New Haven and London: Yale University Press.

PURCHER, J; LEFÈVRE, C (1996). *The Urban Transport Crisis in Europe and North America.* London: MacMillan.

PROOST, S; BRADEN, J (1998). *Climate Change, Transport and Environmental Policy.* Cheltenham: Edward Elgar publishing.

UNITED NATIONS DEVELOPMENTS. Programme (1997). *Human Development Report.* New York and Oxford: Oxford University Press.

UNCHS (1996). *An Urbanising World: global review for human settlements.* Nairobi: Habitat.

VASCONCELOS, E (1996). *Transporte urbano nos países em desenvolvimento: reflexões e propostas.* São Paulo: Ed. Unidas.

VIG, N.; RAFT, M. (1998). *Environmental Policy in the 1990s: Toward a New Agenda.* Whashington: Congressional Quartely Inc.

VIOLA, E. (1997).Globalização, sustentabilidade e governabilidade democrática no Brasil. In: Trindade, A. .A. .e Castro, M. (org.). *A Sociedade democrática no final do Século.* Paralelo 15.Brasília.

------------- (2000). The Global Politics of Climatic Change: How Strong are the Forces Supporting Sustainability. Trabalho apresentado na ISA. Rio de Janeiro.

WALLACE, R. and ALISON,W.(1995). *Contemporary Sociological theory. Continuing the Classical Transition.* Prentice Hall. New Jersey: Englewood Cliffs..

WHITELEGG, J.(1993). *Transport for sustainable future: the case for Europe.* London: Belhaven Ed.

Brazil and Global Governance: the Case of Climate Change

EDUARDRO VIOLA* AND HÉCTOR LEIS**

* University of Brasilia
** Federal University of Santa Catarina,

INTRODUCTION

This paper has four parts. In the first one we discuss the concept of global governance and the differentiation of the world society in relation to the problems of global governance at the beginning of the 21st century: liberals, nationalists, cosmopolitans and neo-feudals. In the second part we discuss the problematic of climate change, one of the most important questions of global governance. In the third part we discuss the Brazilian stand in relation to most of the issues of global governance and show the transition from nationalism to liberalism. In the fourth part we analyze the Brazilian stand in the formation of the climate change regime.

GLOBAL GOVERNANCE AND THE DIFFERENTIATION OF SOCIETY

In the last two decades all the nation-states have increasingly become part of a global world. Obviously, they are not parts of a whole in the same way and/or importance, but all of them depend (and contribute) more and more on the global transformations and flows. From a formal point of view we are still living in an international society of states, but from a real perspective a significant part of the world is going deeper and deeper into a global society with an interdependent and interconnected network of power, capital, people, goods, technology, communications, beliefs, environment, etc. Actually the contemporary world is formed by three worlds: the OECD world formed by countries with well established market economies and democratic regimes; the transitional world formed by countries with heterogeneous status that are in some kind of transition towards a market economy and/or democracy. These range from Brazil or Poland (which are very close to the first world) to Saudi Arabia and China (where there is some market economy, but no democracy) to Russia (where there is some democracy and some market economy); and to the marginal world, where there is no transition toward markets or democracy (Africa, Middle East, Central Asia) and where neo-feudal forces prevail. In this new context it is essential to clarify the emerging concepts of global politics and global governance.

Common sense associates globalization with economy. It seems more difficult to grasp the transformations of politics in a globalized world. Today politics has all kinds of ramifications, inside and outside national frontiers; global politics has increased its sites and actors beyond the states, governments and political parties. The political actions come to society through complex networks of actors and decision-making processes that include, along with "full-time" political actors, a new group of very effective "partial-time" political actors, coming from civil society and the economy. In other words, politics today is something more (or less) than politics yesterday. This is not an effect of globalization; this is globalization itself. Not only politically, but also economically, globalization implies an intensification of interactions among actors, processes and functions of economy, politics and civil society, in all directions.

But globalization, like Janus, has two faces. In one sense, it means the breakdown of barriers between different sectors and elements of the world. The globalizing processes spread readily across national boundaries and are not blocked by cultural or territorial frontiers. Also traditional political hierarchy among actors has been affected. A global political action can be initiated from above or below with equivalent repercussions. A transnational corporation, a big state, a local activist group or even an individual alone can push people, capital, governments and institutions in different parts of the world. Obviously, an action will not imply the same effects in a worldwide scale. Globalization means that any action could reach a global effect, but not that all actions will always have a global effect. Globalization is simultaneously a real and a potential process.

This real and potential power of an action is justified by the complex characteristics of globalization. For the same reason, along with the trend towards integration, globalization presents another opposite trend towards fragmentation and/or inhibition of actors, processes and functions. Rosenau (1997) proposed the concept "fragmigration" to suggest this simultaneity between the integrating and fragmenting trends and processes of the globalized world (many times we find cooperation and conflict among the same actors). In fact, in the global order, order and disorder come together. In part because the world is not any more exclusively centered in the States (and their governments). In the past legal authorities were characterized by the power and regulatory action that they exercised over the social life on a territory. Presently, in the globalized world there are authorities, with and without territory, with power on specific (economic, environmental, political, etc.) issues, in a very extended region (or even the whole world) or with an extended power, on a very local community. The reorganization of the interrelations and power of actors through globalization explains why today the focus is much more on governance than in government.

The old actors tend to be formal, concrete and territorial; the new ones tend to be more informal, less concrete and non-territorial.

In spite of the states' persisting power, these cannot prevent the emergence of other actors in the global arena. Now, an increasing number of intergovernmental organizations (IGOs) and non-governmental organizations (NGOs) – based on the existing states, civil society and multinational corporations – challenge the states. From supranational political entities (like the European Union) and multinational corporations, to local social movements and transnational professional or scientific associations, all participate in global politics. Even those small actors that are against globalization for ideological reasons cannot deny an intense participation in global politics. The size or importance of the national or sub-national actors, or whether they are in favor or against globalization is not relevant; all of them are part of global politics.

To some degree, all politics today are global politics. Political action, as well as political authority, appears in a very wide and diffuse way. This fact explains why global politics and democracy have an intimate relationship. The tremendous democratization of the world in the last two decades is not only an effect of the victory of the US and liberalism in the Cold War. Democratization today is more than a political-ideological proposition; its strong development is a direct consequence of a systemic constitution of a polyarchic mixed actor-system for the global arena (Held et al, 1999). A non-polyarchic actor system could work in a democratic or non-democratic way, but a polyarchic one would only work with democratic actors and procedures. The governance of global politics is incompatible with authoritarianism and/or bureaucratic centralized regimes.

There is no globalization in a world of only authoritarian states. In the same way, in the long run there is small room for authoritarian states in a globalized world. Many authoritarian states have collapsed during the 1990's, presently remaining as failure states that are giving room for the development of powerful neo-feudal forces; some of which are forming a global network of crime. Many criticisms to globalization come from the wrong perception of this phenomenon as a side effect of the expansion of capitalist economy. This blind reductionism prevents the perception of the multiplication of actors and issues that emerge with globalization. Global politics presupposes global governance because, in the present changing political life, governments alone (being democracies or dictatorships) are not capable any more to govern the world. Global governance presupposes democracy because without general accountability, obedience to law and public rational values and arguments, there is no way to understand each other and establish rules and procedures for the solving of conflicts and problems in a complex world with a mixed actor system.

The fact that global governance is based on a polyarchic mixed actor system does not provide support to the societal version of "governance without government". This interpretation led to the overestimation of civil society's actors and utopian perspectives against states' actors and realistic perspectives. It is a mistake to exclude or underestimate the role of governments, as well as that of multinational corporations or NGOs, in governance. Global governance supposes an effectively mixed actors and values system that cannot be confused with ideological approaches. The international regimes are good examples to understand this point. They do not exist in a political vacuum of states. It is true that regimes are elements of a decentralized system of public order that have affected the sovereignty of nation-states. But regimes are firmly tied to states; they cannot exist out of them. There is no doubt that regimes are supported by the international society of states as well as by global society and multinational corporations (Young, 1997). Regimes do not exist on behalf of states, civil society or corporations; they exist to deal with specific and concrete global problems. It does not matter who is officially in charge or who is primarily responsible for the problem. What really matters is to establish correct rules and relationships among all the actors related to the problem, in order to deal efficiently with the solutions in the global public arena. Regimes means governance at work. Governments, corporations and civil society are deeply associated with each other when and where regimes are at work.

One assumption of this paper is that regimes work better when the different actors establish a democratic and realistic relationship. Governance implies democracy in a realistic sense, not only as a form of government, but as a strong commitment to liberal economy and cosmopolitan law. When regimes do not work, most of the times it is because the actor system related to them has established rules without any compromise with the real dynamics of the contemporary world. It is important to insist that democratic relations among actors of a polyarchic system imply the existence of much more voting and elections, rule of law and accountability, and also rational values and actors. There is no democratic rule that could solve the global climate change without a realistic approach capable of enforcing a cosmopolitan law.

The real challenges to global governance (and this especially applies to environmental global governance) are not so much of a technical-formal type, but rather of a realistic communicative character (Leis, 2000). Governance does not exist without negotiation among different actors, with different interests. And there is no negotiation without participation supported on realistic bases and efficient communicative action. Utopian or instrumental proposals have no capacity to create and to operate a regime. To embrace a wide specter of actors, from global to local, including national and departmental authorities, big corporations and small NGOs, a realistic and complex institutional approach is necessary, as well as a rational and complex value approach (which means that the actors should give to each other the right hierarchical position on the issue at stake, as well as looking forward to realistic and non utopian solutions). At this point we might ask if real governance is possible in a world like this, so anarchic and full of particularisms? Certainly, global government would not be possible, but global governance is. It is so since the chance for governance is never singular, but many; each of them depending on the particular issue at stake. There are issues that are easier to grasp from a rational and realistic approach, than others (for example, the ozone problem facilitate this kind of approach, but climate change is more difficult to confront).

In fact, there is no single global governance; there are many global governances, as many as issues at stake. So, global governance is a complex balance of several processes that are going together, sometimes in a converging way, and other in a diverging way. Everything is possible in a globalized world, but global governance, as we understand it here, implies value-actors that are more cosmopolitan than national or neo-feudal, more liberal than utopian and more realistic than formal.

From the beginning of modernity, the rise of the nation-state brought national identity to the center of political life. The most remarkable characteristic of national oriented actors is a fantastic ranking of the national interests. Even moral values are put aside when the national interest is jeopardized from outside or inside the country. Cosmopolitan philosophers (like Kant) have denied that national community could come before moral community. The concern of these philosophers was traditionally presented as related to human kind in first place. But the cosmopolitan approach is more than a substitution of particular values by universal ones. Cosmopolitanism cannot be reduced to a kind of utopian world citizenship or moral law. It is wrong to deduce, from the fact that national identity privileges particular interests, that the cosmopolitan identity gives importance only to universal interests (even in Kant's view, cosmopolitanism does not deny the importance of the nation-state). In fact, this identity seeks the harmonization of universal and particular obligations, through a broadening of political and moral values from a rational perspective (Linklater, 1998).

The liberal modern nation-state does not prevent citizens from acknowledging that they and non-citizens are part of the same (moral, political, economic and ecological) world. Only despotic and/or fundamentalist states prevent that. Cosmopolitanism is not unitarianism, but pluralism; it is not a particular value-interest transformed into a universal one, but many particular value-interests transformed in a whole. National identity demands loyalty to the nation-state; the cosmopolitan identity accepts many loyalties, but these must be rational and work together. From a cosmopolitan point of view, the alien could be our enemy only when they try to impose on us the supremacy of their own particular value-interests. In the 20th century, it was proved in the Second World War that only the liberal nation-states (not the nazi-fascist neither the communist) were committed to maintaining a pluralist international society. The nazi-fascist and communist occupation of Europe did not look for a pluralist relationship (cooperation among free partners), like the American did. The (totalitarian) alternatives to liberal democracies were implicitly in favor of a "national" world government, demanding the recognition of supremacy for one single national authority to organize the world. Historically, the liberal nation-states were the only ones not in favor of a world government. In this sense, the liberal states never were idealistic or utopian, but realistic. Before the explosion of globalization, they believed in the capacity of international society to organize the world; now they believe in global governance. Because the liberal value-actors are realistic, they are now becoming cosmopolitan.

We must insist in this point. Today, the only realistic way to organize (and to accept) the disorder among different nation-states deriving from global relations is adopting a cosmopolitan conception of law and social action, and a liberal vision of political and economic life. To get a balance between universal human rights and international law, on one side, and particular political and economic background of the nation-states, on the other, cosmopolitan-liberal concepts are necessary. To develop only capitalist economy throughout the whole world could be liberal, but not cosmopolitan. After the tremendous breakdown of the Soviet Union, even the remainder communist parties know that there is no alternative to capitalist economy. This does not imply that every country and social group should accept the western vision of the world, but implies that the relationship among people must be guaranteed by international and transnational rule of law.

Far away from cosmopolitan value-actors, in the contemporary world there are neo-feudal forces. We have said that globalization has two faces. One faces integration, the other, fragmentation. In the same way that the dynamics of the first trend push liberalism towards cosmopolitanism, the dynamics of the second trend push nationalism towards neo-feudalism. The fragmentation processes of globalization create conditions to develop and strengthen forces that are dissolving the union among sovereignty, territoriality, nationality and citizenship. These forces are not new in the political framework of the modern world, but their importance is growing faster with globalization. The neo-feudal value-actors produce a system of corruption, crime, overlapping powers and multiple loyalties, inside and outside the nation-states. If cosmopolitan forces are the best support for global governance, the neo-feudal ones are the worst (with the national and liberal forces somewhere in between).

The cosmopolitan value-actors are oriented towards global governance because they accept multiple and compatible loyalties. The contrary is the case of neo-feudal forces. They constitute a system of multiple authorities, but each of them demands exclusive and total loyalty due to their fundamentalism and criminal bias. There is no global governance without the recognition of each particular culture and actors, and such recognition occurs only through the existence of cosmopolitan law. But the neo-feudal forces do not respect any law other than their own "law". It is usual to understand cosmopolitan law as a reference to that law, which transcends the claim of nation-state and tries to protect basic humanitarian values (see Held, 1995). As a consequence, cosmopolitan law sets down basic standards that may contradict the national law and conflict with national governments and neo-feudal forces. But here we mean more than simple humanitarian law; cosmopolitan law implies the establishment of rules of commerce, diplomacy, warfare, etc.

From the perspective of cosmopolitan law, cosmopolitanism cannot be confused with a human rights movement and/or with a hegemonic ideology of western superpowers. In fact, nationalism serves better than cosmopolitanism to hegemonic aspirations. The latter allows subordinate nations, regions and social groups to organize themselves globally, in defense of actual common interests and values. Nationalism (and, much more, neo-feudalism) prevents the use of the interaction capabilities of globalization. It is from a cosmopolitan point of view that a minor actor or a complex issue can obtain new chances or opportunities. Certainly, in a neo-feudal or national political system it would not be possible to deal with climate change. On the contrary, from a cosmopolitan social action, it is possible for indigenous people in the Amazon, for example, to counteract hegemonic forms of globalization, through multidimensional cooperation among civil society, economic corporations and governments (Santos, 1995).

CLIMATE CHANGE AS A CRUCIAL ISSUE IN GLOBAL GOVERNANCE

The existing mechanisms for Global Environmental Governance are very limited for multiple reasons: lack of coordination among mechanisms; overlap of institutions; contradictions between environmental sustainability and social integration creating a strong problem of legitimacy; and supremacy of economic institutions or regimes, like the World Trade Organization, the International Monetary Fund and the World Bank (Haas, Keohane and Levy, 1993; Young, 1997).

UNEP has succeeded to a certain extent in its catalytic role, with notable achievements in the fields of ozone depletion, toxic waste, biodiversity and regional seas. But it has been a failure in the role of coordinating global environmental action and, in the mid 1990's, seemed to lose its way completely: lack of effectiveness, inefficiency and inability to attract qualified staff. However, in recent years, due to the energetic leadership of Topfer, UNEP has regained some momentum, strengthening its role in monitoring and assessing related functions.

More than 200 Multilateral Environmental Agreements (MEA) now exist, forming a central part of the framework of global environmental governance (Brown et al, 2001). Some have been initiated by UNEP, such as Ozone and Biodiversity, while others, as the UN fish stock agreement, had their beginning elsewhere in the system. MEAs vary significantly in nature, from narrowly focused agreements, such as CITES or the Convention on Migratory Species of Wild Animals, to MEAs with broad coordination functions, such as the UNFCCC and the Kyoto Protocol. In general, the more recent MEAs, such as the Montreal Protocol, the Kyoto Protocol and the Cartagena Protocol on Biosafety, have dealt with increasingly important areas of economic activity, contrasting to earlier MEAs, which were much more limited in their coverage, often dealing with the protection of a single species in a limited geographical area (Porter and Brown, 1996).

The current MEAs, as well as those under negotiation, cover many of the key environmental issues, but there still is a lot of room for improvement in their coordination, as well as that of the institutions related to them. Roles, responsibilities and priorities are fragmented and functions are duplicated to an unnecessary extent. The proliferation of MEAs, their Conference of the Parts (COPs) and interim meetings make it difficult for small and poorer countries to become effectively involved in the real decision-making. On the other hand, many MEA regimes have proved themselves dynamic and innovative, not being constrained by existing institutional structures.

The Montreal Protocol is the best example of success in building international regimes for protecting the global environment. The use and production of CFCs has fallen dramatically in developed countries since the beginning of the 1990's and has been closely controlled in other countries, although illegal trade in CFCs does remain a problem. Four factors underpin the overall successful result: the rapid achievement of consensus in the scientific community about the severity of the problem and the efficiency of information exchange within this community; the fact that CFC production is concentrated among a few key players in the private sector; focused leadership by one country in the international negotiations; and the fast response of CFC producers in investing in research and development of alternative technologies that were commercially feasible. The ozone regime is a clear example of a successful coalition between Liberals and Cosmopolitan-Liberals, with the support of many Nationalists (Gehring, 1994; Porter and Brown, 1996; Clark, Van Eijndoven and Jaeger, 1998; Weiss Brown and Jacobson, 1998).

The developing of the Global Environment Facility, since 1991, has implied a conceptual revolution: the definition of incremental cost for protecting the global environment. The operation of the Facility is being slowly consolidated and it has become an efficient mechanism for transferring multilateral resources for specific projects. Although the amount of resources available for disbursement is small compared to needs, the achievements to date have been commendable. Several features have contributed to the success of the Facility's work: a negotiation process in which the developed countries endorsed a democratic decision-making process and developing countries took seriously the need to achieve efficiency in the Facility's operations; the formation of a dynamic *ad hoc* structure to manage funds involving the UNDP, UNEP and the World Bank; and a multi-disciplinary approach to problem analysis and project implementation (Sandler, 1997).

The Biodiversity Convention has had a double impact. On the one hand, it has significantly increased international public awareness about the importance of biological diversity, an area not well known outside a specialist community before the 1990's, and it also introduced the concept of "common concern of humankind." On the other hand, its implementation has been constrained by the attitude of the US concerning the Convention's provisions for the protection of intellectual property rights. This is particularly significant given that two-thirds of the world's scientific and technological capacity in biodiversity research and biotechnology is located in the US. A second major constrain to the implementation of the Biodiversity Convention is the existence of the much more powerful World Trade Organization. Whenever there is conflict between free trade and protection of biodiversity, a coalition of Liberals and Nationalists in the USA and emerging countries assures the victory of the first. Whenever there is conflict between intellectual property rights and rights of traditional people, the WTO will rule in favor of intellectual property rights as defined in the Western world. The Biodiversity Regime implies a diffuse victory of a wide coalition among Liberals and Nationalists. A proposal for strict regulation for protecting biodiversity, coming from Cosmopolitan-Liberals, was defeated.

Efforts to protect biodiversity and to promote its sustainable development are extremely diverse and not always encouraging. Some countries, such as Costa Rica and Australia, have been strongly promoting the merge between their biodiversity and transnational corporations. In these cases parts of the biotechnology cycle (small in Costa Rica, large in Australia) are settled within the national territories. These bring new intensive capital investments and qualified jobs, either through the re-qualification of workers (in the case of biodiversity, taxonomists and para-taxonomists) or knowledge intensive jobs, provided by local universities or coming from foreign countries. Both cases demonstrate how vital is the existence of a strong rule of law for the development of the biodiversity/biotechnology complex. Australia and Costa Rica are tropical forest countries, with vital democratic regimes and weak internationally organized crime.

The Biosafety Protocol to the Biodiversity Convention, signed in January 2000 after four years of difficult negotiation, can be seen as a compromise between the European culture, which is very concerned about transgenic foods, and the American culture, whose more dynamic sector has promoted research and the use of transgenic crops. This division is well reflected in South America, where Argentina follows American positions and Brazil is divided, with consumers in metropolitan areas tending to follow the European position, while farmers and agribusiness are aligned with the Americans.

Tensions between the world trade regime (WTO) and the MEAs have grown over the second part of the 1990's. The WTO's Committee on Trade and Environment was established to analyze the relationship between trade and environmental policy, and to consider whether modifications on trading rules were required. Up to now, the role of this Committee has been marginal, since conventional trade forces dramatically shape the dynamic of WTO.

The United Nations Commission for Sustainable Development (UN CSD), created by the General Assembly in 1992 and established in 1993, has slowly developed and to date has had questionable influence on international relations. The destiny of the UN CSD was connected to the obsolescence of the main bodies of the UN (especially the General Assembly and ECOSOC) and to the failure of reformist efforts that seemed to hold promise when Butros Ghali was appointed secretary-general of the UN, in January 1992. The UN Commission on Global Governance led by Kenneth Carlson, that produced the 1995 report "Our Global Neighborhood", and the Commission on the Fiftieth Anniversary of the UN, led by Paul Kennedy, were excellent initiatives and presented innovative proposals for reforming the UN. Unfortunately, their impact on the organization has been minimal. At the time of the Rio "Earth Summit", optimists envisioned that environmental issues would drive the UN and be the principal catalyst for internal structural reform. What actually happened was the contrary: the inefficiency and inertia of the UN system obstructed the dynamism of environmental issues. Today, it is evident that a mainstreaming of environmental issues into the UN system depends on deep structural reform within the organization, which may require shifting from an international-intergovernmental system to a transnational-transgovernmental system (Viola, 1998). In the absence of strong actors to champion change, however, "business-as-usual" prevails. Countries such as the US, France, the UK, Japan, Russia, China, India, Brazil, Korea and Indonesia strongly oppose any shift to a transnational-transgovernmental system. Others such as Germany, Italy and Canada are reticent, while the Netherlands, Denmark, Sweden, Costa Rica and New Zealand are mildly favorable. Many representatives of transnational corporations oppose the idea due to perceptions on controls over their global operations; meanwhile, international NGOs are generally in favor. The position of the scientific community on this issue is unclear and that of major religious groups is ambiguous. Besides the traditional inefficiency of the UN bureaucracy, their initiatives were affected by a wrong conception of global governance. Most of them are not cosmopolitan-liberals as they should be in order to facilitate the processes of governance. The founding fathers of the UN were nurtured by cosmopolitanism, but their children were slowly transforming this spirit into something more nationalist and more utopian and unrealistic (almost anti-liberal). The UN did not totally lose the original spirit, but it wasted the last decades going wrong, against history. In this scenario, changes are likely to be incremental; they will never match the actual challenge of global governance.

The climate regime, including the United Nations Convention on Climate Change, signed in Rio in June 1992, and the Kyoto Protocol, signed in December 1997, is the most comprehensive and ambitious of all environmental regimes. As such, it links in a very direct way one of the most crucial economic issues, the use of energy and energy efficiency, with the strongest environmental threat (Rayner and Malone, 1998, Soroos, 1997, O'Riordan and Jager, 1996). The Kyoto Protocol was approved with Liberals and Cosmopolitan-Liberals as key supporters. Actually the Cosmopolitan-Liberals were fighting for a much stricter Protocol, but facing the possibility of a total defeat, they moved to support the liberal position.

The importance of the Kyoto Protocol is hard to overstate. Although 150 countries agreed in principle to reduce global warming pollution in Rio in 1992, it was not until Kyoto that they agreed to legally establish binding limits. If ratified by the required number of countries, the Kyoto Protocol will demand that developed countries reduce their combined output of six greenhouse gases by a modest 5 percent quota below 1990 levels, by 2010. This is a small step toward protecting the Earth from the risks of global warming, but a vital one. Emerging (middle-income) countries do not have commitments in the first period of the Protocol, and most of them have stated that they will not assume commitments at least until 2025. This creates a very strong resistance against Kyoto among Liberals and Nationalists in the USA, but also, in a lesser degree, in all developed countries. The impasses on the climate change regime are a product of the weakness of the Liberal/Cosmopolitan-Liberal coalition and its difficulties to cope with a conservative coalition, formed by Nationalists from all over the world and some Liberals in the U.S. and emerging countries.

Some transnational corporations, like British Petroleum Amoco, Shell, DuPont, Volvo, Toyota, and Honda, support the Kyoto Protocol and have committed themselves to the reduction of emissions from their own operations by more than the agreement would require. The Pew Center on Climate Change has been successfully acting as a coordinator of corporations supporting Kyoto. Corporations' opposition to Kyoto is diminishing at an increasing rate over the last year. Although still expressing doubts about the Kyoto agreement itself, Ford Motor and Daimler-Chrysler recently withdrew from the virulent Anti-Kyoto Global Climate Coalition, citing the desire to take a more constructive approach to the global warming problem.

The Kyoto treaty includes some excellent mechanisms to make it less rigid, but in practice it is not easy to implement them. One of these would allow developed countries to buy excess emission permits from Russia, Ukraine and Belarus, which have a large surplus due to the fact that their economies have collapsed and, consequently, emissions are likely to be much lower than their targets. Such transfers could help countries like the United States and Canada to reach their obligations. A second mechanism would allow developed countries to buy credits from developing countries that reduce their emissions through the adoption of Clean Development Mechanisms (CDM). This would stimulate energy intensive corporations to invest in projects that would reduce emissions in developing countries. That is an excellent idea, but working out rules for the CDM is highly complicated and time-consuming.

As a tool for helping in the analyses of policies related to climate change, we elaborated a classification of country positions on the Climate Regime, as of the year 2000 (Viola, 2000). First, we have the High Carbon Intensive Developed Countries (HCIDC), in which the energy matrix is strongly based on fossil fuels (particularly coal and oil) and/or are continental countries with high-energy consumption in transportation (U.S., Canada, Australia, New Zealand). In these countries, society is highly divided in relation to the implementation of the Kyoto Protocol: most Nationalists and some Liberals are strongly against, but there is a sector of Liberals and Cosmopolitan-Liberals that is strongly in favor. In these societies there is strong support for demanding voluntary commitments from emerging countries.

Second, we have the Middle Carbon Intensive Developed Countries (MCIDC), in which the energy matrix is mostly based on renewable resources or nuclear plants and/or have a very efficient system of public transportation (European Union, Japan, Norway, Switzerland). In these societies there is a growing consensus, led by Cosmopolitan-Liberals and Liberals, in favor of the implementation of Kyoto. Also in these societies there is strong support for demanding voluntary commitments from emerging countries. In the negotiations this group divided itself in two: the European Union from one side and the others. Japan, in particular, has been strongly inclined to be allied with the first group due to difficulties in having high reduction commitments. This is derived from the fact that the country had already progressed in energy efficiency during the 1970's and 1980's.

Third, we have the Former Communist Countries (FCC), whose economies have dramatically declined during the 1990's (Russia, Ukraine, Belarus, Romania, Bulgaria). These countries have strong interest in the implementation of the Kyoto Protocol since they can profit from emission credits. The existing consensus in favor of Kyoto in the FCC is much more related to the possibility of obtaining cash in the middle term than to the concern with the country's long term carbon emissions profile. Their strong interests in Kyoto seem to be a paradox, insofar as these countries are nationalists (also with neo-feudal elements inside the power system). But precisely the presence of this kind of opportunism shows the extreme difficulties on producing a long run agreement on this issue.

Fourth, we have the Super Carbon Intensive Emerging Countries (SCIEC), in which the energy matrix is strongly based on fossil fuels and the exportation of oil is more than a half of the total production (Saudi Arabia, Kuwait, United Arabs Emirates, Iran, Venezuela, Mexico, Nigeria). In these societies most people are against Kyoto, though in some of them, like Mexico and Venezuela, there are significant Cosmopolitan-Liberal sectors in favor of Kyoto.

Fifth, we have the Middle Carbon Intensive Emerging Countries (MCIEC), in which the energy matrix is based on fossil fuels and/or have strong carbon emissions derived from land use/land change/ deforestation (China, India, Thailand, Malaysia, South Africa, Turkey, Brazil). In these societies there is strong support to Kyoto as long as no commitments of their countries are demanded, because they expect to benefit from the Clean Development Mechanism. In those societies there is strong rejection to voluntary commitments for emerging countries. In some of these countries (Brazil, India) there are significant Cosmopolitan-Liberal sectors that are in favor of diminishing the importance of carbon in the energy matrix, fight strongly against deforestation, and promote some kind of commitment for emerging countries.

Sixth, we have the Low Carbon Intensive Emerging Countries (LCIEC), in which the energy matrix is mostly based on renewable resources or nuclear plants and/or have an efficient system of public transportation (Poland, Czeck Republic, Hungary, Baltic states, Philippines, Argentina, Chile, Uruguay, Costa Rica). In these societies there is significant support to Kyoto due to the same reasons of that of

HCIEC. In some of these societies (Argentina, Costa Rica) Cosmopolitan-Liberals support voluntary commitment for emerging countries because they suppose that they will be able to manage the rate of increase of carbon emissions and because they think that this will increase international cooperation.

Seventh, we have the Island/Low Continental Countries (ILCC), which are extremely vulnerable to climate change (Fiji, Malta, Jamaica, Barbados, Bangladesh). In these countries there is massive support to Kyoto, since all sectors will be benefited from it, and also for commitments from the emerging countries. Indonesia is in a very contradictory situation because, as an Island-nation, it is highly vulnerable to climate change, but its economy strongly depends on oil exportation.

Eighth, we have the Stagnant Countries (SC), which are very vulnerable to natural disasters due to their high levels of poverty, but do not perceive themselves as immediately threatened as the IC. These countries are strongly in favor of Kyoto as far as they perceive that they will benefit from the Clean Development Mechanism.

Since the signature of the Kyoto Protocol, the Clean Development Mechanism has occupied a prominent position in the emerging and stagnant countries agenda, due to its voluntary character and its capacity to catalyze new sustainable foreign investments. This has strengthened support to the Kyoto Protocol in emerging and poor countries. The combination of developed countries' obligations to domestic reduction of emissions and the CDM produced a shift in many emerging countries' diplomacy, such as Brazil and India, which previously had a militant position against the joint implementation mechanism. For most emerging and poor countries, the CDM should be considered a very important instrument capable of offering increasing emission reductions at a cheap cost and consequently beneficial to everybody as a market mechanism of resource transfer to emerging and poor countries, also capable of establishing cleaner development patterns.

In spite of years of intense negotiations, the Kyoto Protocol is far from being ratified. This could be attributed to the combination of three major factors. A first difficulty is the weak leadership of major greenhouse gases emitting countries and lack of progress in meeting stabilization targets, settled by the Kyoto Protocol, in crucial developed countries. Compared with the baseline of 1990, in 1999 carbon emissions were the following: United States (commitment: 7% lower by 2010) was 14% higher; Canada (commitment: 6% lower by 2010) was 17% higher; United Kingdom (commitment: 12% lower by 2010) was 15% lower; Germany (commitment: 21% lower by 2010) was 17% lower;

The Netherlands (commitment: 6% lower by 2010) was 17% higher; Japan (commitment: 6% lower by 2010) was 4% higher. Among developed countries, only U.K, Germany, Sweden and Denmark were in a trend of compliance with their commitments by 2010. Even countries like The Netherlands and Austria, which have been for decades at the world vanguard in environmental policies, are having strong difficulties in addressing their commitments. The weak leadership of OECD countries should also be explained by the fact that their populations are ambivalent in relation to cosmopolitan positions.

A second factor would be the reluctance of part of the large (and fast growing carbon emissions) emerging countries (China, India, Mexico, Brazil, South Africa, Indonesia, Malaysia, Thailand) to commit to reduction in the future rate of growth of their emissions. The third factor results from the disagreement among the scientific and economic community about respectively, the pace and extent of climate change, and the costs and benefits of alternative response strategies. Anyhow, this disagreement seems to start to narrow after the publication of the 3rd IPCC Report, in January 2001.

Even with the withdrawal of the U.S. and other difficulties to be ratified, Kyoto is quickly becoming a crucial source of soft law. The Kyoto Protocol is already shaping technological development in favor of de-carbonizing the world economy. In the expectations of the scientific community and corporations, wind, solar and hydrogen powers are rapidly moving into the main stream of the energy matrix. According to many analysts' estimations, renewable energy (solar PV, wind, modern biomass, geothermal, etc.) will jump from less than 2% of the world energy matrix, in 2000, to around 20%, by 2020. The recent agreement among the Iceland government, Daimler Chrysler and Shell in order to develop a large-scale test on hydrogen powered cars is a clear demonstration of the shifting expectations. This paper is not the right place to explain the role of emerging technological innovations in global governance. But, especially in the last two decades, we have seen many examples (not only environmental ones) of technological innovations creating very positive conditions for governance (The Montreal Protocol is a good example). In this sense, a good reason to be optimistic is the increasing dependency of these countries on technological innovations. On the other side, pessimist analysts make their argument exemplifying with the fact that the plan for energy development, elaborated under the Bush administration, is mostly based on the expansion of oil, coal and natural gas.

Beyond the specific fate of the Kyoto Protocol, there is little doubt that climate change is the greatest threat to the future of humankind. In the last year the intensity of severe storms and floods in different parts of the world has intensified, probably as a demonstration that we are already undergoing a soft version of climate change: the increase of extreme meteorological phenomena. These events dramatically demonstrate how vulnerable societies are, especially the poorest: stronger than historically recorded floods in China, Mozambique, Venezuela, United Kingdom and India; devastating drought and famine in Central Asia and the Horn of Africa; extremely severe storms in France and Italy; severe drought in the Great Plains and Western United States; glacier retreat in the Andes, the Rockies and the Alps ranges; devastating fires in Indonesia, Australia, the Northern Amazon and Western United States.

The Cosmopolitan-Liberals have already built up strong support from the Liberals in the European Union and Norway. The Cosmopolitan-Liberals are already significant in USA, Canada, Australia, Switzerland, Japan and New Zealand and also in some key emerging countries (Brazil, Argentina, Chile, Mexico, Hungary, and Poland), where they still have not gained the support of Liberals. The ratification of Kyoto and the deepening of the climate regime depend upon the capacity of Liberal-Cosmopolitans to win the support of Liberals in the countries mentioned above, and to increase their presence in other key emerging countries where they are still quite inexpressive (India, Thailand, Malaysia, Philippines, South Africa, Turkey). Also, the fate of the climate regime depends upon some development of Cosmopolitan-Liberals and Liberals in China and Russia.

During the Conferences of the Parts, after the signature of the Kyoto Protocol (Buenos Aires, 1998; Bonn, 1999; and The Hague, 2000), countries formed four large groups of negotiation: the European Union; the Umbrella Group (USA, Canada, Japan, Australia, New Zealand, Norway, Russia and Ukraine); the G77/China; and the Small Island States Alliance. Within the G77/China, there were two sub-group differentiations in some issues: the less developed countries and the oil exporter countries. The most relevant cleavage has been between the European Union and the Umbrella Group. Basically, the first proposes the following: that countries' targets should be accomplished mostly through domestic efforts to reduce emissions; a restricted role for the flexibility mechanisms (emission trade, joint implementation and clean development mechanism); a strict regime of compliance; and against deducting carbon sinks from the national emission. On the other side, the Umbrella Group strongly reacts to these propositions. The G77/China have been strongly divided in those issues. A second relevant cleavage has been between the European Union, the Umbrella group and the Small Island States proposing that middle income developing countries should assume voluntary commitments for reducing the rate of increase of their carbon emissions in the first period of commitments, on the one hand, and the G77/China, on the other. The third cleavage has been between the G77/China and the Small Island States, proposing a large package of free transfer of climate related technology from developed to developing countries, on one side, and the European Union and the Umbrella group, on the other side, proposing a very small package. The failure of the Hague Conference in reaching an agreement was mostly due to the differences between the European Union and the Umbrella Group in relation to the amount of carbon sinks to be computed as deduction from the national emissions.

BRAZIL IN GLOBAL GOVERNANCE

Brazil participated in the foundation of the League of Nations, in 1920, and kept its commitments with the organization until its collapse, in the late 1930's. The Brazilian contribution to the Second World War effort was significant. Unlike other Latin American countries, Brazil dispatched troop units to Europe to participate in combat. Because of the participation in the war effort, Brazil was considered by the U.S. and the U.K. as the crucial country in Latin America for the construction of a New World Order. Brazil participated in the foundation of the Bretton Woods and United Nations systems as an allied nation, having the prerogatives of the victorious. Brazil has been a member of the United Nations, the International Monetary Fund, the World Bank and the General Agreement on Tariffs and Trade (GATT) since the beginning. At no moment since 1945, has Brazil questioned the main role of these institutions. Many times Brazil has been member of the U.N. Security Council and, in some periods, has presided it. Brazil has also been a major player in the development of regional institutions: the Organization of the American States, the Inter-American Military Treaty, signed in Rio in 1948, and the Inter-American Development Bank.

Brazil has been a very active member of the international community during the whole 20th century. During the second half of the 20th century, the Brazilian economy has grown above the world average and, since the 1970's, the country has become one of the ten largest world economies. In spite of the strong economic growth, Brazil has not been capable of overcoming its extremely unequal income distribution. During the whole century, Brazil has always been one of the worst countries in the world in terms of income distribution. The links between Brazil and the rest of the world have been mostly based on the principles of world peace, free trade, cultural diversity and religious freedom. Political democracy has been almost always a stated goal of Brazilian society, though there have been several periods of political authoritarianism. Between the middle 1960's and the late 1980's, there were some sectors of the military/foreign policy establishment that attempted to build up nuclear weapon capabilities as a platform for achieving great power status in the global arena. Between 1960 and 1990, Brazilian society experienced a strong tension derived from the growing of two contradictory economic-political trends: from one side, the development of transnational corporations, private initiative and Western values; from the other side, the expansion of state-owned corporations, bureaucracies and nationalist values (Martins, 1986).

The Brazilian position in the emerging arena of global environment, in the early 1970's, corresponded to its economic profile. The Brazilian stand was based on the principle that the main pollution was poverty; environmental protection should only come after economic development had dramatically increased, in per capita income, to the level of the developed countries (Guimaraes, 1991). At the Environment Conference of Stockholm, in 1972, Brazil and China led the formation of a coalition of Third World countries contrary to recognizing the importance of emergent global environmental problems. Brazil stand in Stockholm was based on three principles: defense of unrestricted national sovereignty on what is related to the use of natural resources; environmental protection should come only after reaching high per capita income; and the burden of paying for the protection of global environment should be an exclusive responsibility of the developed countries (Viola, 1997).

During the 1990's, Brazil has consolidated a democratic regime, expanded the rule of law (though more is still needed), opened its economy to foreign trade and investment and developed a foreign policy based on a new approximation to Western countries. By the end of the 1990's, Brazil continues as an important country in the world arena. Brazil has 6% of the world surface and 2.7% of the world population, growing 1.3% per year; with a density of 20 inhabitants per square kilometer, and 80% of its population urban. Brazilian GDP is around 650 billion dollars, considering exchange rates, and around

1,050 billion dollars, considering purchasing capacity; 4,000 dollars of per capita income, considering exchange rates, and 6,400 dollars of per capita income, considering purchasing capacity; around 1.9 % of the world economic output, considering exchange rates, and around 2.4% of the world economic output, considering purchasing capacity. The proportional value added in the economy is: 8.4% agriculture, 28.8% industry and 62.8% services. The gross domestic investment is 21% of the GDP, the government revenues (federal, state and municipal) are 32% of the GDP and the overall budget deficit is 4% of the GDP (discounting the interest of the public debt, the country has a surplus of almost 3%, considering the GDP). The annual foreign direct investment in the period 1996 - 2000 was 27 billion dollars, a record in Brazilian history and among the highest in the world. In 1999, the public debt was 300 billion dollars and the total debt service amounted to 50 billion dollars. Brazil participates with only 0.9% of the world foreign trade (mostly commodities and intermediate technology manufactured products) and ranks 64th in the Human Development Index. Brazil produces around 2.5% of the world carbon emissions (considering industry, energy, transportation and land use/land change), with a carbon emissions level of more than 2 metric tones per capita. Brazil has 5.5 million square kilometers of forests, with a deforestation rate of 0.5% per year; around 20% of the world's terrestrial biodiversity, and around 15% of the world's fresh water supply, of which only 0.5% is used.

There is some dissonance that at the core of the Brazilian position in the global arena. Income per capita is slightly below the world average, if calculated according to exchange rates, and slightly above the world average, if calculated according to purchasing capacity. Thus in both cases income is unfairly distributed, since the 1% richest sector of the population gets 13.8% of the national income and the 50% poorest sector of the population gets 13.5% of the national income. The economy is still significantly more closed than the world average because the export capacity is low, only 8% of the GNP. Carbon emissions per capita are well below those of developed countries, but above the average of middle income countries. Carbon emission per unit of GNP doubles the world average, well above that of developed countries and slightly above the average of middle income countries. Carbon emissions coming from the modern sector of the economy (industry, transportation, housing and modern agriculture) is very low because the electricity matrix is based on hydropower. Consequently, carbon emissions coming from the modern sector are well below the average of the middle income countries. Carbon emissions coming from land change and traditional land use are extremely high. On this point, modern Brazil is hostage to traditional Brazil. In terms of energy efficiency, modern Brazilian economy is doing much better than most middle income countries, and traditional Brazilian economy is doing much worse than most middle income countries (Fearnside, 1999).

The Cardoso administration produced a paradigmatic shift in Brazilian foreign policy. It strengthened economic ties with the U.S., the EU and Japan, and abandoned previous attempts of having close ties with China, India and Russia (Onis, 2000). Brazil also strongly supports the West on most global issues: Human Rights Protection, Intellectual Property Rights, expansion of the International Monetary Fund, the World Bank and the World Trade Organization, Protection of Women and Reproductive Health, and condemnation of Nuclear Proliferation and Terrorism. However, the Ministry of Foreign Affairs kept some non-Western approaches (continuity in the affiliation to G77 in U.N. forums, some conflict with Argentina because of its military alliance with the U.S.) as a way of looking for wide support for its candidacy to the U.N. Security Council as a new permanent member.

Brazilian participation in the global arena in the 1990's was shaped by four major trends: the acceptance of limits to the principle of National Sovereignty; the clear decision of attempting to become one of the new permanent members of the United Nations Security Council, when it is enlarged (a subject of diffuse consensus); the strong commitment to strengthen the Mercosur, as a way of coping with the integration challenges of globalized economy; and the support of universalism and contractualism in issues like Human Rights, Political Democracy, Reproductive Rights, Women's Rights, Social Equity and the Environment. The four principles are far from the Brazilian foreign policy of the 1970's and 1980's, based on the principle of absolute national sovereignty and the attempt to become a great power in the global arena, including building a strong military force.

The Treaty of Asuncion that created Mercosur, signed in 1991, was developed under Brazilian leadership. The treaty was crucial in two dimensions: it definitively ended the regional rivalry between Brazil and Argentina, by deepening the civilian argument in favor of diminishing the importance of the military on both sides, and promoted rapid increase in the flow of goods, capital, people and information among the four countries (Fauriol and Weintraubb, 1995). Some months after Asuncion, the presidents of Brazil and Argentina signed the treaty that created the Common System for Accounting and Control of Nuclear Materials, submitted to the regulations of the International Atomic Energy Agency. Both Senates rapidly ratified the treaty and, since 1991, the Brazil/Argentina relationship made a turning point, definitively overcoming the profile of nuclear proliferation rivalry that prevailed between the 1960's and the 1980's.

In 1994, after significant efforts for trade convergence, led by Brazilian diplomats, the countries signed the Ouro Preto Protocol and deepened Mercosur, though still falling short of committing to the building up of supranational institutions. Mercosur was strongly shaken during the global financial crisis in the emerging markets in 1997/99, particularly after the devaluation of the Brazilian currency, in January 1999. However, more recently Mercosur has become an excellent means of promoting commitments to benefic public policies in all countries (trade liberalization, fiscal equilibrium, coordination of macroeconomic policies, middle-term prospect of a common currency). In 2000, the Mercosur framework supported the social-economic forces committed with responsible policies against populists forces. During 1996-99, the four Mercosur countries, led by Brazil, negotiated an additional Environmental Protocol that was ready to be signed by the end of 1999. According to this Protocol, the process of economic integration should include, at its core, environmental protection. The Protocol states that Mercosur is committed to clean air, clean water, the appropriate disposal of solid waste, the careful management of hazardous waste, the preservation of biodiversity, the integrity of the ozone layer and the stability of global climate. The negotiated Protocol reflects the higher environmental standards present in the Brazilian economy. Because of apprehension from Argentine and Uruguayan entrepreneurs, in terms of undermining their competitiveness, by early 2001 it has not been signed.

During the final negotiations of the GATT Uruguay round, in 1993, Brazil strongly supported trade liberalization, based on the recent opening of its economy, and was generally allied with USA, Australia, Canada and Argentina against the agriculture protectionism adopted by the European Union and Japan. Also, Brazil strongly supported the foundation of the World Trade Organization, in January 1995. However, Brazil opposes a general initiative raised by the U.S. Vice-President Gore, for starting a new negotiation round based on the settlement of environmental protection clauses in international trade. During the second ministerial meeting of WTO, in Singapore (1996), Brazil had a very defensive stand. It strongly opposed the establishment of labor and environmental standards in international trade. Also, Brazil was not part of an agreement involving developed and some emerging countries for a complete liberalization, in eight years, of trade on information technology products. During the difficult negotiations that preceded the Seattle ministerial meeting, in 1999, Brazil assumed the leadership of the developing countries, strongly confronting developed countries on their trade barriers in relation to farming and industrial products, and opposing labor and environmental conditions, in a proposed new round of trade liberalization.

During the negotiations for the creation of a Free Trade Area of the Americas (FTAA), initiated during the Miami Summit in 1994, Brazil always defended a slow pace: not before 2005 and with no previous partial implementation. The Brazilian position was based on the need to prepare its industry for competition, before completely opening its economy to the American market. During the period between 1994-97, the Brazilian stand implied conflict with the official policy of other Latin-American countries (like Argentina, Uruguay and Chile) and USA, that were in favor of a faster pace. Finally, in 1997, the American Congress denied fast track legislation to Clinton, and the slow pace became a real constraint for the players favorable to fast track. The tensions between Brazil and the rest of the American countries, on what is related to the pace of FTAA, have restarted in 2001when Bush signaled his interest in moving quickly toward the FTAA.

During his short tenure as Minister of Foreign Affairs in 1992-93, Cardoso persuaded the Franco government of the necessity of moving the Space Program from military to civilian control, in order to get highly needed technological exchange for the development of the program. Also following Cardoso's initiative, the Brazilian Senate ratified the Nuclear Free Latin American Zone Treaty (Tlatelolco), in 1994, significantly later than other Latin American countries. Immediately after starting his presidency, in 1995, Cardoso started a series of foreign policy movements that put Brazil in complete alignment with the West. In 1995, Cardoso completed his work in the missile area, signing the Brazilian adhesion to the Missile Technology Control Regime (MTCR). During the 1980's and the early 1990's, the military had strongly opposed the MTCR. Immediately after the signing of MTCR, Brazil was invited to participate in the project for the construction of the multinational space station, to be set in orbit around 2008. Consequently, Brazil became a member of the select Space Club. Finally, a last step in Brazilian credibility as a peaceful member of the world community was taken in 1999, when Brazil overcame some American distrust and signed an agreement with USA, Italy and Ukraine, for launching commercial satellites from the Alcantara base, which offers strong competitive advantages due to its location near the Equator.

Brazil was one of the leading countries in designing the Nuclear Test Ban Treaty, signed in 1995, and supports, with some resistance from the military, the Land Mines Ban Treaty, signed in 1997. It took more than two years of his presidential term for Cardoso to persuade the military and diplomatic establishment of the need to sign the Nuclear Non-Proliferation Treaty. By the time Brazil signed (1997), it was one of the four countries (besides North Korea, Iraq, and Libya) that still had not signed the Treaty. Brazilian diplomacy was very rigid on this issue and continued arguing that the treaty was unfair, as if the world were still in the 1970's. The long opposition to adhesion to the Nuclear Non-Proliferation Treaty is one of the major errors of Brazilian diplomacy in the 20th century. By early 2001, the Brazilian Senate has not yet ratified the Treaty, though there is no more significant resistance to ratification. Also in 1995, Cardoso persuaded the military of the need for better coordination between the U.S. and Brazil on fighting Narco-traffic and, consequently, both countries signed a new cooperation agreement.

The U.S. and Western European countries were very pleased with the new policy, adopted by Cardoso, of breaking up the State monopolies in petroleum and telecommunication sectors. Since 1996, American and European telecommunication corporations have invested very strongly in Brazil. During Clinton's visit to Brazil in 1997, Brazil and USA signed important agreements for cooperation in several areas: higher education, science and technology, environment and drugs. According to the environmental agreement, both countries committed themselves to exchange ideas as deeply as possible in order to reach more common positions on several issues: climate change, depletion of the ozone layer, conservation and sustainable use of biological diversity, deforestation, desertification, ocean pollution and management of hazardous wastes and toxic substances. In order to further their common agenda for the environment, the United States and Brazil stated their intention to hold regular high level consultations, on at least an annual basis, to discuss priority issues of mutual concern in the area of environment and sustainable development. Such consultations should involve the participation of the relevant governmental agencies, from both sides, concerned with environmental protection and sustainable development.

On three of the United Nations deliberations on military intervention during crisis situations in the 1990's, Brazil was reluctant to approve interventions: Iraq, 1991; Haiti, 1994; and Kosovo, 1999. The Brazilian position in relation to the Persian Gulf War was highly costly in terms of the country's credibility, but at least this was the last time that Brazilian military had some influence in the definition of crucial foreign policy issues. A Brazilian contingent participated in the UN observer force that guaranteed the October 1994 elections in Mozambique, and in the UN observer force in Bosnia, in 1995. Brazil has sent peacekeeping military contingents to Angola (1998) and East Timor (1999). In the Angolan case, Brazil participated in the core of the U.N. forces.

In all World Summits related to Human Rights – the Vienna Conference on Human Rights, in 1993; the Cairo Conference on Population and Development, in 1994; and the Beijing Conference on Women Rights, in 1995 - Brazil strongly aligned with Western countries' liberal coalitions, in the promotion of individual rights against State or traditional institutions (Nickel and Viola, 1994). In the Cairo and Beijing Conferences, the Brazilian delegation had significant participation of feminist leaders from the civil society and, consequently, had a leading role in promoting universalistic and liberal causes. Differently from other Latin-American countries, the Brazilian Catholic Church was not successful in shaping the international stand of the country in a conservative direction.

Brazil has supported, since the beginning, all the treaties related to global environment signed throughout the 1990's: the Basel Treaty, for controlling and discouraging international trade on hazardous waste, in 1989; the London Amendment to the Montreal Protocol, establishing technology transfer mechanisms for substituting CFC, in 1990; the Madrid Amendment (1991) to the Antarctic Treaty, extending for fifty years the moratorium to economic activities on that Continent; the Convention on Biodiversity (1992); the creation (1991) and expansion (1993) of the Global Environment Facility; and the Biosafety Protocol (2000). In the development of all these treaties Brazil had middle to low profile participation, with the exception of the Biodiversity Convention.

During the negotiations of the Biodiversity Convention (1990-92), Brazil had a leading role derived from its position as the richest country in the world in biodiversity. One of the most important issues at stake, during negotiation in the Biodiversity Convention, was related to the connections between biodiversity and biotechnology. On one side, the U.S. (where 2/3 of the world's biotech industry is installed) strongly defended the principle of intellectual property rights, according to the conventional definition. On the opposite side, a coalition of countries rich in biodiversity led by Brazil, defended the right to royalties from biotech products derived from biodiversity, for the countries where biodiversity is located. The Convention, in May 1992, implied a partial victory for the coalition of countries led by Brazil, since the convention did not recognize the full principle of intellectual property rights; defining it in a broad sense, giving rights to indigenous people.

During the 1990's, Brazil progressed dramatically towards a deep convergence with Western democratic capitalism: deep market reforms; created, for the first time in history, an economy that is more based on market mechanisms than on State regulations; direct foreign investment from developed countries have been among the highest in the world; independent courts have operated in all dimensions of social/economic life; open public accounts and fiscal responsibility became goals of the national society; corruption in politics has diminished, though continues relatively high; political democracy became more deeply rooted in the social web and the political culture; respect for human rights became core goals of domestic and international public policies; attempts to build up a strong military machine were abandoned and the military have gradually been civilized; public awareness about environmental protection has grown continuously; and the country's foreign policy has achieved, for the first time, some capacity of regional and global leadership, giving substantial foundation to the pretension of becoming a permanent member of the U.N. Security Council. The only area in which there has not been significant transformation is income distribution and its consequences since the early 1980's; growing urban and rural violence. From this point of view, there has been a vicious circle between the difficulties in overcoming the heritage of slavery and the trend of the globalized economy to produce a new digital/information divide. The growth of urban/rural violence in the 1990's has been strongly related to the production, trade and consumption of illegal drugs and other illicit activities, like smuggling in electronic products and trade of endangered species. By the end of the 20th century, Brazil has successfully adapted to globalization from an economic point of view, but highly unequal income distribution and growing social violence and crime are still a major threat to the stability of the society.

The Brazilian position in the global arena is ambivalent, in spite of the significant and positive shifts of the Collor and Cardoso administrations, in the last decade (1990's), to tune in and converge with the liberal policies of OECD countries. A significant part (military, foreign affairs, industry) of the Brazilian state bureaucracy is nationalist (although pragmatic) and traditionally afraid of any kind of cosmopolitanism (always associated with Amazonian fears). The civil society, in spite of the active participation of many Brazilian NGOs in global networks of social action on several issues, defends a diffuse and utopian nationalism (almost anti-liberal). Brazilian native corporations are taking good advantage of globalization, but they continue more nationalist than liberal. In this context, it is hard, to any local authority or citizens, to avoid back and forth movements in essential issues of global governance. The hope for a better Brazilian contribution to global governance depends more upon external (than internal) conditions. The Brazilian will for modernization is a historic certainty; it would be very hard to change this feeling (even for a leftist national government). In the present circumstances, we can expect that the country will follow (even if not very closely) the mainstream of globalization and global governance. Brazilian participation in the climate regime is a good example of its ambivalent position in terms of global governance.

BRAZIL IN THE CLIMATE REGIME

The Brazilian government's position in the Rio-92 Conference was founded on the following: global environmental problems are very important and priority should be given to them by the international community; the cause of global environmental problems has had differentiated historical responsibilities and that should be reflected in the measures for coping with them i.e., rich countries should assume the higher cost. During UNCED negotiations (1990-92), the Brazilian government was progressively retreating from its nationalism (1972-88) and was assuming a liberal position: it had a leading role in writing the Biodiversity Convention; it facilitated negotiations and the agreement in the climate change convention; and it supported funding commitments in relation to Agenda 21, though the nationalist heritage emerged when Brazil supported Malaysia in its opposition to a forest convention (Viola, 1997).

The Brazilian position in UNCED was liberal, with some components of cosmopolitan-liberalism, though inserted in a State structure in which a declining nationalism still prevailed. In spite of its decline, nationalism was still hegemonic in fundamental sectors of the State like the armed forces. The Brazilian liberal stand was a product of four factors: a) the crisis of the protectionism model, during the 1980's, produced a significant public sensitivity towards sustainable development (even when the adhesion to this ideal was very diffuse); b) the young president Collor developed a sensibility towards environment because of his international background; c) the Brazilian energy matrix is mostly renewable, in a context where most of the countries are strongly dependent on fossil or nuclear fuels. In this sense, Brazil is sort of a world vanguard in terms of the development of a new energy matrix, oriented towards sustainability; d) Brazilian society was becoming aware of the Amazon forest (the world's largest biodiversity reservoir) and consequently was moving from supporting deforestation (predominant up to 1988) to supporting preservation and conservation.

Brazil's stand in relation to the climate change regime has been full of internal tensions: cosmopolitan-liberals views in the direction to support a strong climate regime and nationalistic views and fears to dilute the treaty. This tension has been present simultaneously in many Brazilian positions during the whole history of the negotiation. In COP1 (Berlin, 1995) and COP2 (Geneva, 1996) there was some weakening of Brazilian liberalism and some revival of nationalism: Brazil kept a defensive approach, focusing on the historical responsibility of the developed countries, and stood strongly for the right of emerging countries to develop, without considering their proportion on new carbon emissions. (Viola, 2000).

During the 1997 negotiation of the Kyoto Protocol, Brazil had a very important green initiative: the creation of the Clean Development Fund. However, in most of the issues, Brazil had a highly contradictory position derived from its belonging to the G77: it manifested some support to the irresponsible nationalist position presented by emerging countries strongly dependent on fossil fuels (China, India, Indonesia and South Africa), instead of assuming a clearly differentiated position framed on its renewable energy and its vast tropical forests carbon sink. This differentiated and pro-active approach would have implied taking a position between liberal and cosmopolitan-liberal: alliance with the European Union and Island-States on the question of strong emissions reduction; alliance with USA, Canada, Australia, Japan, Russia, Ukraine and Argentina on the question of having a strong role for flexibility mechanisms (tradable emission permits, joint implementation and Clean Development Mechanism) and alliance with G77/China on the question of technology transfer.

Until the beginning of 2000, few actors constituted the Brazilian domestic arena of the climate change regime. However, during the year 2000 there was an increasing interest in climate change coming from academia, NGOs and corporations. Recognizing this change, in June 2000 the government established the Brazilian Forum for Climate Change (with participation of government officials from the federal, state and municipal level, NGOs, academia and business).

The Ministry of Foreign Affairs in coordination with the Ministry of Science and Technology are in charge of the negotiations on the climate regime. Until

2000 there was poor participation of non-governmental organizations, both in decision making and in the process of policy implementation. Some large corporations began to be interested on climate change due to the influence of the Business Council for Sustainable Development. There was also the participation of some scientists, responsible for offering technical backup to Brazilian diplomacy in the multilateral debates. In parallel, there is also an effort of diplomacy in guaranteeing that Brazilian scientists participate in the Inter-governmental Panel on Climate Change. Congress almost does not participate in decision-making of Brazilian policy in environmental regimes. Its function is restricted to the ratification of agreements signed by the Executive, which happens without significant involvement of civil society. Brazilian foreign policies on climate change have been internally consistent in the last years (after the open clashes between environmental and foreign policy officials, at the time of UNCED): restricted number of actors, concentration of decision making in the highest steps of the bureaucracy and good articulation among the agencies inside the bureaucratic structure. This profile could be changing since 2001, due to the newly created multi-stakeholder forum.

In order to achieve emission reductions from developed countries, the Kyoto Protocol allows that, besides the reduction initiatives taken domestically, the developed countries accomplish this goal through coordinated action with emerging and poor countries. This would be performed with the adoption of the Clean Development Mechanism (CDM), adapted from the original Brazilian proposal for a Clean Development Fund. The Brazilian proposal was formally presented during the negotiations of the Protocol (Bonn Meeting, in August 1997) and after several sessions of negotiations it was inserted, with modifications, in the text of the Kyoto Protocol. During the negotiations of the Clean Development Mechanism there was, for the first time in the evolution of the climate regime, a significant convergence between Brazilian and American diplomacy: Brazil moved in the direction of accepting flexibility mechanisms for reducing emissions, and USA moved in the direction of accepting flexibility beyond the Annex 1 countries. The Clean Development Fund had as a goal the possibility of developed countries supplying financial help to Non-Annex 1 countries that were strongly committed to the use of cleaner technologies. Without the punitive character of the Brazilian original proposal, that established penalties for Annex 1 countries that failed to reduce emissions, almost all countries supported CDM. The Brazilian Clean Development Fund was the most cosmopolitan-liberal position ever assumed by Brazilian diplomacy throughout the formation of the Climate Change Regime (not considering as the country's official position the strongly cosmopolitan-liberal approach assumed by Lutzenberger's Ministry of the Environment, in its confrontations with the positions of the Ministry of Foreign Affairs during the 1990-92 Prepcoms).

Since Kyoto, the CDM started to occupy a prominent position on the Brazilian agenda, because of its voluntary character and its capacity to catalyze new sustainable foreign investments. The combination of developed countries' obligations to domestic reduction of emissions and the CDM produced a shift in Brazilian diplomacy that has been previously militant against the joint implementation mechanism. For Brazilian diplomacy, the CDM should be considered only as an auxiliary instrument capable of reducing emissions, at a cheaper cost. Consequently, it would be beneficial to everybody as a market mechanism of resource transfer to emerging and stagnated countries, capable of establishing cleaner development patterns.

In spite of the growing support for the Clean Development Mechanism, most emerging countries have kept their stand against assuming voluntary commitments, consequently increasing the rejection of Kyoto among American senators. Brazil and Argentina are very important emerging countries that are not highly dependent in fossil fuels, consequently finding themselves in a favorable situation for assuming commitments for reducing the rate of increase of their carbon emissions. In COP 4, in Buenos Aires, Argentina proposed that emerging countries should assume voluntary commitments. This was proposed without previous consultation to Brazil (unexpected because of Mercosul), which reacted negatively, following the G77/China position. In October 1999, Argentina formally announced that it was assuming commitments for reducing carbon emissions in 12%, by 2010. This, in comparison to what would be its business as usual emissions profile by that year. According to the announcement, most of the emissions reduction would come from strengthening energy efficiency. There are many doubts about the precision of the Argentine calculations and about its readiness to effectively implement the commitments, but there is consensus that the formal announcement had strong political and symbolic impact. Brazil is in a more difficult situation than Argentina for assuming voluntary commitments: on the one hand, the energy matrix is more based on renewable sources than Argentina, but, on the other, the rate of deforestation and land conversion is much higher than Argentina.

There is always a major opportunity for Brazilian global leadership in the climate change regime. This is dynamically derived from the acceptance of voluntary commitments. The Brazilian stand on voluntary commitments is derived from a combination of vulnerability and ideology. In order to assume voluntary commitments, Brazil would have to shift to a policy of cracking down on deforestation. In order to be significant in the national carbon balance, deforestation in the Amazon should be reduced by around 70% of the present annual rate (from around 0.45% of the Amazonian forest, to around 0.15%). Though there is strong support in public opinion for curbing deforestation, it is difficult to assess how deep support would go if strong confrontation with the coalition of interests supporting deforestation in the Amazon occurs. These Amazonian interests form a powerful network that defends the deforestation status quo (even if they are confronted in other issues): most of the timber industry, big and small landowners, land-less people movements, gold-miners and urban ruling elites. Because of the peculiarity of the Brazilian federative arrangements, those interests are strongly represented in the Brazilian Congress.

A second major implication of a potential shift in Brazilian policy related to voluntary commitments would be the moving away from the G77, in the United Nations arena. There is significant support for this change in Brazilian society, due to the shift from nationalist to liberal economic policy in the 1990's. However, there is a significant sector of the foreign policy decision makers that still support the association to the G77, due to ideological reasons (keeping Brazil as a leader of the South in negotiations with the North) and because they suppose that continuity in G77 would be crucial for getting a Brazilian permanent seat on the Security Council, in case of a reform of the UN structure. The opposite happens with Argentina (country where environmentalism is significantly weaker than in Brazil): its environmental foreign policy is shaped by its non-alliance to the G77 and its commitment to be as convergent as possible with Western countries in any global multilateral forum. Most of the Argentine ruling elite suppose that the alignment with the West would favor their role in the global arena. The Brazilian situation is very dynamic and two driving forces push in the direction of following Argentina in the climate regime standing: because of the continuous trend towards huge foreign direct investment and strong development of the new information economy, Amazonian deforestation is becoming more and more dissonant with the mainstream of the Brazilian economy; and the support for curving deforestation, continuously growing in public opinion, as shown in 2000 during the high profile debate about the new Forestry Law. The Cardoso administration is a very complex and unstable coalition among Liberals, Cosmopolitan-liberals and some Nationalists. The distancing of Nationalists and the strengthening of Cosmopolitan-liberals in the governmental coalition is likely a prerequisite for a shift in the Brazilian stand in favor of voluntary commitments.

BIBLIOGRAPHY

BROWN et al. 2001: State of the world. A Worldwatch Institute report on progress toward a sustainable society. New York, Norton.

CLARK, W. VAN EIJNDOVEN, C. and JAEGER, J. (eds.) 1998 *Learning to manage global environmental risks: a comparative history of social responses to climate change, ozone depletion, and acid rain* Cambridge, MIT Press

FEARNSIDE, P. 1999 "*Forest and global warming mitigation in Brazil*: opportunities in the Brazilian forest sector for responses to global warming under the Clean Development Mechanism" IN: Biomass and Energy # 16

FAURIOL, G., WEINTRAUB, S. 1995 "*U.S. Policy, Brazil and the Southern Cone*" Washington Quaterly 18/3

GUIMARÃES, R. 1991 *The Ecopolitics of Development in the Third World: Politics and Environment in Brazil.* Boulder, Lynne Rienner

GEHRING, T. 1994 *Dynamic international regimes: institutions for international environmental governance.* Frankfurt am main: Peter Lang

HAAS, P., KEOHANE, R., LEVY, M. (eds.) 1993 *Institutions for the Earth:* sources of effective environmental protection. Cambridge: MIT press

HELD, D. (1995) *Democracy and the global order: from the modern state to cosmopolitan governance,* Cambridge: Polity Press.

HELD, d. et al. (1999) *Global transformations*, Stanford University Press.

LEIS, H. (2000) "A politico-communicative model to overcome the impasse of the current politico-technical model for environmental negotiation in Brazil", In: CAVALCANTI, Clovis (ed.) *The Environment, sustainable development and public policies*, London, Elgar.

LINKLATER, A. (1998) *The transformation of political community*, Cambridge, Polity Press.

MARTINS, L. 1986 "*The Liberalization of Authoritarian Rule in Brazil*" In: O'DONNELL, G., P. SCHMITTER, P., WHITEHEAD, L. *Transitions from Authoritarian Rule:* Latin America. Baltimore, The John Hopkins University Press

NICKEL, J., VIOLA, E. 1994 «Integrating Environmentalism and Human Rights» In: *Environmental Ethics*, v. 16, 1994,

ONIS, J. 2000 "*Brazil's New Capitalism*" In: Foreign Affairs 79/3

O'RIORDAN, T., JAGER, J. 1996: *Politics of climate change: a European perspective.* London, Routledge

PORTER, G., BROWN, J., 1996 *Global environmental politics.* Boulder, Westview Press.

RAYNER, S., MALONE, E. (eds.) 1998 *Human choice and climate change.* Colombus, Batelle Press.

ROSENAU, J. (1997) *Along the domestic-foreign frontier* Cambridge University Press.

SANDLER, T. 1997 Global challenges: *an approach to environmental, political, and economic problems.* Cambridge University Press

SANTOS, B. de S. (1995) *Toward a new common sense*, New York, Routledge.

SOROOS, M. 1997: *The endangered atmosphere. Preserving a global commons.* University of South Carolina Press.

VIOLA, E. 1997: "*The environmental movement in Brazil: institutionalization, sustainable development and crisis of governance since 1987*" In: MAC-DONALD, G. , NIELSON, D., STERN, M. (ed)

Latin American environmental policy in international perspective. Boulder, Westview Press.

VIOLA, E. 1998: "*Globalization, environmentalism and new transnational social forces*" In: CHUNG, C., GILLESPIE, B. (org.) Globalization and the environment. Paris, OECD

VIOLA E. 2000 "*The Global Politics of Climate Change: How strong are the forces supporting sustainability*" 10th World Congress IRSA, article # 305

WEISS BROWN, E., JACOBSON, H. (eds.) 1998 *Engaging countries: strengthening compliance with international environmental accords.* Cambridge: MIT Press

YOUNG, O. (1997) "*Global governance: toward a theory of descentralized world order*". Global Governance, Cambridge, MIT Press.

TECHNOLOGICAL CHANGE AND DEVELOPMENT

Energy and Development Strategies within the Context of Global Environmental Changes

MAURICIO TIOMNO TOLMASQUIM AND CLAUDE COHEN
Energy Planning Program (PPE),
Graduate School of Engineering (COPPE),
Federal University of Rio de Janeiro (UFRJ)

INTRODUCTION

As the basis of production activities, energy consumption inevitably causes impacts on the environment. Although dealt with in the past as merely a problem of supplying input for production, threatened by the oil crises of the 1970s and consequent price hikes, during the 1980s energy developed into an issue linked strongly to environmental preservation. Particularly outstanding through international discussions and studies carried out in many different countries is the in-depth expansion of this relationship (Rosa, 1990).

Threats to the environment caused by energy use are located at three levels: local, regional and global.

The resettlement of riverbank communities imposed by flooding land for hydro-power dams and the contamination of areas by radiation due to inadequate protection mechanisms are examples of local impacts caused by power generation activities.

The emissions of SO_2 and NOx by industries and transportation have marked local impacts and cause regional damage as well. While emissions of these gases in large urban conglomeratess has had a negative effect on the life of their residents, at the regional level these environmental hazards consist of environmental acidification affecting lakes, soils and plants.

At the global level, the main threat arises from the danger of global warming, whose severe effects sweep indiscriminately across both the industrialized and the developing countries. The origin of this phenomenon is the greenhouse effect, caused mainly by emissions of carbon dioxide into the atmosphere, which in turn is fueled largely by energy consumption.

Although the industrialized countries are today responsible for more than half the world's CO_2 emissions, the developing countries also have a role to play in countering the greenhouse effect, particularly in terms of a prospective analysis, as any hypothesis of minimal development necessarily results in higher energy consumption in these countries due to powerful demographic dynamics and low income levels among the populace in general.

Furthermore, the environmental problems faced by the developing countries are currently worsened by a situation that is specific to the present day: resources that were once abundant for spurring growth in the developed countries would no longer be sufficient if used by the developing countries in the same proportion, and/or would be far more expensive, due to this situation of increasing shortages.

As clearly indicated by Sachs, (1980 (a) and (b)), it is necessary to move away from the dominant methodology in the study of the energy prospects for the developing countries. If supply-side action is a significant aspect of the problem, this certainly does not cover its entire scope. This does not involve merely replacing more pollutive energy sources by new sources in a classic import substitution situation, through the use of an elasticity coefficient in energy demand, estimated on the basis of past development. This approach curtails the scope of the freedom to adopt a development strategy that is less energy-intensive, muddying discussions of energy issues. To the contrary, attempts should be made to decouple the economic growth rate from the pace at which energy demand is rising. This means that demand-side efforts are broader-ranging: more than just responses at the sectoral level, the energy issue requires a redefinition of the global development strategy (offering a great opportunity for a complete re-think) due to the simple fact that energy is found in all human activities, as a facet of any development process. Consequently, far greater emphasis should be laid on identifying development strategies with a moderate energy consumption profile.

In this article, we reflect on the key factors that determine energy intensity in an economy, seeking to identify the possibilities available to the developing countries for moving ahead along a path that is less energy intensive and highlighting the case of Brazil.

ENERGY CONSUMPTION AND DEVELOPMENT STAGES

According to the United Nations Development Program (UNDP), "human development is the process of expanding individual options". And this can only be brought into effect through expansion of three essential capacities: "individuals should live a long and healthy life, with access to knowledge and the resources required for decent living patterns" (UNDP, 1999). This is the basis for the concept of the Human Development Index (HDI) which is designed to measure the general performance of a country or some other grouping in terms of three aspects: longevity, knowledge and living patterns.

The HDI was introduced relatively recently (since 1990) and is still consolidating its working methods, but it is proving to be a useful tool for raising discussions on development, particularly as it highlights the conditions of well-being of the populace, drawing attention to the needs for continuous data-base monitoring on this matter. However, this does not mean that we should take the HDI as a complete expression of the level of human development in a country, as this would merely replace one type of reductionism by another. Human development is something that is far broader in scope than anything that could be measured by the HDI or any other synthetic rating system for human well-being (Sachs, 1993). The real use of the HDI lies in providing conditions for discussing and reflecting on the issue of development, based on criteria that are of more social interest than those that are usually adopted (Bôa Nova, 1999).

The tables published in the Human Development Index for 1999 assess 174 countries by their respective HDI ratings, calculated from 1997 onwards. First place goes to Canada (HDI rating of 0.932) followed by Norway and the USA (both with an HDI rating of 0.927), while Sierra Leone trails well in the rear at 0.254. Brazil ranks 79th with an HDI rating of 0.739. According to the HDI classification, countries were grouped into three categories: Low Human Development (HDI under 0.5), Medium Human Development (HDI between 0.5 and 0.799) and High Human Development (HDI of over 0.8).

Based on the analysis of the links among the recorded levels of energy consumption and the results of the HDI ratings, Bôa Nova (1999) compared countries in order to assess the correlation between development levels and energy consumption. This report shows that countries with high human development levels consume 4,977 kgEP/capita, while the mid-level countries account for 1,007 kgEP/capita and the lowest absorb only 400 KgEP/capita.

Consequently, at first sight there seems to be a relatively clear link between energy consumption and human development levels. However, the association between energy consumption and human development levels is not as obvious as it may seem initially. In fact, looking more closely at a breakdown of the data, the panorama changes significantly, revealing a significant number of cases that do not fit into these initial generalizations (Bôa Nova, 1999).

Two countries with similar HDI ratings may post significantly different *per capita* energy consumption levels. A classic example is the USA, with the same HDI as Norway, but with a *per capita* energy consumption that is almost twice as high (8,051 kgEP/capita compared to 5,284 kgEP/capita). There are cases where the situation is reversed, with the country consuming more energy posting a lower HDI rating. This is the case with Japan, for instance (HDI of 0,924 and energy consumption of 4,058 kgEP/capita) compared to Singapore (HDI of 0,888 and energy consumption of 7,835 kgEP/capita).

The listing of similar cases could be far longer, although it cannot be denied that energy consumption levels tend to match higher development levels to some extent. However, what is particularly noteworthy is the fact that significant variations from country to country suggests that certain nations are dealing more successfully than others with the task of translating higher average energy consumption patterns into improvements in the quality of life of their peoples. Rising concern over the manner in which the developing countries are using energy resources to boost their growth appears clearly in analyses such as these.

According to Berrah (1983), the links between energy and growth are historically reflected through the development of elasticities, represented in the form of a parabola that initially rises until it reaches saturation point and then begins to drop: the weak elasticities at the start reach values of 1.5 to 2 during the early years of industrialization, then drop back steadily to close to one value. Berrah suggests that a "tunnel should be dug" (as shown by the arrow in Figure 1) in order to avoid the peak of the curve, driving ahead towards elasticity levels higher than one through a less energy-intensive development.

The feasibility of this less energy-intensive development is supported by the empirical work of Martin (1988) on the development of energy intensity for economic activities in the industrialized countries, which upholds the theory that there is no rigid law linking energy consumption to economic growth.

FIGURE 1: ENERGY/GDP ELASTICITY ACCORDING TO THE DEVELOPMENT LEVEL.

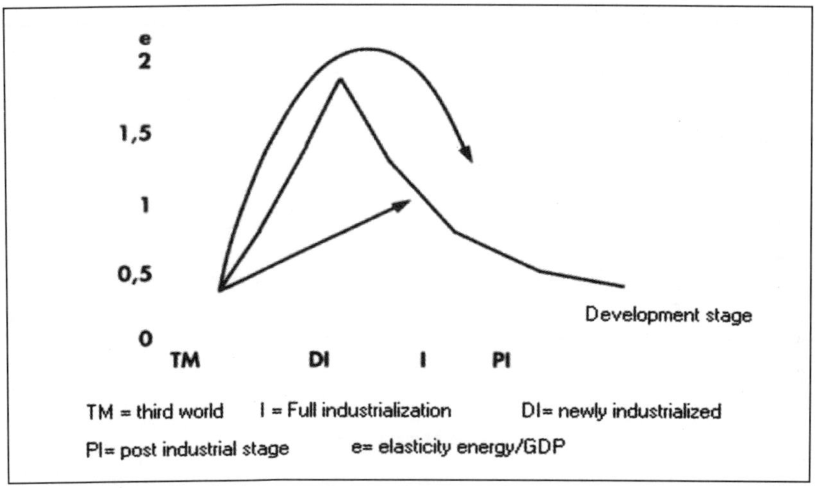

Source: Berrah (1983).

Martin shows the "hill evolution" described above in the link between energy (fuel wood excepted) and GDP. From 1850 onwards, there was a steady rise in GDP energy intensity worldwide up to a date that varies from country to country: 1880 for the UK, 1920 for the USA and Germany, 1929 for France and 1970 for Japan.

According to Tolmasquim (1993), if on the one hand the five countries analyzed by Martin show a "hill evolution" to their energy intensity, on the other it is worthwhile noting that these "hills" peak at different times and are increasingly less pronounced. Energy intensity in the British economy in 1880 is higher than that of the USA economy and the German economy in 1920, which are in turn higher than the energy intensity peak achieved by the French GDP in 1929, which is higher than that of the Japanese GDP in 1970. In other words, due to technical progress and the availability of "more efficient" energy sources, the tip of the curve that reflects national energy intensity evolution is taking place in increasing lower levels.

FIGURE 2: COMPARISON OF GDP ENERGY INTENSITY EVOLUTION

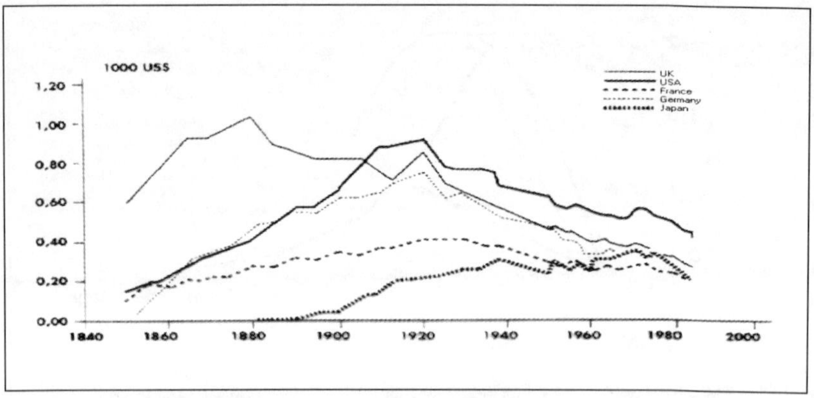

Source: Martin (1988)

In fact, this discussion and the relationship between energy and development is set within a wider-ranging issue that consists of analyzing how far energy, feedstock and economic output, could be decoupled through structural, technological and institutional changes. This means that a materials intensity indicator for the production of goods and services is required. This materials intensity indicator for the economy is calculated on the basis of the amount of materials needed to produce one monetary unit of the GDP. In fact, through dematerialization, one tries to find out what is the absolute or relative reduction in the amount of materials used and the quantity of wastes generated in producing one economic production unit.

Some authors such as Ayres (1989) for instance, have noted that as income rises, consumer preferences tend to shift towards the consumption of services, with the amount of materials built in to these supplies being naturally lower than most goods produced. Furthermore, saturation levels are also noted for the consumption of goods that are intensive in material embodied in their production. In fact, as an economy expands, the need for infrastructure assets such as bridges, railroads and heavy industry shrinks, reducing the consumption of steel, cement and other infra-structure industries materials. Moreover, materials are used more efficiently, and are replaced by lower-cost materials that are lighter and more durable.

The Kuznets Environment Curve (KEC) is a widely-used environmental sustainability indicator that follows this same line of thought. The hypothesis behind the KEC is that the use of resources and pollution tend to reduce as income rises, resulting in an inverted U curve. Empirical analyses seem to show that this relation is confirmed for certain types of pollution (Selden & Song, 1994) and deforestation (Panayotou *apud* Cleveland, 1999). The most usual explanation follows the line of the works mentioned above, showing that energy consumption and growth can be decoupled and expands the analysis to include the assumption of different stages of national development. During the early development stages when income is low, material input remains equally low, as the economy is almost completely based on non-mechanized agriculture. Subsequent industrialization triggers an upsurge in the demand for material input in order to build the basic infrastructure. As development expands, the need for basic infrastructure shrinks, as the demand for services that are less material-intensive rises. Some people argue that the KEC is noted not only within a nation but also can be used for material comparisons among countries, denoting differences in their economic development levels (Bernardini & Galli *apud* Cleveland, 1999). In other words, countries located on the upside of the curve are at earlier development stages than those on its downside.

Despite many empirical results demonstrating the relationship noted in the KEC (Malenbaum, Radcliffe, Brooks & Andrews, *apud* Cleveland, 1999), there are many criticisms for this analysis. More recent works – such as those by Jänicke (*apud* Cleveland, 1999) and de Bruyn & Opschoor (*apud* Cleveland, 1999) – challenge these results. For instance, through an analysis covering nineteen countries from 1966 through 1990, de Bruyn & Opschoor concluded that most developed countries in fact decoupled consumption and growth from 1966 through 1984, although some returned to couple energy and growth during the late 1980s, with energy and material intensities rising in step with the GDP. This results in an N curve, rather than an inverted U.

As no consensus has been reached so far on this matter, which remains open to a wide variety of interpretations, the links between energy and development require a careful approach avoiding a deterministic stance. We analyze below the driving forces shaping the connection levels between energy and economic development.

RATIONAL ENERGY USE AND TECHNOLOGICAL PATTERNS

There is a certain amount of elbow-room in the amount of energy consumed by a country in order to achieve a certain GDP level, depending on technological choices and social and economic options. It is important to distinguish between the development of GDP energy intensity caused by a structural distortion in production and that caused by technical or behavioral changes. Consequently, the effects of specific consumption (meaning content effects) can be distinguished from structural effects.[1] The Content Effects allow us to assess the evolution of a given quantity of energy required to produce one GDP monetary unit during a period, assuming the GDP structure to be *ceteris paribus*. The Structure Effect calculates the energy consumption variations by GDP unit, prompted by an alteration in the structure of this GDP and taking the energy consumption for each sector as being constant. Finally, changes in energy consumption may result from production scales, also known as the Activity Effect.

A reduction in specific energy consumption is the outcome of more rational energy use. Three types of policies can be distinguished, all designed to reduce the energy content of different types of human activities:

- Upgrading production and consumption systems through more effective management and more rational production of goods and services;
- Eliminating waste such as direct and indirect energy losses through regulating equipment and reusing production wastes;
- Investment in the conceptualization and use of more efficient production processes and items of equipment.

Consequently, although the expansion of energy supply is necessary for economic development, this may well be far less important when an effective conservation strategy is in place. Works such as those by Goldemberg et al (1988), Geller et al (1998) and Mielnik & Goldemberg (2000) have indicated the potential for a style of development with a low energy profile through technological development that enhances energy efficiency among the developing countries in general, particularly in Brazil. Mielnik & Goldemberg (2000) argue that from the 1970s onwards, the developed nations decoupled rising energy consumption from GDP growth through efficient energy use and production, thanks to clean and efficient sources and technologies, that developing countries could benefit from.

In Brazil for instance, a series of papers has attempted to demonstrate the potential for more efficient energy use. The three types of policies described above may be found in almost all the studies, with modifications concerning mainly the main focus of the case study, that is to say, focusing on the equipment, on the energy source, on the consumption sector, or on more efficient energy generation.

The work of Delgado & Tolmasquim (2001) for instance fits into the first case study category, analyzing the feasibility of introducing more efficient electric motors into Brazilian industry. The results presented by this analysis indicate that the present average prices for electricity do not hamper the introduction of more efficient technologies. Solely from the price standpoint, it would consequently make sense to bring into operation almost all the high-efficiency engines found on the market.

The article of Geller *et al* (1998) is a good example of an analysis focused on a specific source of energy: electricity. This article shows that there is tremendous potential for reducing the use of electricity through boosting end-use efficiency in Brazil. Many efficient technologies for refrigerators, air-conditioning units and engines are currently produced in Brazil. Although the adoption of these efficiency-enhancing measures is expanding, they have penetrated the potential market by less than 5% in most cases.

A study by Soares & Tolmasquim (2000) focuses on an industrial sector that consumes ample amounts of energy and accounts for a large portion of CO_2 emissions: the cement industry. In 1996, the Brazilian cement industry consumed 4.3% of the energy demand of the industrial sector, contributing over 24 Mton of CO_2. The authors present the prospects for energy conservation in the Brazilian cement industry until 2015, based on the introduction of new production technologies for this sector, as well as the use of waste-based fuels, cogeneration, the use of cementitious materials and other measures, using an energy demand simulation technical and economic model. In fact, in all scenarios, the drop in energy consumption (from 10.9 to 24.7 Mtep) and CO_2 emissions (from 200 to 400 Mton) ushered in by energy conservation measures is quite significant in the Brazilian cement industry.

Finally, a series of works that analyze the prospects for more efficient energy generation may be listed. Good examples are the recent technical and economic feasibility studies for cogeneration (Szklo *et al* (2000), Soares *et al* (2000), Tolmasquim, Szklo & Soares (2001), and Szklo & Tolmasquim (2001)).

Despite the undeniable potential for energy conservation, its limits as a palliative measure for curbing the steady increase in energy consumption cannot be ignored, and has been amply demonstrated by authors such as Georgescu-Roegen (1977), as well as Ayres & Kneese (1969). All of them affirm that a closed physical system should meet mass conservation conditions and that economic growth consequently increases both the extraction of environmental resources as well as the volumes of waste materials dumped back into the environment. One of the main theoretical results of these analyses is that the optimum path would lead the economy to a stationary status. Georgescu-Roegen (1971) shows the implications of entropy and the Second Law of Thermodynamics for economic analysis, describing the difference between primary production factors (energy and materials) and the agents (capital and labor) that transform these factors into goods and services. The agents are produced and sustained by the flow of energy and materials that enter the production process as high-quality, low-entropy feedstock, leaving it as low-quality and high-entropy wastes. Consequently, according to this author, there is a constraint on the substitution undertaken by the production agents (capital and labor) between low quality stocks and energy flows and input materials tapped from the environment.

Consequently, thermodynamics offers us an idea of the constraints: minimum quantities of materials and energy are required to produce one unit of a product that technical modifications cannot supplant. In sectors that depend heavily on the processing and/or fabrication of materials, technical modifications are subject to diminishing returns of scale when approaching the thermodynamics minimum. In fact, Georgescu-Roegen argues that materials and energy are irreversibly dissipated and that consequently the notion of total recycling is impossible. This observation prompted him to postulate the "Fourth Law of Dynamics or the Law of Entropy" stating that due to material dissipation and the decreasing quality of the resources used, materials could become more crucial than energy itself.[2]

Although criticized by economists and physicists (Cleveland, 1999), there is an aspect of his argument that is particularly noteworthy, even though his theoretical development may be subject to disagreement, which is how efficient technology can be in infinitely extending the planetary support capacity. An indicator of the scale of anthropic participation in global material cycles is the concept of **industrial metabolism** (Ayres *apud* Cleveland, 1999) which shows the human role in material cycles at differing spatial scales.

The industrial metabolism studies analyze the lifecycles of materials or nutrients. The flow of energy and materials in economic systems differs from the natural metabolism of the Earth on one basic point: most natural cycles are closed, while industrial cycles are open. In other words, the industrial system does not recycle its nutrients. To the contrary, the industrial cycle absorbs high-grade input material (fossil fuels, for instance) tapped from the Earth, and returns them in a degraded form. Very often these leaks in materials cycle also degrade air and water quality with consequent adverse effects on human health as well as other species.

A broad-ranging analysis of the issues underlying the industrial metabolism has obvious implications on environmental policies. If vast quantities of materials with deleterious effects on human health and ecosystems are mobilized, this mobilization must be reduced in some way. To a certain extent this is possible through energy conservation and recycling, thanks to technological improvements, but as they are limited, this type of action is insufficient to solve the problem.

CHANGES IN THE PRODUCTION STRUCTURE AND INTERNATIONAL TRADE

Another important determining factor in the energy intensity of an economy is the composition of the GDP. Since the 1970s, the developing countries have been beefing up their position as the producers of energy-intensive semi-finished goods, resulting in higher energy intensity for their industrial structures, while the share held by industry in the GDP of the industrialized countries has shrunk, losing ground to the services sector.

This behavior by the industrial structure is clearly reflected in the work of Jung *et al* (2000). The authors selected a group of thirty countries and split this sample into three sections (Table 1). The first consists of countries that are members of the Organization for Economic Cooperation and Development (OECD) rated under Code 1 - 10; the second group (Code 11 -20) consists of OECD countries and recently-industrialized nations. The third group (Code 21 - 30) consists mainly of the developing countries.

TABLE 1 - ANNUAL GDP GROWTH RATES AND SHARES HELD BY INDUSTRY

Country	Code No.	Annual GDP Growth Rate (%)	Shares held by Industry 1970	Shares held by Industry 1991
Australia	1	3.0	24	14
Austria	2	2.9	34	27
Belgium	3	2.5	32	23
Canada	4	3.5	20	17
France	5	2.7	27	22
Sweden	6	1.8	25	20
Denmark	7	2.1	19	18
Japan	8	4.3	36	28
UK	9	2.1	29	19
USA	10	2.5	25	18
Mexico	11	4.0	23	22
Korea	12	8.7	21	29
Portugal	13	3.7	32	27
Greece	14	3.1	16	17
Singapore	15	8.0	20	29
Hungry	16	2.6	36	21
Thailand	17	7.4	16	28
Brazil	18	4.8	25	20
Peru	19	1.5	19	25
Malaysia	20	7.0	15	28
Venezuela	21	2.8	16	20
Uruguay	22	1.7	19	23
South Africa	23	2.3	22	24
Philippines	24	3.6	24	25
India	25	4.3	14	18
Tunisia	26	5.8	8	17
Sri Lanka	27	3.2	14	18
Pakistan	28	5.5	15	18
Kenya	29	5.1	9	12
Zambia	30	1.0	19	25

Source: Jung et al. (2000)

From 1970 through 1991, the countries in the first group all posted a reduction in the share held by the industrial sector in the GDP. The conduct of the countries in the second group was mixed, while the share held by the industrial sector in the GDP for the countries in the third group expanded.

These different types of behavior are set against a backdrop that is redefining the conditions of the international economic order, characterized by a rising trend in the industrialized countries that is shifting low-priority industrial activities to regions where labor and raw materials are less costly. As shown by Tolmasquim (1991), while Brazil's II National Development Plan (1974) rated capital goods, heavy electronics and semi-finished goods as high priority, Japan opted to redirect its growth towards high-technology sectors, after more than two decades focused on heavy industry. This decision was prompted by problems - such as industrial pollution levels - that were becoming a matter of much concern, while also foreseeing constraints on energy supplies.

There are cases where the alterations in the GDP structure may limit positive results in terms of gains in energy efficiency. This is actually the case for many developing countries, as their industrial structures are energy-intensive. In other words, Content Effect measures are limited, as the Structure Effect locks developing countries into production structures that specialize in energy-intensive products and relegates them to the unenviable condition of price-takers. This downgrades the terms of exchange resulting from the monetary devaluation of the output of these countries, prompting them to produce more in order to offset this devaluation. According to Furtado (1998) the prices of the manufactured products exported by the developing countries such as Brazil rose 12% in nominal terms (in dollars) during the 1980s, while the prices of manufactured products exported by the industrialized countries rose 35%. This phenomenon affects the industrial output of the developing countries that depend on importing capital goods and to an increasing extent of sophisticated technology in order to compete on the international market. These countries are constrained to produce more and more in order to counterbalance the dropping values of their manufactured products compared to variations in the prices of the products that they import from the industrialized countries. In this case, it should be noted that hardening up of the production structure not only limits gains in energy efficiency, but may also prompt an Activity Effect that outstrips the Content Effect. In fact, the results of Tolmasquim, Cohen and Szklo (2001) corroborate this statement quite clearly.

This redefinition of the global production structure has a significant effect on greenhouses gases emissions in each country. In fact, the CO_2 emissions assigned to each country are based on domestic emissions. However, these emissions do not necessarily represent the sole contribution by each country to increasing greenhouses gases in the atmosphere. Many different products with relatively large amounts of carbon embodied in their production are exported and imported by these countries.

Brazil offers a good example of this. When calculating the direct and indirect energy embodied in Brazilian imports and exports through a feedstock/product matrix, Machado (2000) found values of around 813 PJ and 13.5 MtC for exports in 1995, and 679 PJ and 9.9 MtC for imports that same year.

In parallel, Tolmasquim, Schaeffer & Machado (2000), estimate the effects of direct energy and direct CO_2 emissions. Although including only the direct effects in these calculations, the study by Tolmasquim, Schaeffer & Machado (2000) indicates an appreciable difference between the amount of energy that the country is using (and the carbon that it is emitting) in order to export its products and what it is not using domestically because goods are imported rather than processed within its own territory. In fact, some 6.6% of the final energy consumed in the industrial sector and some 7.1% of the carbon emitted by this same sector are the outcome of international trade.

In the case of Brazil, the energy intensity of the economy and its carbon emission levels may be artificially high due to the significant amounts of energy and carbon embodied in its international trade transactions. That is why it would be interesting to re-think the National Carbon Emissions Inventory System, ensuring that it also reflects the emissions arising from international trade.

SPATIAL AND URBAN ORGANIZATION: METROPOLITAN TRANSPORTATION STRUCTURES

The spatial configuration of national economies and their transportation structures are two other elements that shape GDP energy intensity. Rising urban concentration brings in its wake an increase in the energy costs required to build and maintain the infrastructure of major urban centers.

Cities have been studied by social scientists, urban planners and have even been discussed by Greek philosophers such as Aristotle in his *Politics*. Urban centers are analyzed by Castells (1983) on the basis of an approach using natural systems, treating the city as an organism linked to the environment through two-way input / output flows. Although anthropic effects on the environment had already been noted long before, it was only from the 1960s onwards that the extent of these impacts could be perceived as a matter of much concern in terms of sustaining life on the Planet. Many studies of cities have focused on the individual systems that sustain them: transportation, water supplies, sewage disposal, food and other aspects have been covered by analyses of the environmental impacts of the production and consumption cycles that underpin these cities (pollution, greenhouse gases emissions, waste materials, and land use). Studies are currently underway at centers such as the Institute for Global Environmental Strategies (IGES) and the Asian Urban Information Center in Kobe (Vellinga & Herb, 1999), through comparative analyses of cities in Asia where the social changes prompted by industrial and technological development have been striking. These studies show that consumption patterns are changing dramatically, with *per capita* number of motorized vehicles rising appreciably. These works share the common factor of treating cities as complex organisms slotted into local, regional and global systems, with modifications to one section of the system affecting other parts of the whole.

Therefore, one of the basic requirements of studies focused on cities is the movement of people, raw materials, manufactured products, water, food and wastes, reflected through transportation. In most countries, transportation systems necessarily depend on burning some type of fossil fuel, which directly links human transportation requirements to modifications in the carbon cycle. Some 25% to 30% of anthropogenic carbon dioxide emissions are caused directly by burning fossil fuels for transportation purposes (Vellinga & Herb, 1999). Most cities depend on global transportation systems, whose pollutive effects extend over the entire Planet as the particulates and pollutants produced by burning fossil fuels and biomass are carried over long distances.

Issues focused on restructuring transportation systems are based on three standpoints: technology, institutions and space, all of which are intrinsically linked. From the technological standpoint, vehicles can be upgraded, aiming at lower energy consumption and less pollution through more efficient engines, lower gas emissions, lighter bodywork, and the use of alternative sources such as fuel cells, solar energy or renewable sources of energy like alcohol. Information technology can also help through reducing the length and number of trips.

In institutional terms, incentives could be introduced to the use of certain modes of transportation rather than others, as there are considerable differences in energy intensities among different modes of transportation. A good example would be greater encouragement to use mass transportation facilities rather than individual options, for instance, but measures of this type must necessarily take into account the relevant spatial structure.

In terms of space, cities could be built in a manner that ensures most trips can be made efficiently on foot or by public transportation. This is the case with many major European centers such as Paris, Amsterdam, Milan and Berlin, where personal automobiles are more of a problem than a privilege. But this does not apply to many cities in the USA (other than New York) such as Los Angeles or Miami, for instance.

Even in countries where industrial civilization is more advanced and with economic systems that are very similar, differences appear in energy consumed by the transportation sector (Darmstader, Dunkerley & Alterman, 1978).[3] Some factors can explain this: the variation in the volumes of passengers and goods transported for the same GDP level (passenger/kilometers and tons/kilometers transported); differences in the share-out of demand among the various types of transportation available: mass or individual for passengers, road, rail, water or air for cargo and passengers, and in the case of individual transportation, the frequency of use (number of trips) of the automobile.

The selection of the mode of transportation is shaped by a series of factors such as the extent of the transportation infrastructure (roads and highways, railroads, waterways, airports) and fuel prices. As this infrastructure requires regular maintenance and in many countries this is Public Sector's responsibility, the transportation infrastructure of a country depends on the availability of capital for the different types of transportation or even the demand for each of them. For instance, in the USA, due to low automobile use costs, the low population density of most metropolitan areas and long distances between urban centers, individual transportation is preferred over mass facilities. Fuel prices have more effects on passenger transportation than cargo shipments, but may also have significant impacts on the energy intensity of transports because consumers and producers tend to pay more attention to vehicle efficiency and the way of using it when fuel prices are high or rising (Schipper *et al*, 1992). This means that countries with high gasoline prices such as France, Japan and Germany are far less transportation and energy intensive than countries with low gasoline prices such as the USA, Mexico and Venezuela.

Similarly, the quality of the transportation infrastructure affects the type of transportation selected, as well as the energy intensity. In other words, roads and highways that are poorly maintained, in poor condition or constantly grid-locked will tend to increase the energy intensity of the vehicles traveling along them. In fact, lengthening traffic jams in major Brazilian cities are due to heavy demand for automobile road-space that is fifty to a hundred times higher than the one required for mass transportation for a typical home-to-work trip (Poole & Moreira, 1991). Moreover, even with significant improvements in vehicle efficiency, private cars consume far more energy than mass transportation facilities, in relative terms. Additionally, Rosa & Tolmasquim (1993) indicate that despite of a substantial increase in the energy efficiency of equipment in the developing countries, more efficient automobiles and household appliances in these countries keep pace with life-styles that result in higher energy consumption levels than previously. This means that although automobile engines are more efficient, the average size of the vehicles is larger, they are used more frequently and driven at higher average speeds.

Furthermore, the traditional response to increased motorization of the populace is to build more roads, which is a shortsighted and expensive solution, as it shows a marked trend towards saturation. The problem is worsened by population growth and urban expansion. These factors together trigger an explosive increase in road space, with average trip differences also lengthening, putting pressure on energy consumption. Part of this process is inevitable, but its progress can be at least minimized through a mass transportation expansion strategy, such as the one implemented in Curitiba, Southern Brazil (Lerner *apud* Poole & Moreira, 1991). As related by Poole *et al* (1998), Curitiba demonstrates the possibilities of reducing energy consumption in the transportation sector. The measures introduced include urban planning and the introduction of backbones and easy ticket purchase facilities in order to increase average bus speeds along special bus lanes. Although energy was not the main reason prompting the introduction of these measures, which were rather designed to thin out traffic jams and reduce local pollution, vehicle fuel consumption levels dropped, compared to other towns of a similar size. Despite the success of this program, it has not been widely deployed in other towns, as local lifestyles must adapt to certain priorities in order to expand the use of mass transportation facilities. This shows the importance of the patterns of consumption followed by the population.

CONSUMPTION PATTERNS OR LIFESTYLES AND THEIR EFFECTS ON ENERGY AND DEVELOPMENT

There seems to be intuitive agreement on the fact that most people in the North and the elites in the South consume more (Princen, 1999). However, the question of modifying consumption patterns is crucial, but it is difficult to approach because of its undeniably deep-rooted personal aspects. Furthermore, another matter that makes discussions of consumption a cause for discomfort is the fact that in analytical terms this issue slips away from the dominant rationalist paradigm. This is quite clear when observed from the standpoint of efficiency. Gains in efficiency can become ways of avoiding the problem, dissimulated means of masking social costs in space and time, particularly when political borders and generations are not respected. This may divert attention from the real problem, meaning issues of scale and ecological functioning. For instance, in the USA automobiles have become considerably more efficient since the oil crises of the 1970s, while the country as a whole remains just as dependent on imported oil and has not reduced its CO_2 emissions. Moreover, any mention of modifications to consumption patterns in the USA is still very difficult, if not practically impossible.

Therefore, the "modern lifestyle" inherent to current market economies is proving increasingly expansive, mobile and harmful to the environment. Many technical options seem to be available for boosting energy and material efficiency, as well as reducing the amounts and variety of wastes produced. But the environmental advantages of these many technical options are lost through scale effects and the Rebound Effect.[4] For instance, although many modes of transportation (bicycle, automobile, train, boat and aircraft) have boosted travel speeds appreciably, the time spent commuting every day has remained almost the same for many decades (Vlek, *et al*, 1993), due to lifestyles that are more dependent on mobility.

Energy economists have been studying the Rebound Effect since 1980, based on an article by Khazzoom (1980) that was widely discussed during that decade due to the hike in oil prices. The Rebound Effect indicates that enhanced efficiency achieved through modern technologies does not result in reduction in the consumption of resources (or energy) in the same proportion, as technological improvements ratchet up behavioral responses. In some cases, an increase in efficiency may even result in an increase in energy consumption.[5]

Even when the improvement in efficiency results in a drop in energy consumption, the total estimated reduction is rarely achieved. Transportation and housing are the areas where the Rebound Effect is noted most strongly, due to enhanced energy efficiency (Biesiot & Noorman, 1999).

Greening and Greene (1997) propose a terminology that distinguishes three types of Rebound Effects: rising use of an energy service prompted by a price reduction through enhanced efficiency; there is also the situation where energy service costs drop while the prices of other products remain constant, meaning that consumers have more income available to spend on other products and services (in other words, a type of income effect) that also require energy supplies and consequently increase total energy consumption in areas not directly related to the enhanced energy efficiency (indirect Rebound Effect); and finally, changes in feedstock prices for production and alterations in the demand for feedstock prompted by substitutions and income effects that spread throughout the economy and result in adjustments in the supply and demand for all sectors affecting the general equilibrium. Although this phenomenon is controversial and challenged by some authors (Lovins, 1988),[6] the Rebound Effect brings up a relevant issue: to what extent the effects of using certain more efficient technologies has a concrete impact on the dematerialization of energy consumption and production.

Sanne (1998) discusses "overconsumption" in terms of three distinct components: wastefulness, sumptuousness and affluence. Wastefulness is expressed in using more resources than effectively required; sumptuousness refers to elitist consumption patterns affordable by only a limited number of people through what are known as "position goods"[7] (Hirsch, 1976); and affluence refers to general consumption levels that may be too high compared to the available resources, or compared to some established international standard. Linttot (1998) also affirms that there is a "relative income" that consumers look for, rather than a higher absolute income. Consumers consequently seek a specific income level *compared to other consumers*. This highlights the importance of analyzing position goods.

The issue is that these goods have the specific characteristic of causing "social congestion": a type of shortage expressed through the fact that the "good things in life are restricted not only by physical constraints but also on absorption ceilings for their use" (Hirsch, 1976). Position goods became more important, but individual attempts to enhance well-being through them do not result in any widespread improvement in well-being as a whole.

Consequently, the argument put forward by Hirch (1976) is that consumption generally occurs within a social context where factors such as socialization, habits and imitation are very important and buttressed by the media in establishing social consumption patterns.

The Norwegian Royal Ministry of the Environment (1998) adds five reasons for transforming current consumption patterns, particularly in the wealthy countries: "Sustainable consumption would reduce environmental damage and ensure the supply of essential goods and services to the poor, making a positive contribution to human health and fostering economic efficiency while upgrading the quality of life of individuals". Given the long-established focus on economic growth found among current societies, this view of sustainable consumption assumes that more stress is being laid on the dimensions of human development, in addition to social and environmental aspects.

Along these same lines, Baudrillard (1998) stresses that many consumers all over the world "do their duty to consume" in order to save jobs and support the free market system, while Durning (1992) refers to this as the "myth of consumption, or societal decline". He claims that this is based on the "call to consume or perish" which is deployed as a justification by stating that if there were no consumption there would be no production and consequently no employment or income. In this case the environmental variable is eliminated from the argument, with no stress laid on the fact that a high level of material consumption is not necessarily a vital condition for generating employment or eradicating poverty. High material consumption and unemployment can in fact exist quite well together, as has always been the case.

In fact, the view that economic behavior reflects the impact of dynamic aspects and the "life process" of the culture was already found in Veblen, a critic of the analytical tradition of the neo-classic economy (Rima, 1987). Veblen saw the main dynamic influences on human behavior as deriving from changes in technology, namely: "(...) the methods for dealing with the material means of life". The technological activities of man would reflect his instinct to produce things with his hands, his idle curiosity and his paternal inclination. Consequently, Veblen saw an essential dichotomy in all human behavior that on the one hand reflected the impact of dynamic technology while on the other static ceremonial influences derived from institutions prevailed. Ideally, according to Veblen, economic science should consequently be reshaped on a Darwinian structure. This would accommodate the essentially revolutionary character of consumer conduct (Rima, 1987). According to Veblen (1934) consumption patterns are less the outcome of the rational calculation of marginal gains in losses than the result of habit, and consumption patterns of others in society that encourage the "emulative" exhibition and "conspicuous" consumption. This is the source of basic dichotomy that characterizes consumer conduct. Society seeks to emulate the behavioral patterns of the consumer as established by the idle rich. But these patterns do not meet the needs of man arising from his instinct to work with his hands, to be deeply involved in useful activities that foster well-being. Consequently, the technological inclination of man is skewed by a culture that is slanted towards wasteful consumption and ostentation. "The effect (of conspicuous waste) on the serious activities of man consequently guides him solely towards acquiring as much wealth as possible, disapproving of work that does not bring in any pecuniary gain." (Veblen, 1934).

Few Brazilian authors have undertaken any in-depth analyses of consumption patterns. One of the earliest studies on energy and social classes in Brazil was produced by Bôa Nova (1985), analyzing energy consumption within the nation's social structure. This author covered issues such as the energy cost of the development policy introduced in Brazil, the strategies implemented after the oil crises of the 1970s, and the distribution of energy consumption levels among different social groups.

Furtado (1998) related the issue of consumption patterns to the macro economic dynamic of nations through their links to national saving, and on the bottom line (like Veblen, to a certain extent) to cultural factors. According to Furtado (1998), "the adoption by the dominant classes of the consumption patterns of countries with accumulation levels that are far higher than our own explains high income concentration and the persistence of social heterogeneity, as well as the manner of inclusion in international trade. The independent variable is, on the bottom line, the flow of innovations in consumption patterns that spreads through countries with high income levels. The counterpart of cultural mimetism is found in the income concentration levels found in Brazil. In order to eliminate the effects of this perverse cultural imperative, it would be necessary to alter consumption patterns within the framework of a broad-ranging social policy while at the same time boosting savings substantially and clamping down consumption by high-income groups". In fact, according to Furtado, what is noted is that in the marginalized sectors of the developing countries, the penetration of technical progress is initially circumscribed to consumption patterns, with limited effects on modernizing lifestyles among only a few segments of the population. This gives rise to the problem of the marked propensity for consumption among the modernized segments of society, growing into structures that hamper social and economic dynamics, such as large-scale land ownership and corporativism, resulting in lost savings potential through abusive types of consumption that siphon off funds abroad.

The core issue here is that current consumption patterns in the developed countries *cannot* be generalized at the global level in economic terms because the North has achieved its current levels through exclusion: most of the world's population lacks access to these patterns and consequently they are only "momentarily sustainable" as there is no way of extending them. The Report prepared by the United Nations Development Program (UNDP, 1999) mentions "compulsive consumption" and "conspicuous consumption "[8] as reasons for the excessive consumption of the North. On the other hand, considering environmental aspects, these patterns *should not* become more widespread, as this consumption has proven predatory at current technological levels: the speed at which natural resources are depleted in order to perpetuate these patterns is far higher than the reproductive and regenerative capacities of the environment.

The quest for better technology – to the detriment of an analysis taking into account both the production and consumption sides on a complementary basis- has produced unsatisfactory results. On the one hand, looking at technical and economic aspects, machinery is efficient from the physical and energy standpoint but is associated with wasteful habits that cancel out some of these gains in efficiency, while on the other, opulent societies are extremely exclusive in terms of social aspects. In fact, any mention of consumption patterns or levels means stepping "outside the paradigm". This requires moving away from the production standpoint and focusing rather on analytical issues that are hard to handle through conventional tools such as price determination, cost benefit analyses and even life-cycle studies.

Finally, globalization and international cultural and consumption patterns have had sweeping effects on the elites in the developing countries, with well-known social consequences: transplanted technologies that are highly capital and energy intensive and are the only available options for the joint production of consumer goods that shifts the labor force and worsens inequalities in income distribution to an even greater extent. Consequently, the imitative growth model of the developing countries has it its core a structural crisis element worsened by easily-available low-cost energy, leading to the replacement of production factors that were abundant locally by others that must be imported.

CONCLUSION

Countless criticisms have been fired at the abusive use that is often made of the GDP and its variations for measuring economic soundness and even levels of social well-being, in the absence of any other summary indicator that is easy to use. Dupuy & Robert (1976) affirm that GDP growth is not analogous to the growth of a child (meaning the state growth variable) and its achievement of adult status (an image frequently used to describe the process to be followed by the underdeveloped "young nations" on their path to development).[9] This is clearly a flow variable that is always destroyed and reproduced over again. Furthermore, as argued by Furtado (1998), "there will always be some ambiguity as long as any attempts are made to reduce the expenditures of different groups to a single denominator, within a non-egalitarian society. When consumer expenditures are amalgamated, yet another effective demand variable is sought for an economic system, rather than a social well-being rating scheme".

Similarly, energy consumption levels depend on the systems used to measure its production, processing and end-use (such as meeting energy requirements) that in turn depend on the purposes for which this energy is intended (What is it used for? To manufacture what? To meet which needs?). These queries finally lead to the need to define the beneficiaries of energy consumption (Who uses it?) and on whom do decisions depend that are required in the course of the process (Who decides what? To what extent? For what purpose?). Consequently, close links between energy use and development styles show that there is not necessarily only one path leading to an ideal level of well-being for society.

In the case of the Northern countries, the need to reshape the development models of industrial societies - based on a standard of civilization that grew up from technological and industrial foundations - has been clearly reaching its limits from both the environmental and social standpoints. This is also curtailing widerspread levels of well-being that are needed to expand *per capita* energy use.

In the specific case of the developing countries, it is important to note that the position commonly encountered is to rate an energy conservation strategy as low priority for application in these countries (as their *per capita* energy consumption is still extremely limited), although this is in fact a dangerous simplification. From the standpoint of allocating available investment funding, this in fact indicates that higher unit investments channeled to boosting energy supplies place the objectives of the energy sector in direct competition with urgent demand for the social and economic development of these countries over the next few years. It is quite clear that the main problem of the developing countries lies in boosting the low consumption levels of most of their inhabitants, and to do so, expansion of the economic surplus assigned to energy investments is to a certain extent inevitable. However, in all countries -including those in the South- it is saved energy that is generally cheaper, creates more stable jobs and is more environmentally friendly (Sunkel, 1979).

In terms of global impacts for an energy conservation strategy, the decisive contribution must today come from the industrialized countries (basically the USA), due to vast disparities in consumption. Looking at longer-term prospects and in view of the obvious need to boost the level of meeting the energy requirements of the developing countries, the adoption of a broad-ranging energy conservation policy is of vital importance, not only in order to reduce primary energy generation holding steady useful energy consumption levels, but also to build up a style of development based on lower useful energy demand profiles through deploying alternative solutions in fields such as transportation and spatial arrangements, as well as new technologies to cut energy consumption by the industrial sector.

Another aspect of this same issue is that behind the low average energy consumption levels of the developing countries lies a marked disparity between two co-existing situations. On the one hand, vast masses of people cannot achieve energy consumption levels that meet their basic needs in terms of food, healthcare, housing and medication (UNDP, 1999); while on the other local elites absorb significant portions of global income in order to underpin lifestyles that are comparable in absolute terms with their peers in Europe and North America. This results in energy wastage levels that are also comparable with those of societies in the North, not only in terms of direct consumption (air-conditioning, household appliances, and automobiles) but also indirectly through the energy needed to produce these goods and the supply of luxury services.

Finally, poverty and environmental degradation are closely interconnected. Although poverty imposes heavy stresses on the environment, the main causes of the deterioration of the global environment are production and consumption patterns that are not sustainable, particularly in the industrialized countries, worsening poverty and exacerbating imbalances. In order to usher in sustainable development, production sector efficiency is needed, in parallel to reshaping consumption patterns. While optimizing the use of resources and minimizing the output of waste materials, spatial arrangements and urban structures can ensure that appropriate targets are both established and achieved.

NOTES

1. *Introduced initially by Darmstader (1978).*

2. *Georgescu-Roegen (apud Costa, 1996) considers that the Law of Entropy is applicable not only to energy transformation but also to qualitative changes in macroscopic material. He postulates the IV Law of Dynamics as follows: 1) non-available material may not be recycled; 2) a closed system (that does not swap material with the environment) cannot operate indefinitely at a steady rate. "This Law states for material what the Law of Entropy states for energy. One difference is that in an isolated system, instead of a trend towards thermodynamics death when all energy becomes unavailable, there is a trend towards chaos when all material energy becomes unavailable. Consequently, we should avoid speaking of material entropy as a measurable entity. There is a measurable entropy for energy, because energy is homogeneous, in contrast to material which is heterogeneous, as shown by the Mendeleev Periodic Table. Material dissipation factors vary widely from one substance to another and consequently we cannot include all material dissipation under a single formula - which does not mean that this dissipation does not occur or that we cannot speak of a general degradation of energy material from an available form to another that is unavailable."(Georgescu-Roegen, apud Costa, 1996).*

3. *In general, as a country develops, energy consumption rises in the transportation and industrial sectors, to the detriment of the residential sector. For lower income levels, the residential sector is the major energy consumer (some 50% of total consumption) while at higher GDP/per capita levels, the share held by this sector drops to 20% of total consumption, while the percentage absorbed by industry and transportation increases.*

4. *A special issue of Energy Policy (6-7, Volume 28, 2000) is devoted to this topic.*

5. *This effect is also known as "backfire".*

6. *In fact, Lovins does not question the existence of the Rebound Effect, but rather its effective impact, which he affirms is negligible, as the demand for energy services tends to be inelastic. However, Binswanger (2001), relativizes the position of Lovins, explaining that the impacts which he claims are negligible can be rated at this only if analyzed through a single service model (of energy). When taking into account the effects of enhanced efficiency on other services in the economy, the Rebound Effect may be quite significant, if the replaceability of services is high and the demand for energy services rises with income. This is even more true when the efficiency reduces the time needed to perform/obtain some service (faster modes of transportation, for instance) as this tends to result in wider-spread use of this service. When opportunity time costs are higher (wage rate) and energy prices are lower, energy improvements will tend to pump up energy use to a greater extent.*

7. *This term defines goods that are typically consumed by an elite through defensive costs that rise systematically, keeping pace with income. These are assets such as automobiles and second homes, whose contribution to the well-being of each person drops as other people acquire them.*

8. *This term was initially used by Veblen (1934) and has become part of standard economic terminology, describing the type of consumption behavior associated with the idle rich.*

9. *This is in fact quite inadequate because, as noted by Abdalla (apud La Rovere, 1989) in contrast to human beings, societies do not follow a single preprogrammed life-cycle.*

BIBLIOGRAPHY

AYRES, R. U. (1989). Industrial Metabolism. In: CLEVELAND, C. (1999). *Ecological-Economic Underpinnings of Industrial Transformation. Industrial Transformation Research Directions*, Draft for discussion at the IHDP-IT Open Science Meeting, Amsterdam, IHDP, 1999.

AYRES, R. U. & KNEESE, A.V. (1969). Production, consumption and externalities. In: CLEVELAND, C. (1999). *Ecological-Economic Underpinnings of Industrial Transformation. Industrial Transformation Research Directions*, Draft for discussion at the IHDP-IT Open Science Meeting, Amsterdam, IHDP, 1999.

BAUDRILLARD, J. (1970). *La société de consommation*, Folio Essais, Ed. Denoël, 1970

BERRAH, N.E. (1983). Energie et Développement: l'effet tunnel. *Revue de l'Énergie*, Paris, n.356, Aug./Sep. 1983.

BIESIOT, W. & NOORMAN, K.J. (1999). Energy requirements of household consumption: a case study of The Netherlands. *Ecological Economics*, n 28, p. 367-383, 1999

BINSWANGER, M. (2001). Technological progress and sustainable development: what about the Rebound Effect?. *Ecological Economics*, n 36, p. 119-132, 2001.

BÔA NOVA, A.C. (1985) *Energia e Classes Sociais no Brasil.* Ed. Loyola, 1985.

------ (1999). Níveis de Consumo Energético e Índices de Desenvolvimento Humano. In: *Anais Do VIII Congresso Brasileiro De Energia,* 8. *Anais.* Rio de Janeiro, SBPE/COPPE/UFRJ/Clube de Engenharia, 1999. v.1

CASTELLS, M. (1983). *The City and the Grassroots: A Cross-Cultural Theory of the Urban Social Movements.* Berkeley, University of California Press, 1983.

CLEVELAND, C. (1999). *Ecological-Economic Underpinnings of Industrial Transformation. Industrial Transformation Research Directions*, Draft for discussion at the IHDP-IT Open Science Meeting, Amsterdam, IHDP, 1999.

COSTA, M.M. (1996) *Energia, Informação e Entropia: uma Visão desde a Física até os Sistemas Industriais*, MSc Dissertation., Rio de Janeiro, PPE/COPPE/UFRJ.

DARMSTADER, J., DUNKERLEY, J., ALTERMAN, J. (1978). International variations in energy use: Findings from a comparative study. *Annual Review of Energy*, USA, 1978.

DELGADO, M.A.P. & TOLMASQUIM, M.T. (2001) «Analysis of a Feasible Tariff Policy for the Introduction of High-Efficiency Electric Motors in Brazil», *Energy Conversion and Management*, accepted in March 2001.

DUPUY, J.P., ROBERT, J., (1976). *La trahison de l'Opulence.* Paris : Universitaires de France, 1976.

DURNING, ªT., (1992). *How Much is enough? The consumer society and the future of the Earth*, London: Earthscan, 1992.

FURTADO, C., (1998). *O Capitalismo Global.* São Paulo, Ed. Paz e Terra, 1998.

GELLER, H., JANNUZZI G. M., SCHAEFFER, R., & TOLMASQUIM, M.T., (1998). The Efficient Use of Electricity in Brazil. *Energy Policy*, UK, v. 26, nº1, 1998.

GEORGESCU-ROEGEN, N., 1971, *The Entropy Law and the Economic Process.* Cambridge, MA, Harvard University Press, 1971.

GEORGESCU-ROEGEN, N., (1977). Matter matters, too. In: WILSON, K.D. (ed.), *Prospects for Growth: Changing Expectations for the Future.* New York, Praeger, 1977.

GOLDEMBERG, J., JOHANSSON, T. B., REDDY, A.K.N. , WILLIAMS, R.H. (1988), *Energy for a Sustainable World.* New Delhi, Wiley Eastern, 1988.

GREENING, L.A. & GREENE, D.L., (1997). Energy use, technical efficiency, and the Rebound Effect: a review of the literature. Report to the *Office of Policy Analysis and International Affairs*, U.S. Department of Energy, December, Washington, D.C.

HIRSCH, F., (1976). *The social limits to growth*, Cambridge, Harvard University Press, 1976.

JUNG, T.Y., LA ROVERE, E.L., GAJ, H., SHUKLA, P.R., ZHOU, D., (2000). Structural Changes in Developing Countries and Their Implication for Energy-Related CO_2 Emissions », *Technological Forecasting and Social Change*, 63, 111-136.

KHAZZOOM, D., (1980). Economic Implications of mandated efficiency patterns for household appliances. *Energy Journal*, 1, p.21-40, 1980.

LA ROVERE, E L., (1989). Energia e estilos de desenvolvimento: os exercícios de avaliação das perspectivas energéticas do Terceiro Mundo - um diagnóstico inadequado e suas causas. In: *Energia e Desenvolvimento: a política Energética no Brasil.* Montevideo: ROSTLAC/ FINEP/PNUD/UNESCO - BRA 82/004 Project, 1989.

LOVINS, A.B., (1988). Energy saving from more efficient appliances: another view. *Energy Journal*, 9, p.155-162, 1988.

MACHADO, G.V., (2000). Energy use, CO_2 emissions and Foreign trade: an I-O approach applied to the Brazilian case. *Paper presented at XIII International Conference on Input-Output Techniques*, 21-25 August 2000, Macerata, Italy.

MARTIN, J.M.L., (1988). Intensité Énergétique de l'Activité Économique dans les Pays Industrialisés: les Évolutions de Très longues Périodes Livrent-elles des Enseignements Utiles?. *Économies et Sociétés*, France, 1988.

MIELNIK, O., GOLDENBERG, J. (2000), Converging to a common pattern of energy use in developing and industrialized countries, *Energy Policy*, UK, v.. 28, 2000.

NORWEGIAN ROYAL MINISTRY OF ENVIRONMENT, (1998). *Consumption in a sustainable world*, A. Soolgard, PO Box 8013 Dep, N-0030 Oslo, Norway.

POOLE A., HOLLANDA J. B., TOLMASQUIM, M. T., (1998). *Energy conservation and greenhouse gas emissions in Brazil,* report for project "General Description of Measures Taken to Implement the Framework Convention of the United Nations on Climate Change", Instituto Nacional de Eficiência Energética, COPPE/UFRJ, for the Ministry of Science and Technology (Project BRA/95/G-31), Brazil.

POOLE, A. & MOREIRA, J.R., (1991). Energia e meio Ambiente - Subsídio Técnico para a Conferência das Nações Unidas sobre o Meio-Ambiente e o Desenvolvimento - UNCED 92. Science and Technology Bureau. (final version, mimeo, 6/6/91).

PRINCEN, T., (1999). Consumption and the Environment: some conceptual issues. *Ecological Economics*, 31, p.347-363, 1999.

RIMA, I.H., (1987). *História do Pensamento Econômico*, São Paulo, Ed. Atlas, 1987.

ROSA, L.P , TOLMASQUIM, M.T., (1993). An analytical model to compare energy-efficiency indices and CO_2 emissions in developed and developing countries. *Energy Policy*, UK, v. 21 n.3, p.276-283, 1993.

ROSA, L.P, (1990). Tendência mundial de consumo, produção e conservação de energia - o meio-ambiente e os avanços tecnológicos. *Revista Brasileira de Energia*, Rio de Janeiro, v.1 n.3, 1990.

SACHS, I., (1980a). *La ville et son profil énergétique.* Paris, CIRED, EHESS, 1980.

SACHS, I., (1980b). *Estrategia de desarrollo con Perfil Energético Moderado.* Paris, CIRED, EHESS, 1980.

SACHS, I., (1993). *Estratégias de transição para o século XXI - desenvolvimento e meio-ambiente.* São Paulo, Studio Nobel/FUNDAP, 1993.

SANNE, C., (1998). *Cutting consumption: How to understand and reduce overconsumption in wealthy countries*, Paper presented at Ecological Economics Conference, Geneva, March 1998.

SCHIPPER, L., MEYERS, S., HOWARTH, R. B., STEINER, R., (1992). *Energy Efficiency and Human Activity*: past trends, future prospects. Cambridge, GB, Cambridge University Press, 1992

SELDEN, T., SONG, D. (1994). Environmental quality and development: is there a Kuznets curve for air pollution emissions?. *Environmental Economics Management*, 27, p.147-162, 1994.

SOARES, J.B. & TOLMASQUIM, M.T., (2000). Energy efficiency and reduction of CO_2 emission through 2015: the Brazilian cement industry, *Mitigation and Adaptation Strategies for Global Change*, Dordrecht, Boston, London., 5, p.297-318, 2000.

SUNKEL, O. (1979). *Estilos de desarrollo y medio ambiente en América Latina: Una interpretación global.* Santiago, CEPAL/PROY.2/R.50, 1979.

SZKLO, A.S. , TOLMASQUIM, M.T., (2001). Strategic cogeneration: Fresh horizons for the development of cogeneration in Brazil, *Applied Energy*, Oxford, 69, p.257-268, 2001.

SZKLO, A.S., SOARES, J.B. & TOLMASQUIM, M.T., (2000). Economic potential of natural gas-fired cogeneration plant in Brazil: Two case studies", *Applied Energy*, Oxford, 67, p.245-263, 2000.

TOLMASQUIM, M. T., (1991). Modes d'adaptation aux chocs pétroliers: Brésil, France et Japon. *Revue de l'Énergie*, Paris, 433, p.626-642, 1991.

TOLMASQUIM, M. T., (1993). Meio Ambiente, Eficiência Energética e Progresso Técnico, *Cadernos de Eficiência Energética*, Rio de Janeiro,INEE, 1993.

TOLMASQUIM, M. T., COHEN, C., SZKLO, A.S., 2001, "CO_2 Emissions in the Brazilian industrial sector according to the Integrated Energy Planning Model (IEPM)", *Energy Policy*, UK, 29, p.641-651, 2001.

TOLMASQUIM, M.T., SCHAEFFER, R. & MACHADO, G. (2000) *"Energy and Carbon Embodied in the Internatiotional Trade in Brazil"*, Paper prepared under the aegis of the Project entitled "General Description of Steps Taken to Implement the United Nations Framework Convention on Climate Change" (Project: BRA/95/G-31), Brazilian Ministry of Science and Technology.

TOLMASQUIM, M. T., SZKLO, A.S. & SOARES, J.B., (2001). Economic Potential of Natural Gas-Fired Cogeneration Plant at Malls in Rio de Janeiro. *Energy Conservation and Management*, 6, 42, p.663-674, 2001.

UNDP (United Nations Development Programme), (1999). *Human Development Report 1999*, Oxford, Oxford University Press, 1999.

VEBLEN, T., (1934). The Theory of the Leisure Class, Nova York, Modern Library, 1934.

VELLINGA, P. , HERB, N. (editors), (1999). *Industrial Transformation Science Plan*, IHDP (International Human Dimensions Programme on Global Environmental Change) Report nº 12, Bonn, Germany.

VLEK, C., HENDRICKX, L. , STEG, L., (1993). A Social dilemma's analysis of motorized-transport problems and six general strategies for social behaviour change. In: EMCT: *Transport policy and global warming*, European Conference of Ministers of Transport, Paris, OECD, 1993.

Science and Technology on the Threshold of the New Century

HENRIQUE RATTNER
School of Economics and Management, University of São Paulo, and Brazilian Association of Leadership Development

THE HISTORICAL CONTEXT AT THE END OF THE CENTURY

When discussing national policies for science and technology, scientists approach the issue as if it functioned in a vacuum, independently of political dynamics and economic problems. In order to understand and occasionally raise questions about the aims and trends of national policies for science and technology, it is essential to approach this issue within its historical context, identifying the ties by which it is closely dependent on economic and financial policies, both national and international.

The country's economy has experienced considerable turmoil in recent years whose consequences for the Brazilian people in the areas of employment, income and savings, as well as education, health, and science and technology sectors, are far from over. The discussion of guidelines and policies to minimize the effects of this crisis should be preceded by an analysis and assessment of the economic policies which, if not a direct cause of this turmoil, certainly created the conditions for it. But the Brazilian economic and financial crisis has not been restricted to the national territory only. To a greater or lesser extent, wealthy and poor societies alike were affected, showing the harsh reality of a globalized planet and its communication networks that provide instant feedback and response.

The global financial system greatly influences all that is constructed, planned and decided within national boundaries. It is therefore essential that any sectorial discussion (for instance, on the prospects for science and technology in the country) begin with a diagnosis - even if hypothetical - of the directions and trends of the world system, and by extension and in the light of these, of the Brazilian society.

Science has become a hegemonic ideology for reconstructing reality, arrogating to itself the role of the sole arbiter of truth. But despite all the truths produced, the problems afflicting mankind - famine, poverty, ignorance, violence and injustice - are still unresolved. Will more scientific and technological information change governmental policies in the areas of health, education, sanitation and, above all, job generation?

We cannot ignore the fact that the main problems affecting the global society remain unresolved, challenging scientists to come up with adequate responses and proposals. Having shown very limited explanatory power and even less predictive ability regarding the cataclysmic phenomena which plague the lives of millions of human beings, science runs a serious risk of losing its hegemonic position to religious fundamentalist dogmas, which would undoubtedly be a step backwards in human evolution.

In the days of the declining feudal order, natural science was censored and persecuted both by lay and religious authorities, because it dared to contest and transform the conventional conceptions of the world and mankind. In this period of decline and crisis of the capitalist order, critical social science has come to be vilified and at times persecuted by the authoritarian State and the scientific establishment. The former sees social criticism as a threat to political stability or to the status quo, while scientists, who are either co-opted by or allied with the structures of power, deny its scientific rigor in content and methodology. Because it adopts a radically critical posture towards the destructive practices employed in the exploitation of natural resources and the labor force, the knowledge produced by critical social science has either been ignored or rejected as unscientific and therefore of insufficient value and significance to be useful in the formulation of public policies. As a consequence, social critique has almost totally disappeared from scientific discourse, both inside and outside the University. The plea for ethics is offered as a Band-Aid solution for social conflicts, while it functions as a non-critical, individualistic doctrine in relation to the system of capitalist exploitation. "Ethical" corporations and foundations, totally submissive to official policies, are proposed as alternatives to critique and proposals for transformative action.

In spite of the pretended "human dimensions", there is no mention of the social actors, with different visions, interests and cognitive structures. Society is not the homogeneous entity subsumed in the official discourse. Different groups with different needs, wants and aspirations fight for scarce resources. Any proposal for "change" has to start with the identification of the contending actors, their discourses and ideologies by which they try to legitimate claims and policies. This argument evidences the relevance of the political relationship and power structure that conform the decision-making processes. In this sense, human dimension is primarily determined by the political process by which priorities are established and scarce resources are allocated.

The next step should be the description and analysis of the global context, the trends and its driving forces. To mention just a few relevant issues, we may observe the dynamics of integration, the creation of common markets and regional associations, and the fragmentation of the former Soviet Union and ex-Yugoslavia and many other separatist movements, all over the world. From fewer than fifty nations present at the foundation of the United Nations in 1945, today the number of member-states reaches almost two hundred. The asymmetric relationship within and between countries determined by the concentration of wealth, power and access to information is at the roots of polarization and exclusion. The dominant factor in this perverse dynamics is the uncontrolled flow of investments from the rich to the poor countries and back to the international centers of finance. All the "emergent" (formerly so-called "underdeveloped") countries are literally broke, asking for alms from IMF, IBRD, IBD and private bankers. As there is "no free lunch" in financial help from these organizations, imposed policies like ESAP have increased the external, internal and most of all, the social debt of the poor countries. Probably, the most serious problem in contemporary societies is the growing number of unemployed people, considered "unemployable" because of their precarious qualification. Besides the phenomenal growth of the informal sector, the most serious consequence is the spread of social anomie, even in the metropolitan areas of the rich countries, a retrogression to the state of homo homini lupus.

Do we imply that by more efficient science and technology policies one could improve the existential conditions of the deprived 1,2 billion people living below the poverty line? Another crucial question is: to whom do we present our research projects and the eventual findings? To national governments? To the headquarters of transnational corporations? Or, to the international finance organizations? Those are exactly the actors responsible for the present state of calamity, the degradation of the environment and of human beings' deplorable existence, most interested in keeping the status quo. Are we unable (or unwilling) to think in terms of alternatives for the victims?

TOWARDS A SYSTEMIC AND INTER-DISCIPLINARY VISION

We postulate that the conventional disciplinary and departmental approaches no longer work in this highly dynamic and complex arena. The challenges of building a sustainable world demand the creation of a complex framework of reference. whereby economic, political and environmental dynamics are closely interconnected and dependent on one another, and all of them on the whole system.

Therefore, it is not very fruitful to study each problematic area or sector in isolation, and later try to describe or interpret the behavior of the system as a whole. A more effective way to begin this task would be to think about society (or the world) as an elaborate system whose elements are inextricably intertwined, even if the models currently available for this analysis are not perfect.

An awareness of the complexity of the world around us, and our attempts to explain the multiple phenomena and problems caused by the actors and their interactions, require the efforts of interdisciplinary groups capable of considering and focusing beyond their own area of specialized research, the total dynamics of the system.

The progress of science in the last centuries, while helping us improve our understanding of reality, also produced an enormous paradox: while intellectual and scientific processes tended to continually reduce the barriers established by fragmented disciplines, they have led to increasing specialization and fragmentation in the academic area. According to many scientists, this trend has resulted in a sweeping crisis in all fields of knowledge, on account of the isolation and self-sufficient environments in which each discipline develops.

Epistemological and methodological approaches – the way we observe, analyze and interpret reality, are of fundamental importance. The world around us is not fixed and immutable. We change the world through a dialectic process of observing and interpreting "facts" in new and different ways. If we perceive phenomena in the world in a different way, we create a new "Gestalt" which in turn, allows us to perceive different relationships between the parts and the system. The process of scientific observation depends on a set of premises, which are developed by scientists in a particular social, cultural and political context.

The theoretical background – this set of premises - will necessarily influence the analysis and interpretation of the "data" collected, which affects the meanings attributed to it as well as the resulting actions. From a methodological point of view, the formulation of premises is essential to decide what the pertinent facts are and the norms and techniques appropriate to empirical research. Moreover, the premises help us to identify the problematic themes, a step which precedes observation and analysis. Our perception is influenced by reality, at the same time as reality is also influenced by our perception. Therefore, our learning occurs only through successive approximations to 'truth", in a permanent process of change.

Those researchers who try to explain natural or social phenomena in accordance with their academic background and experience select some variables or factors considered to play a determining role in the behavior of the object(s) of study. Such selection is not based only on a detailed literature review. Along with their specific theoretical assumptions, scientists also have beliefs and values that inevitably affect the object under study and the methodology used.

Therefore, the scientific method and its products - knowledge, theories, technology, etc - should never be placed as completely objective or neutral, because scientists, like any human being, cannot expect to be above and beyond personal feelings, interests, beliefs and passion.

The same reasoning applies to technology – whether state-of-the-art, intermediate or appropriate - which should be placed within the historical, cultural and social context of its introduction, assimilation and use. To reduce the solutions of environmental and developmental problems to mere "rational choices" amongst the best techniques available, while ignoring the political forces at play, would be naiveté or mystification, serving only to help maintain the status quo.

Scientists try to explain the real world in terms of statements subject to confirmation or rejection, thereby producing "true" knowledge. By demanding empirical evidence to decide what is true or false, science has contributed to the progress of knowledge (as distinct from wisdom), and to the gradual decline of superstition and irrational authoritarian behavior.

The world held great expectations that science would in time turn out to be an agent of emancipation for humanity, transforming authoritarian religious and political regimes into more rational and democratic societies. Such expectations, however, have not been fulfilled. Science and technology have also produced a wide range of negative effects on the lives of human beings, such as toxic substances, radioactivity, sophisticated weaponry, and total alienation of the working class.

SCIENCE AND PUBLIC POLICY

Science has traditionally been defined as theory, regarding reality as if it were an entity detached from human action and thought. Modern science, however, aims at exerting control over natural and social reality, which is to be explained by theory and transformed by technology. Contemporary science has surpassed the dreams of previous generations, not only in the description and interpretation of reality, but also in its intellectual construction and transformation, as well as experimentally, in laboratories and research centers. Advanced technologies – computation, genetic engineering, automation and nuclear power – demonstrate the huge potential for humanity to control nature, culture and society.

The triumph of natural science over critical thought in regards to society is not just a coincidence. The basic arguments of science have played an important role in promoting capitalism by providing "objective" models and paradigms which were crucial in legitimizing existing social relations, despite all their contradictions and injustices.

When we see contemporary societies struggling with economic and social crises, unemployment, fear and violence, political infighting, ethnic and religious conflicts, the possibilities for peaceful cooperation and solidarity amongst and within nations, in accordance with ethical and moral principles, seem more and more unattainable.

The widespread aspirations for democracy, justice and equity are being frustrated by the neo-liberal ideology, which postulates a free market and competition as the structuring principles of life in society. Consequently, the violence of economic actions and transactions and their devastating effects at the individual and collective level, are considered legitimate and necessary, with no regard for ethics or social values. Thus, neo-liberal ideology not only tries to justify individualistic-narcissistic behaviors, but also in calling for and supporting accelerated technological innovations and their impact on production, trade and capital mobility at a global scale, contributing decisively to the expansion of globalization within and amongst societies. The concentration of wealth and power, information and easy access to goods and services is paralleled by the swelling ranks of unemployed, excluded, marginalized and alienated, dividing humankind into a privileged minority and a vast majority completely deprived of any rights to life.

Is it possible to escape from this framework when defining programs and research policies and guidelines? Or are technological and scientific research priorities unaffected by the process of questioning relations of power, the role of the State and the struggle to conquer citizenship?

Scientists who do not take political positions are able to coexist with, along with injustice, the lack of national, industrial, and cultural development plans as well as science and technology programs, which should establish society's goals and priorities.

Neither researchers of the most renowned institutes, nor the authorities responsible for agencies and ministries provide us with clear indications of national priorities for their sectors and departments. As there is no definition on the part of the MCT (Ministry of Science and Technology), CNPq (National Council for Scientific and Technological Development), and MCIT (Ministry of Commerce, Industry and Tourism), each area tries to maintain its position in the allocation of ever more scarce budgetary resources, while failing to create a favorable climate for innovation. Nobody dares to question this deplorable lack of pro-active and innovative policies in the field of Science and Technology, which is a reflection of what happens in the financial-economic area, where the government clumsily tries to follow to the letter the recommendations of the International Monetary Fund and the World Bank, even when those multilateral organizations give unmistakable signs of the necessity to change the course of policies, particularly after the financial disaster in Southeast Asia.

In the end, who should be in charge of defining national priorities for science and technology policies, taking into account the great variety of perceptions, interests, life experience and expectations of the population, in a society where the benefits of an urban-industrial civilization are so unequally distributed?

The assumption that it is possible to know the dynamics and possible results of the policies supported by economists and technocrats is subject to criticism, since the concrete behavior of the different social actors has been very different from the "rational" standards postulated by the several areas of scientific research.

There is growing evidence of situations in which priority-setting is subject to pressure, confrontation and imposition according to the bargaining power of the different interest groups and social actors.

Judging by concrete actions of the government, in providing support to banks and subsidies to foreign companies to establish themselves in the country, or on the other hand, the way issues of wages and social welfare are handled, it is clear that the priorities of the technocracy have very little to do with social development and the goals represented by the five fingers of the president's hand during election campaigns.

The biggest cuts in the government's budget have taken place in public health, education, housing and transportation. What will the implication of this be for science and technology policy, and above all, for scientists engaged in research and development?

SCIENCE AND THE ENVIRONMENT

Contrary to the official discourse supported by the scientific establishment, there is no positive correlation between important achievements in scientific and technological research and the ranking of a given country in terms of social indicators. In spite of a reasonable scientific infrastructure (universities and research institutes), in terms of human development indicators Brazil remains well behind several countries with inferior development of science and technology.

The question of the role of science and technology in a society afflicted by tremendous social problems has been systematically avoided by the establishment, scientists and politicians alike. During the last few decades, public opinion has been nourished by the myth of the "trickle-down effect", from more research and development to economic prosperity and social welfare. However, as harsh reality evidences, the nature of our social problems does not require sophisticated high-tech solutions but more rational uses of existing "appropriate" technologies and policies that endeavor the reduction of waste and conspicuous consumption. Another important factor in human development is raising the level of education and knowledge of the whole population by incorporating millions of children still excluded from an adequate schooling system. How can a society progress without the inclusion of all its population? From the foregoing discussion one may infer that science and technology are not politically neutral or non-ethical issues. On the contrary, equipments and work processes as well as their organization and management are inextricably linked with social productive relations. In each historically, spatially and socially determined context, the material forms of technology represent a given combination of different levels of centralized economic and political power, and the countervailing aspirations of the producers for more autonomy and self-management. Hence, technological practices reflect the political contradictions between the dynamics of the economy (concentration and centralization of capital) and the opposite trends of the political system, towards democracy and self-management. This dialectical tension establishes the limits of science and technology as instruments of social change. Technological research and development, the innovations and their incorporation into the productive system obey primarily economic and political criteria. To affirm the belief in the possible change of the balance of power towards more equity and social justice derived from conventional science and technology policies, sounds naive or deliberate mystification. Ultimately, social and economic development, including science and technology, does not only depend on the volume of resources available, but on whom controls and uses them, and for which objectives, plans and values are they being used.

An unequivocal demonstration of the biased way and methods adopted by the official science discourses is revealed by an analysis of the discussions of environmental problems at international meetings and conferences on climate change and correlated phenomena. To avoid reducing emissions at home, representatives of the rich countries, on the evidence of dubious scientific results, propose the most complex ways and mechanisms in order to escape the obligation of adopting a clean and rational climate policy, based on a systemic and interdisciplinary framework. As advised by their scientists, governments consider climate policy as emission abatement and control. However, there is an urgent need to redesign the energy and transportation sectors, as well as industrial production to fight air pollution and traffic congestion. Instead of a climate policy based on the negative stance of emission reduction, we need to advance with positive proposals of industrial transformation, abandoning the narrow piecemeal approach, to be replaced by a systemic global change framework.

A different methodology is required when exploring social, ethical and behavioral fundamentals of human well-being regarded as a priority issue. An important assumption is the extreme relevance of intra – and intergenerational distribution, defending an ethical position rather than scientific neutrality. Current approaches by economists based on natural sciences information seem too limited.

Rethinking the role of science and technology as proposed by Funtowicz and Rawetz (1992, 1993) will change our perception of the problems and help us to develop new ways of thinking about the world around us and its challenges.

The greenhouse gas accumulation is but one of the many symptoms of irrationality in our highly inequitable world, where 20% of the population consume 80% of its natural resources including energy. Other negative manifestations are the ozone layer depletion, the pollution of rivers and oceans, the ever-increasing number of hazardous chemical and nuclear waste deposits which impact negatively on the natural and human environment. These problems cannot be fixed by technological solutions alone.

The unequal distribution of income and of productive assets imposes many constraints on poor countries development policies. The crops to be grown, the energy sources to be exploited, the ways the land is to be used etc, are no more decided by national authorities, but by external forces. In dealing with the problem of ghg (green house gas) emissions, rich countries are less worried than in the case of sulphuric dioxide (SO?). However, the rise of global temperature due to climate change will hurt badly poor countries in the southern hemisphere. One meter rise of the sea level will displace many millions of people and submerge much of land all over the world, while the building of walls to protect vulnerable zones near the sea may involve unbearably high costs for the poor countries.

So far, negotiations about climate change have produced poor results as they are held between unequal partners. The representatives of poor countries are inferior in numbers at the conference tables and generally lack the access to relevant information and negotiation skills. Therefore, it is hard to reach agreement about the levels of CO_2 concentration, which represent risks to the populations' health. Present models emphasize cost minimization for the rich, but not risk minimization for the poor. If poor countries will sell their "rights" to pollute, how much will they be able to emit in the future, to support their industrialization policies? The almost endless negotiations from one conference to another represent objectively a serious delay in taking adequate measures, thereby worsening the situation of inequity to a point of no return.

By including sinks in the CDM (Clean Development Mechanisms), rich countries are probably imposing the worst possible way in dealing responsibly with their obligations to reduce emissions. There are several reasons for not including sinks into the CDM, once a systemic approach is adopted. The still controversial question of the preservation of biodiversity related with GMOs (genetically modified organisms), the land rights of indigenous people in the poor countries, living in areas coveted by so-called development projects (for instance, the Parana-Paraguay waterway crossing the Pantanal region), the uncertainty about the estimates of the storage capacity of carbon by biological sinks and, most of all, the transience of carbon sequestration in the light of unpredictable and uncontrollable factors of human and natural behavior inducing climate change.

The adoption of the precautionary principle and a strong commitment to regional and international cooperation would be the first steps towards a cleaner and safer environment.

Independently from the negotiations results, each country should be held responsible for its own emissions to be checked and evaluated by an independent international committee. The trading of quotas - euphemistically called "clean development mechanisms" may improve efficiency, but certainly not equity among and between nations. CDM proposes to mobilize private investments for poor countries to be able to provide for cleaner development, based on capital and technology flows. Again, negotiations generally held on a bilateral basis between unequal partners do not ensure that a "good" price will be obtained by the poor countries. Furthermore, without concomitant transfer of technology, any deal involving the concession of pollution "rights" will certainly be onerous for the weaker partner.

SCIENCE AT THE SERVICE OF POWER

There is an increasingly pressing need for a systemic and integrated approach, which considers at the same time economic variables and processes of social transformation in all of their dimensions. Although the disciplinary approach may appear to be necessary for scientific and technological progress, perceiving the complexity of the world we live in, as well as explaining phenomena and problems caused by multiple factors and their interactions, require an intellectual exercise that is not confined to a single discipline. To deal with complex systemic relationships, we need a non-linear and reflexive frame of reference in which each variable or element is connected to others, receiving and transmitting impacts of different scales and intensity, structuring the dynamics of the system as a whole.

But it was economics that most assiduously sought to harness itself to the canons of the natural sciences, emphasizing mathematical and statistical procedures in analyses and interpretations. Both economics and the natural sciences adopt methods guided by "functionalism" and "reductionism". Functionalism is limited to analyzing the form of observable behavior and its functions, relegating questions about the "meaning" or the essence of the object of investigation to the realm of metaphysics, ideology or subjective opinion. When functionalism is applied to the problems of humanity, instrumental reason and formal rationality, the targets of criticism of the Frankfurt School, are unavoidable. But if the ends justify the means, where shall we find justification for the ends themselves? Supporters of reductionist science endeavor to reduce higher-order objects and forms to simpler, lower-order objects and forms. Many economists and natural scientists agree that cultural and social phenomena can be explained by economic or biological processes and concepts. Within this approach, thought, conscience and morals, as well as modes of social interaction, can and should be reduced to neurobiological brain activity.

Of course, it is not just purely logical arguments underlying functionalist-reductionist reasoning. Its ideological underpinnings affirm the rationality of the system, which should be efficient (competitive) and functional. Individuals, on the other hand, might be misfits, unskilled, aggressive or lazy. Therefore, policies of intervention and change, when admitted as necessary by neo-liberal supporters of the 'status quo', must be centered on individuals who should be educated, trained and guided, and never on the system which is perfect, balanced and rational!

However, the critique of instrumental rationality of functionalism and reductionist science must not be based on a return to religious or mystical appeals that try to resuscitate medieval obscurantism, aggravating the de-politicization and civic passivity of marginalized populations.

Critical social theory, which situates the role of scientific and economic theories in its proper historical context – closely related to the rise and expansion of the capitalist system – examines them in the light of a given historical subjectivity, decoding their axioms, concepts and categories. Its job is to unveil the connections between capitalism and natural science, while pointing out their negative effects. However, the social critique of the dominant paradigm in the sciences requires overcoming scientists' political "illiteracy" and a parallel politicians' scientific illiteracy. The inflexibility of specialists cloistered in their disciplines and departments does not contribute much to promote the establishment of a dialogue aimed at the construction of a new, interdisciplinary and systemic paradigm.

Curiously, the university has resisted and continues resisting innovative ideas in the production and diffusion of knowledge, thus dramatically reducing its own capacity to effectively contribute to solving social problems. Such behavior is rooted in political linkages and commitments with power - the centralizing State introduced by Napoleon as part of his reforms early in the 19th century. In order to modernize education, the emperor entrusted his minister with developing a body of scholars or instructors showing a marked "*esprit de corps*" to teach knowledge and techniques, while subordinating "what to do" to political logic – in other words, to the interests of the state or hegemonic groups. The result has been the cooptation of the educational apparatus, especially higher education, to the dynamics of social reproduction of the system, educating and training its administrative and management personnel. In fact, to this day the École Normal Supérieure and the Collège de France have systematically staffed France's public administration and political institutions, a model that has been copied by many other countries. The so-called autonomy of universities has been under pressure due to the control of its budget by the State. To transform this external dependence into autonomy would require transforming the State-civil society relationship, leading the people, through mounting and intense political participation, to assume their role as social protagonists.

To this end, it is necessary to counterbalance the pyramidal logic of State power with the demand for political autonomy emanating from civil society. Through effective experiments in autonomy, the false belief in the linkage between efficiency and centralized decision-making processes will be overcome. To deny the role of the State and public policy in planning and allotting resources is unrealistic and ideological. But to submit social and productive organizations generated by civil society to an overarching State would mean to corrupt the process of development towards a pluralist and participatory democracy. The alternative of resorting to the market's patronage has not been very effective in stimulating the production of knowledge by scholars and researchers. Fragmented science at the service of utilitarianism does not lead to a meaningful intellectual production nor concern with the fate and suffering of humanity. By transferring the function of neutralizing the power of intellectual innovation to the universities, the elites protect themselves against "subversive" ideologies. It should be the role of scientists to identify development options, not as a monolithic project controlled by the State or a hegemonic group, but as the result of a multiplicity of individual and collective projects empowering and reinforcing one another.

In the present context, however, scientific doctrines and models, rather than contributing to the understanding of economic and social dynamics, have been transformed into an apology at the service of financial capital and the dominant elites. The university's lack of intellectual production does not bother the elites, as long as it provides the function of pragmatic professional training, more and more questionable in the light of rapid changes in the technical and social division of labor. Any advancement which occurs results more from vertical specialization than from interaction among the different theoretical bodies.

Scientific theories deal with collections of interrelated phenomena that display their own particular dynamics, seeking to explain them and to predict possible trends and events.

Economics, as an effort to systematize knowledge, has from the beginning been closely linked to power relations in society. While mercantilism exalted gains in gold and precious metals in the interest of monarchs, the Physiocrats postulated land ownership as the supreme value. Classical liberalism favored the advancement of the industrial revolution, giving prominence to the role of the capitalist entrepreneur, while Marxism developed the doctrine of dialectic materialism in support of the struggle of the working class. Finally, in the first half of this century, J. M. Keynes tried to rehabilitate the crises-ridden capitalist system, by invoking the mentor role of the State in the process of economic development. All those theories and more than two hundred years of economic history cannot disguise the obvious perplexity of economists when challenged to explain the current crises in the financial markets and their impact upon the productive systems. The inability of politicians and their economist advisors (the "king's wise men") may be attributed to their individualistic assumptions, empirical-inductive rationality and linear causality, whereby the world functions according to individual options, always dictated by rational choices based on individual interests, and not by objective processes that condition and limit the field of action of social actors. The dynamics of capital accumulation and reproduction in the era of worldwide instant communication are carried out more easily and quickly through speculative financial investments than through industrial or agricultural production.

SCIENCE IN CRISIS – SYSTEM IN CRISIS

In light of the above arguments, why not regard the current historical juncture as transitory, one of many stages in human development, and therefore subject to being supplanted by more rational and humane organizational arrangements for life in society?

The widespread crisis that today affects the capitalist system has made it clear that theories about the social benefits of globalization and the free market have no scientific basis and are not supported empirically. As a matter of fact, as is frequently the case with ideologies, we have just another act of faith masquerading as science.

The alleged greater productive efficiency is invalidated by the social inefficacy of fragmented and compartmentalized thinking and organization of work, which extends the logic of machines and their mechanisms to human life, in a manner completely contrary to nature. Mechanistic and quantitative procedures ignore or obscure the affective and creative dimensions of human beings and are thus ineffective in the face of complex social problems. Hence a challenge for the university: to retake its trailblazing, humanistic role in the generation of knowledge, embracing complexity in cognitive terms and in structural transformation, by means of organizational and institutional innovations.

Science may be regarded as a privileged means of producing knowledge, within the social division of labor. But like any other human activity, science is a product of society, and scientists, despite their high-minded discourse of objectivity and neutrality, do not differ in concrete social behavior from other actors, individuals or groups. Thus the hope that science would become an agent for human liberation, transforming the rulers of authoritarian and clerical regimes into enlightened, democratic champions of the public cause, has failed to materialize.

It is not simply a matter of judging the conduct of our rulers against the canons of science. Undoubtedly, science helps us to explain and interpret reality through the formulation of theories and models, granting plausibility and even reliability to its propositions. Technology, for its part, steers us to useful and effective action, instrumentally relating means to ends. But technology, particularly its high-tech sector, geared towards and for the market, nurtures and supports the existing structure of power, judged as necessary or inevitable, and therefore, legitimated by science, regardless of the injustice and deprivation resulting from its uses.

It is clear that science and technology (or scientists and technologists) do not succeed in sidestepping ethical considerations or crossroads in the confrontation with political and economic problems of society. Instead of accepting the limits of the " possible", it is our task to seek out what is "desirable" and what might be a concrete alternative for conscious and liberating human action.

Consciousness presupposes "meanings" of content that cannot be derived from function nor reduced to the form and sequence of some neurobiological processes. A school similar to the reductionist science prevailing in natural sciences has gained followers in the schools of economics taught in universities and popularized by the media. From Malthusian laws, to the social Darwinist doctrine of "survival of the fittest" to the supposed genetic predisposition for criminality (Lombroso) and the stigma of poverty (Herrnstein), society has been biologized or reduced to the "natural" mechanics of price oscillations, supply and demand, GDP expansion or reduction, unemployment, etc.

In spite of the possible good intentions and subjective beliefs of their authors and followers, the objective consequences of the functionalist-reductionist combination are highly destructive. Both, economics and the natural sciences, although successful in the manipulation of man and nature, have not brought about improvement in the quality of life for the majority of the world population. Whereas the economy (or economic theories and assumptions) leads to recurrent cycles of expansion and crises, aggravating polarization and exclusion, natural science provides us with artifacts, which have ambiguous or even destructive effects. This is not a matter, however, of erroneous utilization of the products of scientific and technological research. The system's perverse effects are deeply rooted in the procedures, axioms and categories of economics and natural science.

To expand information and knowledge will not be enough as long as professional practices and public policies remain closed to considerations of ethics, justice and equality. Unless they result in new management practices and behavioral changes, all the research, libraries and data banks in the world will fail to resolve the intricate problems of our society on the threshold of the third millennium.

Ultimately, to act ethically means to challenge the culture of submission and the fear of violence imposed by the elites, and to question theories which legitimate market rationality, competition and (spurious) productivity as universal values, while ignoring the inhuman treatment inflicted on the excluded and the subjugated.

THE HUMAN DIMENSIONS OF SUSTAINABILITY

The basic assumptions that guided our analyses are:

a. our society is a social construction -almost everything has been produced by human beings and therefore, can be transformed and reconstructed by social actors;

b. due to the diversity of social actors in our complex and contradictory societies, changes mean tensions and conflicts, as individuals and groups have different interests, worldviews and cognitive structures;

c. although we believe that "Man makes History", social action does not occur in accordance with the free arbiter principle. We are bound to and largely conditioned by the work of generations who preceded us. Therefore, knowledge of our cultural history is fundamental for transforming social action;

d. the complexity and dynamics of social relations cannot be apprehended through linear, Cartesian reasoning. Complex organizations and processes ought to be analyzed by systemic interdisciplinary teamwork - a requirement still far away from current teaching and research practices.

Current debates about sustainability, by resisting the incorporation of the great challenges faced by humanity, tend to become rhetorical and innocuous. To the question "what kind of society do we want to build"?, the ecological discourse is not sufficiently relevant. We need answers to the problems of unemployment, the lack of participatory democracy, the constant violations of Human Rights, the degradation of the quality of life of the majority of the world's population, particularly the deprived and excluded.

The globalization of the economy and the withdrawal of the State from its historical responsibilities to provide basic services to the low-income population has resulted in poor health and educational services, deficient housing and transportation for the historically underprivileged classes. Yet, the reduction until the total elimination of social inequalities has become a primal condition for achieving sustainability in all its dimensions – social, cultural, economic, political, environmental and ethical – because poverty and environmental degradation are closely related phenomena in our society.

Increasing integration of national economies into an internationalized market generates more material wealth and stimulates commercial transactions. Paradoxically, globalization also generates more unemployment, poverty and exclusion, which increase in rhythm and intensity faster than the accumulation of wealth. Affluence and waste on one side, and poverty on the other are increasingly polarized and concentrated in our society, turning into the major obstacles of the process of sustainable (durable) development.

The world's present development path is not sustainable. Climate change, population growth, loss of biological and cultural diversity, rampant poverty and inequality tend to increase the vulnerability of human life and of the planetary ecosystems. We need a better understanding of the complex and dynamic interactions between society and nature, which goes far beyond conventional approaches of formulating and testing hypotheses, because of the non-linearity, complexity and multi-loop feedback nature of the observed processes. An alternative approach will emphasize the need to analyze and predict the behavior of self-organizing systems, while assessing the roles of various social actors, with different and often contradictory perceptions and aspirations.

The traditional scientific method following sequential analytical phases of conceptualizing the problem, formulating hypotheses, collecting data, building theories and applying the results, will have to be blended with new practices of social learning which incorporate elements of collective action, adaptive management , innovative public policies and social experiments. This means that scientists concerned about the future of humanity will endeavor to work with all kind of social groups to better understand how they elaborate the construction of knowledge and collective behavior practices. The next step would be to evaluate how governments are responding to increasing pressures for participatory democracy and the universal quest for citizens' rights.

In concrete terms, this would result in continuous efforts to reconnect science and scientists to the political endeavor of promoting sustainable development. The future of science, on the threshold of the century and the millennium, is linked with the universal aspiration toward the conquest of democracy, the full obedience to the Charter of Human Rights and a global effort to erase poverty from the surface of the planet.

To achieve these goals requires us to step down from the "ivory tower" of scientific neutrality. However, scientists, like other human beings, have their preferences, values and ideologies, which determine behavior in real life situations, including the research profession. Unfortunately, many scientists have been seduced by the holders of economic power or by the proximity to political rulers. Consequently, they have renounced the mission of being the critical conscience of society, hereby converting Reason into an instrument of oppression. The majority of scientists have lost the courage to question the values of the ruling system and seek to avoid social problems as if those would not exist in our society. When pressed to take a position, their judgment is of a rather conservative nature and in opposition to structural changes.

Social asymmetry is being increasingly recognized as the main obstacle to the building of sustainable societies. Therefore, in defining our research agenda for a sustainable world, the problems of power relationship, the role of the State, the conquest of participatory democracy and the universal rights of citizenship ought to be given absolute priority.

BIBLIOGRAPHY

BERGER, P. L.; LUCKMAN, T. (1966) *The Social Construction of Reality.* Garden City: Doubleday & Co.

BNDES (1999) *Efeito Estufa e a Convenção sobre a Mudança do Clima.* Rio de Janeiro.

CHAUÍ, M. (1994) Ética e Universidade, *Ciência Hoje*, São Paulo, v.18, n.102.

FEARNSIDE, P.M. (2000) Greenhouse gas emissions from land use change in Brazil's Amazon region. In: LAL, R.; KIMBLE, J.M.; STEWART, B.A. *Global Climate Change and Tropical Ecosystems*, Fl. CRC Press, Boca Raton.

FUNTOWICZ, S.O.; RAWETZ, J.R. (1992) The good, the true and the postmodern, *Futures*, v.24, n.10, p.963-976.

FUNTOWICZ, S.O.; J.R.RAWETZ (1993) Science for the post-normal age, *Futures*, v.25, n.7, p.739-755.

GOODSTEIN, E.S. (1999) *Economics and the Environment*, New York: Prentice Hall.

KUHN, T. S. (1977) *The Essential Tension*. Chicago: University of Chicago Press.

LEWONTIN, R. (1999) *The Triple Helix:* Gene, organism and the environment. Cambridge: Harvard University Press.

NOORGARD, R.B. (1994) *Development Betrayed:* the end of progress and coevolutionary revisioning of the future. London: Routledge.

PARIKH, I. (1998) *Linking technology transfer with clean development mechanism* (CDM): a developing country perspective. Paper presented at Columbia University, New York.

RATTNER, H. (1988) Science, Ethics and Ideology. In: *Liderança para uma Sociedade Sustentável.* São Paulo: Editora Nobel.

RATTNER, H. (1999) Sustentabilidade : uma visão humanista. *Ambiente e Sociedade*, São Paulo, v.2, n.5, p.233-240.

SINGER, P. (2000) *A Darwinian Left:* Politics, evolution and cooperation. New Haven: Yale University Press.

TESH, S. N. (1988) *Hidden Arguments* Rutgers, New Brunswick: The State University.

WEBER, M. (1968) Objectivity in Social Sciences. In: BRODBECK, M. (Ed.) *Readings in the Philosophy of the Social Sciences.* New York: MacMillan.

WYNNE, B. (1992) Uncertainty and environmental learning: reconceiving science and policy in the preventive paradigm. In: *Global Environmental Change*, jun., p.11-127.

BEHAVIORAL CHANGE

How Green are Brazilians? Environmental Values, Attitudes and Behavior in Brazil

Solange Simões
Federal University of Minas Gerais and
Institute For Social Research/ University of Michigan

INTRODUCTION

The assertion that the measurement and analysis of public opinion has become a central feature of liberal-pluralist democracies is hardly disputable nowadays. Opinion polls are increasingly used by political parties, newspapers, radio, TV and business organizations to profile issues and influence decision-making. On a similar fashion, survey methodology (the academic, more theoretical and analytically oriented counterpart of commercial public opinion polling) has sought to play a role in the production of scientific knowledge, as well as in the planning and realization of social change.

In the last decades survey methodology has sought to produce knowledge relevant for diversified areas of prevailing social and political issues such as voting, social stratification, racial relations or crime. One could have expected, then, that in the eighties, when global environmental problems became a central issue in social life and in social science, survey research would be called upon to produce scientifically based knowledge relevant for policy making. The challenge that has been posed is to achieve useful knowledge through description and understanding of the human role in causing global environmental change and the consequences of these changes for society (Simões and Stycos, 1995). This challenge has several major components.

The first one resides in the task of measuring the *global* dimension of environmental problems. A crucial issue we should bear in mind is that the environment is a multidimensional concept, and very often the *local*, the *national* and the *global* are presented to respondents as distinct dimensions, although it is rarely investigated how clear-cut, if at all, these dimensions are for the respondents.

Many of the issues which gave rise to questionnaire items in surveys of the mass public are either local environmental problems (which are the ones survey research is better equipped to measure, but are not necessarily directly related to biosphere changes) or are global questions (such as global warming, loss of biodiversity, depletion of the ozone layer) so far removed from the daily life of individuals that the questions might make no sense to large segments of the population, bringing into question the validity of the measures.

We could distinguish then two different approaches to the measurement of individual's perception and behavior concerning global change. The first one directly asks the population about their individual perceptions of the salience, importance, causes and consequences of, for instance, global warming and loss of biodiversity. A second approach argues that, in thinking about the human dimensions of global environmental change, energy use is one of the solution/amelioration via changes in individual values and behavior. Both in terms of the resources consumed or transformed to produce energy and in terms of the pollution or degradation, which results from energy production, this is a major factor. In global environmental terms, one is talking, for instance, about global warming that results from increases in carbon dioxide levels in the atmosphere or biodiversity loss, which is related to forest burning, among other changes (Simões, S. and Hogan, D., 1997).

According to this second approach, one should be measuring *attitudes* (transformations in values relative to consumption patterns, support for public policy related to individual and public transportation); *consumption behavior* (energy conservation through more efficient or less use of fossil fuels; re-using and recycling materials, which reduces the energy necessary for production; or transportation) and *political behavior* related to those issues. This last rationale was only more recently developed by the Global Environmental Survey, but in most surveys we can find a combination of direct abstract questioning about global environmental problems and various attitudinal/behavioral items related to energy use.

The second challenge we face is to understand the growing "greening of the world" – the widespread concern with environmental issues throughout the world and across the various social segments in the last decades. The question being posed is: *what is being measured? Overstatement of concern or deep-rooted attitudes and behavioral change?*

In fact, in this chapter we are concerned with both how broad environmental concern among Brazilians is, and how *deep* it is. In considering the breadth of concern we look into whether it is felt by a majority or minority, and to what degree, and among whom/ what social segments). In examining the depth of the Brazilian mass public's views, we address questions such as the implication of environmental concern in relation to behavior. This chapter also seeks to present and discuss Brazilians' environmental attitudes and behavior from an international comparative approach. In order to do that, we will draw on data from various national and international surveys (see section below).

ARE WE ALL GREEN? VALUES AND ATTITUDES VERSUS BEHAVIOR

In this section we will briefly introduce some of the major findings on Brazilians' concerns and behaviors, with a special focus on the inconsistencies one can find between these two dimensions of environmentalism.

We will illustrate our arguments with data from three major international surveys that included Brazil (Health of the Planet-1992[1]; World Values Surveys – 91 and 95[2]; and the Global Environmental Survey – 1997[3]) and four surveys conducted in Brazil (USAID Survey – 1991[4]; What Brazilians Think about Ecology Survey – 1992[5]; What Brazilians Think about the Environment, Development and Sustainability Survey – 1997[6]; the Rio das Velhas Basin Survey – 1997[7]). Given the choice of an international comparative approach for this chapter, I also draw on findings from an important international survey that did not include Brazil – the International Social Survey Program's Research into Environmental attitudes and Perception Report.[8]

It should be noted that these surveys were conducted in two different points in time. Several of them have been carried out in the early nineties, especially in the period immediately prior or after the Rio Conference. Two of them (WBT-EDS and Rio das Velhas Basin Survey) were conducted in the late nineties (1997) – this should allow us to look at convergence of findings, and continuities and discontinuities in the environmental attitudes and behavior among Brazilians.

A common survey finding in the last decade was the existence of a widespread concern with the environment. Earlier studies (e.g. Lipsey, 1977; Mc Evoy III, 1972) had showen that environmental concern was particularly prominent among higher-educated, higher-income, and younger people. However, more recent studies indicate that environ-mental concern is no longer a mainly middle-class issue, but transcends social-demo-graphic borders (Ester et al., 1993; Nelissen, 1992; Olsen et al., 1992). Furthermore, it even transcends traditional North-South and East-West divisions (Dunlap et al., 1993; Ester and Mandemaker, 1994) – developing as well as developed nations share widespread environmental concern. This sounded to many as a rather puzzling finding since, as Dunlap (1993) pointed out, both *conventional wisdom and theoretical knowledge* would lead us to expect low levels of con-cern for the environment in the so called developing nations. Accordingly, Brazilian researchers (Crespo and Leitao, 1993, pp.235) were rather puzzled by their own 1992 survey findings:

> *"Such findings would support the view that Brazilians have a van-guard environmental awareness, being at the same time ready to stop the uncontrolled exploitation of natural resources as well as to make sacrifices to preserve nature. Such disposition would place Brazilians in the same level of environmental concern as the one of the advanced societies, the ones that have already met the demands set by the realm of necessity or modernity, able to dedicate themselves to explore the open possibilities of the realm of freedom or post-modernity".*

In this text I seek to look into these unexpected (implausible to some) findings on Brazilian environmentalism, by tackling both the *breadth* and the *depth* of environmentalism in Brazil. This section introduces the description and analysis of the **breadth** of environmental attitudes in Brazil.

A first way most surveys measure concern is by asking about the importance or the seriousness of environmental problems.

Survey research findings have repeatedly found that Brazilians consider the state of the environment to be a very serious issue. In the early nineties, nearly the majority (49%) rate environmental issues as "*very serious*" in the country (HPS). Likewise, measuring the importance of environmental problems in Brazil, the USIA Survey found that majorities of Brazilians (62%) considered them to be "*very important*". In the WBTES-92, the respondents were asked how *much* they were interested in the environment, question to which a majority (51%) replied by saying that they were "very interested". Similar findings held true for the HPS – a majority in Brazil (53%) said they were *personally* a "great deal" concerned about the environment.

Looking at **environmental quality**, only 3% of Brazilians rated, overall, the quality of their environment as "very good" (USIA Survey). In its turn, in order to measure the respondent's perception of environmental quality, the HPS distinguished among the *community, national and world dimensions*. Brazilians followed the trend applicable to most countries – the more distant the environment being rated, the more negatively it was viewed. Moreover, a majority of Brazilians (76%) considered environmental problems "a great deal" of a threat to their own and *their family's health* (USIA Survey). In its turn, the HPS introduced a distinction between perception of effects to "*respondents own health ten years ago; own health now and the health of their children and grandchildren over next 25 years.*" Brazilians followed the pattern of less concern in the past and growing concern about the future.

When looking at the above results one should bear in mind, however, that despite considering the environment an important or serious issue with implications for the family well-being, the environment has not appeared as a *salient* issue for Brazilians. This was the case in 1992 and in 1997 as well. In the HPS, only 2%, and in the USIA Survey 0%, of Brazilian respondents *volunteered* environmental problems as the most important problem in the country. In the more recent surveys, the environment was still only spontaneously mentioned by less than 1%, in both the Rio das Velhas Survey and the WBTEDSS-97.

In order to interpret the above findings we first need to bear in mind the distinction between importance and salience of issues. Furthermore, when the concept of the environment is presented in a generic and abstract form and compared to other broad issues such as health and education, it is not perceived as an important issue. When we change the frame of reference from the country to the city/local, and the environmental problems are presented in their more local and concrete dimensions, such as water pollution and sanitation, they appear as more relevant to respondents (this is the case in both the Rio das Velhas Basin Survey and WBTEDSS-97). If respondents are given the chance of spontaneously naming the most important environmental problem, national or local problems, issues such as water pollution and sanitation appear as central.

In my view this can be seen as an indication of the need to further investigate the environment as a multidimensional concept and explore people's views for different frames of reference (local, national, and global, for instance) and different dimensions (attitudes and behavior). The Rio das Velhas Basin Survey findings seem to indicate that people's own definitions can be widely varied ("everything" for 17%; "life" for 13%; "quality of life" for 9%; "source of life" for 7%; "natural resources reserve" for 7%; "health" for 7%; noting that one-fourth of the respondents could not provide an answer). We would like to argue that in all cases, but especially in the case of developing countries (given higher levels of illiteracy and lack of information), it is necessary to make clear which dimension of this broad multidimensional concept people might have in mind when answering attitude items.

It is also worth mentioning that, as measured in the HPS, Brazil was among the developing nations with lowest levels of concern (together with Uruguay 3%, Philippines 2%, Nigeria, Poland and Hungary 1% each). Although we had sharp variations in the level of concern *within* each block of nations, it might be meaningful that only 4 develop-ing nations (out of 12) had more than 10% volunteering environ-ment as *the most important problem*, whereas only 3 developed nations (out of 12) had less than 10% doing so. But Brazil and Mexico, both developing nations, stood in polar positions – one (Mexico – 29%) ranked among the ones where the environment is a *salient* issue and the other (Brazil) where it is very rarely volunteered. Economic problems appeared as "*the most salient*" issue in most societies, developed or develop-ing. What is challenging here is to understand why the "environment" appears or not as a salient issue – although the data shows that it was more likely to be volunteered in most of the developed countries, we are left with the problem of understanding why in societies such as Norway (2%) and Great Britain (3%) so few volunteered the environment as the most important problem, and why in developing countries like Mexico (29%), India (21%), Chile (20%) and Turkey (18%) it appeared as a salient issue. My belief is that trying to explain these somewhat unexpected findings we might gain insight into the other major fac-tors conditioning salience of issue. As various survey researchers have repeatedly argued, we do need *contextual data* to make sense of these variations.

Among a variety of environmental con-cerns, at the *national* level Brazilians are mostly concerned with water pollution (35% in the USIA Survey; 38% in the WBTE-92, 26% in the WBTEDS-97), deforestation (30% in the USIA Survey, 46% in the WBTE-92, 45% in the WBTEDS-97), and air pol-lution (29% in the USIA Survey; 18% in the WBTE-92, 12% in the WBTEDS-97). When it comes to the *local* dimension, sanitation and garbage collection are major issues in the earlier HPS as well as in the more recent WBTEDS-97 and Rio das Velhas Basin surveys. It stands out that those issues are mentioned spontaneoulsy in the WBTEDS-97 and appear as salient (together with water pollution) in the lists presented to respondents by HPS and the Rio das Velhas Basin Survey. It is also noteworthy that these findings converge with international findings (Health of the Planet, 1992; Environmental Monitor,1997), which shows that the mass public in the so-called developing countries choose water pollution as the most pressing problem, whereas the mass public in the developed countries are mostly concerned about air pollution.

Similarly to the perception of problems in the national dimension, when it comes to the global dimension *deforestation* and *water pollution* are again viewed as the most important problems by Brazilians in 92 (WBTE-92) as well as in 97 (both in the Rio das Velhas Basin Survey and WBTEDS-97). Among the global environmental problems, nearly one third of Brazilians also single out loss of biodiversity and depletion of the ozone layer, in 92 as well as in 97 (WBTE-

92 and WBTEDS-97). Smaller percentages mention global warming as one of the most important global problems (17% in the WBTE-92 and 19% in the WBTEDS-97). As *the* most important global problem, global warming is singled out by only 4% in the Rio das Velhas Basin Survey. In fact, most Brazilians say they have never heard about global warming (over 50%), or biodiversity (nearly 70%) (WBTMADS-97).

However, even when we move from the "cost-free" measures of environmental concern to the trade-offs, such as the dilemma between environmental protection vs economic development, majorities in Brazil adopt a pro-environmental position. This is a relatively consistent finding through time and various surveys (64% – USIA Survey, 71% – HPS, 47% – Rio das Velhas Survey). Furthermore, compared to residents of the high-income nations, Dunlap (1993) argues that Brazilians are as likely to give the environment priority over the economy. Even when confronted with the scenario of having to make personal sacrifices to protect the environment, a majority of Brazilians (53%) say they are willing to pay higher prices to protect the environment (HPS). Brazil is among the 17 nations (including low-income ones) out of 24 where majorities in 1992 indicated a willingness to pay higher prices for increased environmental protection (HPS).

> *How have the researchers "interpreted" such findings? Dunlap et al (1993 and 1995) sounded straightforward and confident in the conclusion that "that such higher proportions of citizens of so many nations, including many countries with low standards of living, express a willingness to pay higher prices if it will improve the environment is perhaps the strongest evidence we found of world-wide concern for environmental quality.*" (Dunlap, 1995)

In his turn, enthusiastically quoting the USIA Survey findings to support his analysis of worldwide concern for the environment, Worcester (1993, pp.14) sees as *unsurprising* that by margins of more than two to one, in developing countries such as Mexico (66% to 30%) and Brazil (64% to 24%) people would *give priority to the protection of the environment over economic growth*. This finding was unsurprising, in his view, under the circumstances – the fact that majorities of Brazilians (75%) and Mexicans (62%) believed environmental problems to be a great deal of a threat to their own and their family's health.

The WBTE-92 researchers, however, were at first clearly awestricken by similar findings in their 1992 Brazilian survey. They found quite puzzling that a majority of Brazilians (57%) strongly disagreed with the statement "*a comfortable life brought about by progress is more important than preserving nature*". However, they believed that these amazing findings were taken even fur-ther in the answers to the question "*Would you accept living with more pollution if this means more jobs? Do agree or disagree?*" A majority (64%) strongly disagreed and only 17% strongly agreed. This question was replicated in the 1997 wave and, again, practically the same majority of Brazilians (65%) adopted a pro-environmental stance.

Such results and their interpretation have been met, however, by considerable skepticism among those who stress the pervasive huge gap between attitudes and behavior. We should therefore try to address the question posed earlier on: *does environmental concern translate into pro-environmental action and support for public policies?*

Moving on from the measurement of attitudes to the more methodologically challenging issue of behavior (or at least *reported* behavior), two dimensions of environmental behavior can be distinguished: consumer and political behavior. When looking at the consumer behavior dimension, the HPS found that *avoiding environmentally harmful* products was the most widely practiced of the three indicators of behaviors (the others were *participation in organizations* and *vote*), with respondents in 18 of the 24 nations reporting having done that in the past year. As Dunlap (1993, pp.33) pointed out, "*not surprisingly, respondents in wealthy nations are most likely to report having done this, since they are most likely to have choices among products and to be able to afford choosing on more than price alone*". Even that being the case, it is noteworthy that in the HPS Brazil appeared as the country with the smallest percentage of green consumers (26%), only matched by India (27%).

The WBTES-92 and 97 waves measured *willingness* to adopt behavior, which is rigorously an attitudinal measure rather than a behavioral one. In both waves, Brazilians tended to choose recycling (59% and 72%) and reducing energy and gas consumption (38% and 41%) as measures *they would be willing to adopt*. The mass public did not appear as willing to pay more for food without chemical products (14% and 11%) or to buy energy efficient home appliances (5% in the 1997 wave).

The other dimension of behavior, political behavior, was measured, in 1992 by the HPS, by looking at participation in a group or organization that works to protect the environment. The proportion of Brazilian activists was unsurprising low- only 4%. In 1997 the WBTEDS, though, fifty percent of Brazilians say they are *willing* to join an environmental group, 55% willing to do volunteer work for NGOs, and a smaller percentage (26%) willing to make a financial contribution.

The Rio das Velhas Basin Survey helps to throw light on this issue by measuring both willingness and actual (reported) behavior. As one would have expected, the results suggest there is a substantial gap between willingness and actual behavior. Nearly 90% of the respondents declare they have never participated in environmental groups activities, educational campaigns, attended meetings with government officials or done volunteer work to clean for the environment. But willingness is relatively high for joining environmental groups, participating in meetings, doing volunteer work, and especially for taking part in educational campaigns.

The findings on willingness to join an environmental group should also be contrasted with *knowing* one. In the Rio das Velhas Basin Survey less than 1% spontaneously cited an NGO. Even when presented with a list of NGOs and governmental agencies in the WBTE-92, a large majority (82%) declared never having heard of any group or organization. Greenpeace and some other few NGOs reached 1%.

Still in the political dimension of behavior, as far as *electoral politics* is concerned, Brazil is among those countries with lowest percentages reporting "*having voted or worked for candidates because of their position on environmental issues*" (9% of Brazilians – HPS).

Since the Stockholm Conference in 1972, international agreements signed by nations worldwide have attributed fundamental importance to the communities in the definition of environmental public policies. Chapter 36 of the Agenda 21 is basically dedicated to the preparation of communities to take responsibility, individually and collectively, over development and environmental issues. In several of the international documents there are provisions and operational guidelines meant to place value not just in collective participation but also in the local spheres of government, closer to the environmental questions related to the citizen's quality of life. Support for decentralization in environmental protection can be traced in the findings of the two waves of the WBTEDS survey (with high levels of adherence to the view that the municipal government and the individuals should have main responsibility for the solution of environmental problems).

On a similar fashion, when it comes to the perception of who should have the most important role in environmental protection, the mass public in the Rio das Velhas Basin Survey appears equally divided in relation to the federal government (24%), the local government (24%) and the individual citizen (24%).

The HPS found in 1992 that, worldwide, *national governments* were most likely to be seen as having primary responsibility for protecting the environment. However, Latin-American countries ranked highest among those nations whose *citizen groups* were more likely to be seen as having responsibility. This was the case for a majority in Brazil (60%) and was the highest percent-age among all nations surveyed. For Brazilians, in 92, the *government* came second (26%) and *business* last and much lower (12%); the USIA Survey found that a plurality (43%) believed the *government* should have the primary responsibility, edging individual and citizen groups by a narrow margin (36%). Significantly fewer said that business and industry should have prin-cipal responsibility (17%). Likewise, in the WBTES-92 and the WBTEDS-97 the *government* came first (51% and 43%, respectively), followed by "each one of us" (39% and 36%) and the municipal government (33% in both waves); again businessmen ranked very low (12% in both waves).

This support for decentralization of environmental policies and the reliance on individual action seemed to some analysts to indicate a deeper level of pro-environmentalism. But is that the case? We found out that pro-environmentalism is broad (widespread concern), but how deep (behavior change) is it? One might think that before jumping to any conclusion we must not overlook some "inconsistencies" in our findings:

- High levels of environmental concern, **but** poor knowledge of environmental issues

- Environmental problems do not appear as salient in relation to other national issues, but majorities prioritize environmental protection in a trade-off with economic development

- Willingness to join NGOs, but very poor knowledge of groups and very low reported activist behavior

- Values and environmental attitudes do not correlate with environmental behavior (nor with willingness to adopt behavior) (see section below)

- The proportion of people holding pro-environmental values is higher in the South-Southeast, but willingness to act is higher in the North-Northeast regions (see below).

This gap between values/attitudes and environmental behavior is not, however, a problem particular to Brazil. This happens to be one of the central issues analyzed in most international survey literature, as we shall see in the next section.

THE STATE OF THE ART: *WHAT ARE WE MEASURING? SUPERFICIAL CONCERN OR DEEP-ROOTED ATTITUDES AND BEHAVIORAL CHANGE?*

Survey analysts from several disciplines (sociology, political science and social psychology) have been seeking description, understanding and explanation of individual responses to environmental change through the adoption of an equation comprising the following terms: values, beliefs, attitudes, knowledge and behavior about environmental issues. Different approaches have laid varying emphasis on description or explanation, on the study of attitude or behavior or on the links between attitudes and behavior. Fewer survey researchers have tried to integrate contextual data and external objective factors in the equation. What have the major findings been? If one investigates "the state of the art", one finds a number of major surveys that tap growing public concern about the environment in many countries (Dunlap et al, 1993; Inglehart, 1990 and 1993; Worcester, 1993; Witherspoon and Molher, 1995). Findings confirm other social science reports of a growth in public interest in, and awareness of, environmental issues. Furthermore, there is growing consensus among analysts that environmental concern is found among all social strata, races, educational levels; among the people of developed as well as developing nations, and so on. Nevertheless there is considerable controversy about the depth and consistency of public concern. To the questions ***"how relevant is widespread concern for action?"*** or ***"how does support translate into support for public policies and behavior?"*** the above-cited survey reports present diverging conclusions.

Some reports (Dunlap, 1993 and 1995; and Worcester, 1993 and 1995) highlight that pro-environmental public sentiments are strong enough to affect such actual behavior as those of voters and consumers whereas others (Witherspoon, 1994) cast doubt on the likely short-term political repercussion of the growth in public concern and suggest that much public concern about the environment is (still) relatively superficial. Some believe that their findings challenge both conventional wisdom and theoretical knowledge, whereas others take a skeptical look at the possible meaning and implications of the greening of the world.

Competing (but often convergent) theoretical approaches have been used in the description and explanation of the general public responses to environmental change and environmentalism. Nevertheless, one could argue that the **terms in the equations** used have been the same: **values, beliefs, attitudes,**

knowledge and behavior. The integration of **external barriers** (or external objective constraints), such as levels of pollution or institutional contexts with survey data have been done by fewer analysts, but there seems to be growing consensus about the need to add them to the research agendas and explanatory models. The several approaches certainly differ in the explanatory weight they give to the various factors; in the degrees of complexity of the operationalization of concepts and in the sophistication of data analysis procedures. Drawing on the more fruitful analytical approaches one can point to the alternative (sometimes complementary) ways researchers have tried to explain support for policies and pro-environmental behavior (consumer and political behaviors). Three major factors have been awarded central explanatory power: values, beliefs and knowledge and context/external barriers.

VALUES X ATTITUDES

To what extent is support for public policies and environmental behavior conditioned by values (post-materialist, socio-altruist and biocentric)?

Several approaches have tried to distinguish between **egocentric x socio-altruistic/homocentric x ecocentric x post-materialist values** as the set of values that weigh more in the explanation of pro-environmental attitudes and behavior. Stern and Dietz (1994) and Inglehart (1997) are among those who try to argue theoretically and demonstrate empirically the weight of values in relation to other attitudinal and socio-demographic variables. Inglehart argues that *within* given countries, people with "post-materialist" values – emphasizing self-expression and the quality of life – are much more apt to give high priority to protecting the environment (and are much more likely to be active members of environmentalist groups), than those with "materialist" values – emphasizing economic and physical security above all. Furthermore, he claims that his analysis shows this holds true at the national level: *countries* that have relatively post-materialistic publics rank relatively high in their readiness to make financial sacrifices for the sake of environmental protection. Stern and colleagues (Stern and Dietz, 1994; Stern and Gardner, 1996) have begun to look for direct evidence of links between value clusters (homocentic/socioaltruistic, ecocentric, biospheric-altruistic, egocentric) and individuals' willingness to take *political action* for environmental protection. The cluster of self-transcendent values (or biospheric-altruistic) was found to be strongly predictive of people's self-reported willingness to take actions such as boycotting the products of a company that pollutes, signing a petition for tougher environmental laws, and refusing to invest in or work for a polluting company. Furthermore, they argue that values can affect pro-environmental action both directly and indirectly, through beliefs about consequences.

In the WBTEDS-97, the central focus of the survey is in the measurement of values and attitudes related to environmentalism and sustainability. We should note, however, that in the Brazilian study, unlike Inglehart's and Stern's, but more similarly to Dunlap's, the central focus lies on environmental world views and attitudes rather than on general values (such as altruism and materialism), which, as Stern and Inglehart hypothesize, should be more strongly correlated with willingness to act and behavior. Notwithstanding, the WBTEDS-97 was innovative in its attempt to build a scale of pro-sustainability values and attitudes. The scale, according to Ignacio Cano's analysis (1998), includes 12 items tackling the following aspects: notion of scarcity and the exhaustion of natural resources; concern with the environment; acknowledgement of the limits of science in solving environmental problems; opposing a utilitarian approach to nature; prioritizing environmental protection in relation to economic well being and jobs; attributing responsibility to the present and not to the future generation; favoring environmental education; considering the use of alternative products to chemical fertilizers and pesticides.

When using the scale for the analysis of the breadth of environmentalism, on the one hand, we should point out that looking at the sociodemographic variables, there is no segment that opposes environmentalism and sustainability. These findings converge with those of international studies. On the other hand, however, Ignácio Cano's analysis (1998) shows how the pro-sustainability scale is correlated with some sociodemographic segments, such as age group, education, income, residence in municipalities with over 100 thousand inhabitants and in the more developed regions (south-southeast). The variables with most impact are education and income, in this order. Through statistical regression, Cano sought to isolate the elements that had an independent impact on pro-sustainability and to build a theoretical model to explain it. The conclusion was that pro-sustainability values in Brazil seem to fundamentally depend on education.

However, when the analyst sought to correlate pro-sustainability values/attitudes with willingness to act (Cano, 1998), the hypothesis that the more pro-sustainability individuals would be the more willing to take action was not confirmed. The data shows that those who are more supportive of sustainability are not more willing to help environmental groups or participate in the solution of local environmental problems than others. Assessing these findings from the theoretical framework developed by international research, we should point out that the correlation between attitudes and behavior is often weak. This is a point tested and again confirmed by GOES (see Aoyagi-Usui, 2001; Ester and Vinken, 2001; and Simões and Hogan, 2001), as well as by the Rio das Velhas Survey (Simões and Porto, 1998). The theoretical hypothesis that Inglehart and Stern put forward is exactly that general values, rather than environmental attitudes, would be strong predictors of behavior. Although not very impressive, the GOES surveys in the Netherlands and Japan show correlations between Stern's socio-altruistic/biocentric values and behavior. In the Rio das Velhas Survey, however, the socio-altruistic/biocentristic values only appear to correlate with consumer behavior (Simões and Hogan, 1998).

In Inglehart's case, his data shows that in the advanced societies post-materialists are twice as likely to support environmental protection than materialists are and nowhere materialists favor environmental protection more strongly than post-materialists. More recently GOES findings for Netherlands and Japan show that among a set of independent sociodemographic, attitudinal and values variables, post-materialism proves to be the best predictor of behavior, especially political action. In the case of Brazil and other lower income countries, Inglehart's data (1995), however, show weak correlations between post-materialism and environmental protection. Likewise, the Rio das Velhas Survey, using some of the GOES behavior measures, did not find an association between post-materialist values and support for public policies or behavior among the population of the Rio das Velhas Basin (Simoes and Hogan, 1997).

As in the case of the WBTEDS-97, we are still left with the challenge to understand the absence of correlation in Brazil between pro-sustainability values/attitudes and willingness to act (or values and behavior in the Rio das Velhas Basin Survey case).

KNOWLEDGE AND BELIEFS ABOUT CAUSES AND EFFECTS OF ENVIRONMENTAL PROBLEMS

To what extent is the adoption of environmental behavior conditioned by rationality (perception of risks)?

In Stern and colleagues' (1994 and 1996) formulation, besides values, another key to people's response to any environmental problem are their *specific beliefs about the consequences of environmental problems to the things they value.* In their explanatory model, beliefs mediate between values and attitudes. They hypothesize that value orientations may affect beliefs about the consequences of attitude objects for the things an individual values and thus have consequences for that individual's attitudes and behavior (Stern and Dietz, 1994).

Moving from the "beliefs" to the issue of "knowledge" and information, the Report Into Environmental Attitudes and Perception (REAP report, Witherspoon and Molher, 1995, pp.88) shows that "*scientific knowledge probably leads people to adopt a less apocalyptic view of environmental problems and a less romantic view of nature, but it is also positively associated with environmental concern and activism, once we discount people's general values*". The policy implication that follows in the report is that public information campaigns about environmental issues may be a useful starting point in an attempt to create an awareness of the problems but, on the evidence presented, neither awareness nor understanding is sufficient to bring about support for green policies.

In their turn, Stern and Gardner (1996) challenge the common beliefs about the role of education and information. They assert that education can change attitudes and beliefs, but warn that many barriers, both within individuals (values and ethics) and in their socio and economic environments, can keep pro-environmental attitudes from being expressed into action. Furthermore, education is not seen as likely to work if it clashes with people's basic ethics or values.

Turning to Brazil, we should look at Cano's analysis of the relation between information and sustainability in the WBTEDS-97 survey. As he hypothesized, sustainability values seem to fundamentally depend on educational level; these two variables are strongly correlated – the more knowledgeable, the more supportive individuals are; whereas, the poorly informed ones tend to be more reticent. This finding of the Brazilian survey clearly converges with the international findings mentioned above. In terms of the relation between knowledge and environmental behavior (or willingness to act in the WBTEDS-97 survey), the low association is even more accentuated. Cano (1998) points out that on the one hand, all individuals who know the environmental issues are clearly supporters of sustainability. On the other hand, however, the data show that it is possible to hold those values without knowledge of the environmental agenda. Nevertheless, contrary to what was expected, the individuals who show more concern with sustainability are neither more willing to take part in the solution of environmental problems nor more willing to help environmental groups.

Likewise, the Rio das Velhas Survey shows that the urban and more educated segments are more likely to show concern in a series of attitudinal questions (ranging from perception of problems to prioritizing environmental protection). Furthermore, in a regression analysis testing values/attitudinal and sociodemographic variables as independent variables explaining pro-environmental behavior, one of the few associations found is between educational level and political behavior (Simoes and Hogan, 1997).

THE STRATEGIC ENVIRONMENTS: CONTEXT AND EXTERNAL BARRIERS

To what extent is support for public policy and adoption of environmental behavior conditioned by external barriers (physical environment, institutional arrangements, level of socioeconomic development)? What are the strategic environments in which individuals make their pro-environmental choices?

External barriers (or external objective factors) as opposed to internal factors (values, beliefs and attitudes).

The integration of contextual data to survey datasets is a particularly relevant development of survey methodology and bears great analytical potential, especially in the case of environmental research, given that human perception and behaviors seem to be strongly conditioned by the physical environment and by the social and political structure of opportunities.

Inglehart (1995) has successfully introduced variables such as severity of air and water pollution in the analysis of mass support for environmental protection, whereas Witherspoon (1994) has insightfully pointed to institutional arrangements. According to Inglehart (1995), the claim that public support for environmental protection is shaped by subjective cultural factors, only tells part of the story. As one would expect, mass support for environmental protection tends to be greatest in countries that have relatively severe objective problems (as indicated by levels of air and water pollution). Inglehart asserts that objective problems and subjective predisposition are both involved in support for protecting the environment; the available evidence indicating that both of these factors are about equally important. Stern (1995) argues that structural conditions and other barriers often keep values and beliefs from being enacted as behavior, and when it occurs it will require changes in the structure of society and in people's funds of knowledge for changed values and beliefs to be carried out into action.

The WBTEDS-97 contained some contextual measures (urban or rural residence; clean water supply, sewage, garbage collection, street lights, among others). As Plinio Dentzien (1998) shows us, the contrast among Brazilian regions is clearly portrayed in the lack of basic services in the households where the interviews for the WBTEDS-97 were taken. According to Dentzien, if we move beyond the "naturalist" definition that respondents give to the concept of the environment and reanalyze the answers to the open question about problems in their neighborhood, and add to the items subjectively recognized as belonging to the environment other aspects that are objectively related to the environment (electricity, clean water, sewage and green areas), the 4% who spontaneously mentioned "the environment" as the most important problem in their neighborhood would jump to 33%. Moreover, as one would have expected, given the lack of access to services in the north, northeast and central regions, the inclusion of the objective elements inverts the ranking of the regions in the mention of environmental problems: northeast – 44%, north – 44%, central – 33% and south-southeast – 29%.

The *political* contexts in which environmentalism is likely to take place have also been integrated to the survey analysis. Analyzing findings for the REAP report (Britain, Netherlands, Ireland, West Germany, East Germany), Witherspoon (1995) makes the case:

> *"It does not seem to be coincidental that the most consistent green activism and the most consistent support for policies which attempt to protect the environment are to be found in the social democratic countries in our sample, which tend to be based on some greater substantive consensus on collective goals...Of course, public policies do not simply arise from individual attitudes; they may also follow from an institutional context which places environmental action higher*

on the public agenda than is currently the case in some countries. So the public may be more activist and more consistent in Germany and the Netherlands because institutional structures act to constrain people's views, nudging them towards more active and consistent responses to their environmental fears." (1995, pp.89)

This institutional contextual element that Witherspoon points out bears similarities with the analysis Bruce Chadwick makes about the support for the public policy of "*rodízio de carros*" (alternating the use of private cars by establishing control over which days of the week license plates ending in odd or even numbers, for instance, can be on the roads) in the two WBTES-92 and WBTEDSS-97 waves. Chadwick (1998) shows that the support for *rodízio* increased in the country as a whole for the period 92-97, but much less than in São Paulo (23,4% greater than the rest of the country). Chadwick suggests that the interpretation for this finding resides in the fact that São Paulo implemented the *rodízio* policy in 1996, six months prior to the conduction of the 97 survey wave. Furthermore, in Chadwick's words

"*the rodízio in São Paulo originated from a top-down policy...in fact, although the Environmental Secretary had invested massively in public relations and educational campaigns, by the time of the 97 wave, the mass public in Sao Paulo had already had enough time to move beyond the novelty of the system and feel the impact of the rodízio in their daily lives. It follows that we can assert that the people in São Paulo were able to adapt their lives to the rodízio and even increase their support to the policies – much more than the rest of the country.*" (1992, pp 6)

Last, but not least, I would like to go back to the issue of the multidimensionality of the concept of the environment and the implications of this multidimensionality for the understanding of the supposedly respondent or data "inconsistencies". Based on a survey of Great Britain, Witherspoon and Martin (1992) show that environmentalism is a multidimensional concept, which means that an individual's environmentalism varies according to the dimensions under investigation. For instance, in the British case, the analysis of the relation between education and the various dimensions of environmentalism (global environmentalism versus pollution-centered environmentalism versus anti-nuclear versus green consumption environmentalism), only the green consumer behavior scale follows the hypothesis, with green consumer's behavior increasing steadily with educational levels (Witherspoon and Martin, 1992, pp.10).

In the WBTEDS-97 survey we have findings with similar analytical implications. The multidimensionality of environmentalism can be highlighted when we see, for instance, that while those more interested in doing volunteer work to help with neighborhood environmental problems are younger, with secondary education, in the North/Northeast regions, live in small towns (up to 20,000 inhabitants), those more willing to sort garbage for collection are the higher educated, living in the state capitals or cities with over 100,000 inhabitants. Plínio Dentizien also shows that while the environmental values/attitudes are more present in the more developed South-Southeast regions, the perception of local environmental problems (such as sanitation) is higher in the impoverished Northeast. In Dentizien's (1998) words:

"The first conclusion to arrive at from this analysis is concerned with the scope of the concept of the environmental question among Brazilians. There is not a single pattern of response to the various questions; they are extremely varied. In other words: those who opt for a pro-environmental orientation in one dimension (factor) have about the same chances of those who opt for the anti-environmental orientation in that dimension to opt for a pro-environmental orientation in another dimension. This implies that, in more general, clear-cut and drastic terms, Brazilians do not perceive, like academics and other opinion makers, a single environmental question, which would unfold into multiple aspects, but diverse questions, more or less specific, vaguely related to the environment."

FINAL COMMENTS

When we compare the findings of research conducted in Brazil with findings for other countries, we realize that there is a considerable convergence among them, whether we are looking at the empirical findings or the analytical procedures.

First of all, it is worth noting that what one could assume to be particularly frequent in Brazil is often similar to what happens in other countries. This is certainly the case for a combination of significant environmental awareness linked to a religious respect in regard to nature, but without association with behavior – which is not at all particular to Brazil, as Brazilian conventional wisdom would be likely to assume. The conclusions from the REAP report (Witherspoon and

Molher, 1995) for five advanced nations from the European Community (the Netherlands, Germany, Great-Britain, Ireland and Italy) certainly will sound familiar to Brazilians when the investigators point out:

> "*The research conducted as part of the REAP project shows that majorities in five nations are aware of, and concerned about, environmental issues. However, the research also shows that the standard survey-based measures of concern are likely to be misleading in suggesting there is greater depth, commitment, and consistency than in fact is the case in people's environmental views. In many countries there is a quasi-religious, quasi-romantic concern about the environment, with human intervention being seen as inherently disruptive. This goes hand in hand, however, with an unwillingness to face some of the hard policy choices that a move towards sustainability, much less a recreation of some Arcadian nature unsullied by human contact, would require*" (p.88).

Although seeing the above assertions as hardly disputable, I would like to raise the possibility that the opposition between some researchers' *enthusiasm* for worldwide high levels of concern for the environment versus other researchers' skepticism about the quality of the measures so far developed might well be missing one important aspect. Since the existence of widespread (although still superficial) *concern* over the environment is highly plausible, one could well argue that the problem might not *centrally* revolve around the validity and reliability of the measures of concern developed, but in *how to link concern to behavior.* Here certainly lies one of the most chal-lenging gaps to be bridged by social sciences in general, not just survey research. Moving beyond (although intertwined with) issues of research design and measures, and the need for development of theoretical explanatory models, I believe such apparently inconsistent findings point to the need to construct in our research instruments the *likely scenarios* where so-cial actors make their choices. By further integrating survey and con-textual data, one could develop a better grasp of the *strategic environments* where individual choices are made and, by doing so, construct more useful data for both international comparative analysis and domestic policy making. Environmental surveys conducted in Brazil (and most other places) have overemphasized the measures of concern and done very little in terms of measuring changes in behavior, general values and political and physical contexts, although one must acknowledge that some of the research we commented in this chapter has made pioneering attempts to develop such models. I would like to argue that in order to develop explanatory models to fill in the gap between attitudes and behavior – besides the political and physical contexts – we should be also looking at universal values (such as altruism, post-materialism, biocentrism, sociocentrism, etc.) rather than over focusing on pro-environmental attitudes (that so far have proved to be weakly correlated with pro-environmental behavior, in Brazil and elsewhere). We should also bear in mind that before tackling the challenge to explain environmental behavior, we need to *measure* behavior, in both the consumer and political dimensions. More urgently, we need measures of individual behavior that most impact *global* environmental change (such as energy consumption, for instance).

Another point I would like to raise in these final comments is that the debate about environmentalism and socio-economic development might be overlooking a central and basic plain factor in the issue of development: developing societies are more and more integrated in an international economic and political system and context, and this fact alone should keep these societies from merely repeating the exact same developmen-tal stages undergone by the currently advanced countries. At the time that we, in developing societies, are *still* struggling with basic material needs, environ-mental problems *already* are a major *global* issue. Foreign and in-ternational opinion and organization (especially by middle class ac-tivists, foreign and domestic ones) bring about this "*post-modern*" con-cern to the still "modernizing *societies*". It follows that in our analyses Brazil should be contrasted not only to the developed northern countries, but to less developed southern nations as well. Despite commonality of the huge social inequalities, high levels of il-literacy, centrality of the need for a fairer distribution of in-come that place Brazil among the developing countries, for comparative analysis we cannot fail to point out and try to under-stand the implications of other relevant aspects of that society. Unlike many southern nations, Brazil is an industrialized *emerging* country with over two thirds of the population living in urban areas known for over-crowding, noise, air and water pollution, lack of sanitation and housing linked to severe health conditions. Furthermore, given the development of industrialization (and post-industrialization), we are also confronted with complex class structures in many ways, especially in the case of a few privileged and highly educated social segments, more comparable to the ones of the so-called developed world.[9] From this argument, it follows that one major reservation to Dunlap's analysis of the HPS is that it is limited to *comparison between two polar blocs of countries*. What could be said about variations *internal to a bloc of nations*? Moreover, what could be found about variations *within* countries? Not to mention the insights we might gain by further comparing social groupings and classes *across* countries.

Moreover, from the standpoint of the analysis of *global* environmental change, it might be crucial that cross-country comparative analysis takes into account not only the level of economic development but the country's contribution to greenhouse gas emissions as well. A typology such as the one developed by Viola (2000)[10] could be used for sampling and for analytical procedures more geared to the understanding of the human role in the *global* dimension of environmental change.

Disentangling the various dimensions of global change might not be an easy task though. As we have seen earlier on, for the Brazilian mass public, deforestation, as a *national* problem, comes first (according to the findings of WBTEDS mass public surveys of 1992 and 1997) (Samira, C., 1988a). As a *local* problem, though, water pollution/sanitation tends to be the most important one in the public's perception (this is a finding of both the WBTEDS and the Rio das Velhas Basin surveys). Taking into account inputs from social science research other than environmental surveys, we should distinguish three major dimensions when we talk about deforestation as perceived by Brazilians (Simões and Viola, 2001). The first one is the local, when people living in areas of traditional agriculture – scattered in the North, Northeast and Center-west regions – and in the tropical forest frontiers – mostly in the Northern region – suffer from increased air pollution derived from deforestation. The second one is the national dimension related to the cutting down of the rain forest, depleting the country's natural resources, increasingly perceived not just as timber but also as biodiversity – raw material for developing new medicines and food. The third one is the global dimension related to the global carbon balance. Although most Brazilians do have diffuse/confused ideas about the carbon balance (Dunlap, 1995), it could be argued that they perceive deforestation as somehow affecting it.

Another important related issue arising from the survey findings revolves around the several possible understandings the public might have of the notion "*environment*". What do they have in mind when asked about *environmental* problems? The format of the question (open ended or list of alternatives) and the dimension tackled (national or local) seem to prompt different understandings of the concept. Deforestation appears as more salient in spontaneous answers about national problems, whereas water pollution/sanitation is mostly picked out from lists of local environmental problems. It could be argued that in the "collective consciousness" deforestation links more directly to the abstract notion "environment", but given a broad concrete definition, the public will see Brazilian environmental problems, beyond the deforestation issues, also closely linked to urbanization and quality of life in the cities. In fact, studies carried out in Latin America, by social scientists using methodology other than survey research, have suggested that the environment in such societies is closely associated with health, sanitation, education and housing (Hogan, 1992). In a way, we could say that *traditional social demands* have been "relabeled" *environmental issues* (Martinez-Alier and Hershberg, 1992).

Survey research has not yet explored people's own definitions in order to find out which dimensions of the broad concept of environment people might have in mind when answering attitude items. The ap-proach to the environment as a *multi-dimensional concept* (Witherspoon and Martin, 1992) should be developed in order to explore different definitions thought of for different problems and by different social segments.

Last but not least, the concept of global warming, which has been forcefully leaving the realms of scientific uncertainty, does not appear high on the list of the Brazilian mass public concern. Nevertheless, the perception of urgency of related issues (or causes of global warming) – such as deforestation and air pollution in Brazil – might indirectly contribute to commitment to policies meant to reduce greenhouse gas emissions. It also might imply that national issues might be seen as not so far removed from the global ones. Indeed there is still a lot to be learned about the ways the local, the national and the global dimensions of the environment are articulated by individuals and how that conditions the perception and willingness to tackle *global* environmental change.

NOTES

1. *The Health of the Planet Survey was conducted in 24 nations in 1992 by the Gallup International Institute (Dunlap, 1993; Dunlap, Gallup and Gallup, 1993). The Health of The Planet (HPS) survey was coordinated by the George H. Gallup International Institute. It was conducted in 24 nations covering a wide range of geographical regions and economic levels. For each nation, the Gallup affiliate employed its standard probability sampling procedure for obtaining a representative national sample. Each affiliate translated the original questionnaire into the appropriate language and the Gal-lup International Institute had the translation back-translated to guarantee comparability. Face-to-face interviewers were con-ducted between January and March of 1992. It was a pioneering move toward scientific coverage of the topic. Although the survey was mainly descriptive, it contained items relevant to policy issues and explored attitude formation.*

2. *The World Values Surveys have been carried out in more than 50 societies containing 70 percent of the world's population. They provide a longitudinal data base which with to analyze processes of cultural change that are reshaping orientations toward religion, politics, work, economic growth, the role of the state in the economy, child rearing, sexual norms, gender roles and the environment. The WVS has produced significant findings concerning the relationship between environmental attitudes and behavior, although it covers environmental issues only briefly, as one of many topics.*

3. *The Global Environmental Survey has two components: a survey of the mass publics and a survey of the decisionmakers. GOES was initially launched by the Human Dimensions of Global Environmental Change Programme with the challenge to build on the state of the art in environmental survey. Given the focus on global environmental change, GOES sought to develop measures of behavior related to energy comsumption affecting global warming. Another central goal was to contribute to building explanatory models exploring the relationships between values, attitudes and pro-environmental behavior.*

4. *The United States Information Agency (USIA) survey – This public opinion survey was based on per-sonal interviews with 2000 adults -- 18 years of age and older --in Brazilian cities of 10,000 population or more. The poll was conducted between November 9 and November 16, 1991.The questions were written by the USIA Office of Research. The poll was then conducted for USIA by IBOPE of Sao Paulo. The translation of the questionnaire was reviewed by the Research staff in Washington and by USIS Brasilia. The sample is representative of the adult (18 and older) national population in cities of 10,000 population or more. The contractor selected the sample through a modified probability technique (probability proportional to size) with cities and census tracts as the primary sampling units. Households were then randomly selected and respondents within the household were selected by quotas.*

5. *The What Brazilians Think of Ecology Survey (WBTES) was conceived and designed by a multidisciplinary team of researchers from Museu de Astronomia e Ciências Afins (MAST) and ISER. The field work was conducted by a private polling institute (IBOPE) which interviewed a* *sample of 3650 Brazilians in the first trimester of 1992. This was a pioneering national survey of the mass public's perception of environmental problems in Brazil.*

6. *The What Brazilians Think about the Environment, Development and Sustainability survey, carried out in partnership with MASP (CNPq) Museum of Astronomy and ISER, conducted a public opinion survey representing the population over 16 with 2,000 households interviews. Together with the 1992 wave, this constitutes the largest and most complete dataset in the country regarding what the Brazilian population thinks about environmental themes related to development.*

7. *The "Environmental Perception and Behavior Among the Das Velhas River Basin Survey" – was conducted in 1997 and interviewed 1800 respondents, in urban and rural areas, selected through probabilistic sampling in the Das Velhas River Basin, located in the central region of Minas Gerais and comprising the Belo Horizonte Metropolitan Area and 51 other cities up and down stream. The survey replicated several questions from the GOES mass public questionnaire. The study was comissioned by the Fundação Estadual do Meio Ambiente and was a component of PROSAM (Programa de Saneamento Ambiental das Bacias do Arrudas e do Onça). The fieldwork was conducted by Sensus Data World.*

8. *In 1993 an environmental module was included in the International Social Survey Program (ISSP), which covered 21 countries, mainly advanced industrial societies. This module provides a rapid overview of global environ-mental concerns. An additional module concerning Research into Environmental Attitudes and Perceptions (REAP) was added to the ISSP survey in five West European countries, including some behavioral and policy items. It focused, though, mainly on issues of concern in advanced industrial countries.*

9. *The critical assessment of the questionnaires and reports of the above and other cross-national survey research about the environment would immediately pose issues of validity and reliability of the measures designed. The presentation of the above findings can certainly lead to us raising a number of problems related to question wording, question sequence and format (open-ended vs close-ended; scales vs ranking). Moreover, I would just like to highlight here a very crucial problem in survey methodology, which can be particularly stressed in cross-national surveys of countries in diversified levels of socio-economic development. Adding to the overall issue of social desirability, that leads to distortion of findings, we also have the problems posed by the different levels of education and information of respondents. Besides measuring over-statement of concern we might well be measuring the opinion of "those who do not have an opinion". It is revealing to compare the percentages of "don't know" answers to some open-ended versus some closed questions. This methodological problem can be argued as a possible source of distortion when the Brazilian researchers found in the WBTE-92, for instance, 47% DK answers in an open-ended question on national environmental problems but only 12% DK answers in a closed question on global issues. Unfortunately, environmental survey reports in general more often than not have failed to present the percentages of "don't knows", which should lead readers to increasingly critical questioning of interpretations and inferences drawn.*

My own contention here is that the developed/developing distinction should consider a more sophisticated understanding of the heterogeneous social structures of some emerging countries – measures designed in western industrialized societies might work among the more educated and urban segments of the population of developing countries such as Brazil but we should not ignore the mounting problems of reliability, validity, relevance and comparability of the measures of global environmental problems in face of higher levels of illiteracy and the rural background of large segments of respondents in developing societies.

10. *According to Viola (2000), the arena of the climate regime is very complex because the interests of the countries are differentiated according to both their level of economic development and carbon emissions. See the chapter by Viola and Leis in this book for a discussion of eight major groups of countries in the climate regime.*

BIBLIOGRAPHY

AOYAGI-USUI, M. (2001-forthcoming). *New Images of Old Concepts: Environmental Values and Behaviors of the Japanese Public.* In: SIMÕES, S., ESTER,. P., VINKEN, H.,

AOYAGI-USUI, M (ed.) *The Global Environmental Survey : An International Comparative Study Of Cultural Change And Environmental Issues Among Mass Publics and Elite".*

BERK,, R. (1994). Contextual Data for Surveys. Paper presented at GOES Workshop, Oslo, Norway, may.

CANO, I. (1998). *Ambientalismo no Brasil: Os Valores Pro-sustentabilidade.* Rio de Janeiro: ISER

CHADWICK, B. P. (1998). *Opiniao Publica e Desenvolvimento Sustentavel: de 1992 a 1997.* Rio de Janeiro: ISER

CRESPO, S. , LEITÃO, P.. (1993). *O Que o Brasileiro Pensa da Ecologia.* Rio de Janeiro: Editora Brasil America.

CRESPO, S. (1997). *O que o Brasileiro Pensa do Meio Ambiente, Desenvolvimento e Sustentabilidade.* Rio de Janeiro: ISER

DENTIZEN, P. (1998). *Variações regionais do ambientalismo no Brasil.* Rio de Janeiro: ISER.

DUNLAP, E.R., GALLUP, G.H. , GALLUP A.M. (1993). *The Health of the Planet Survey, a preliminary report on at-titudes toward the environment and economic growth measured by surveys of citizens in 24 nations to date.* Princeton: The George H. Gallup International Institute.

DUNLAP, E.R., GALLUP, G.H. , GALLUP A.M.. (1995). Global Environmental Concern - Results from an International Public Opinion Survey. *Environment*, v.. 35, November.

ESTER, Peter, HALMAN, Loek, SEUREN, Brigitte. (1993). "Environmental concern and offering willingness in Europe and North America." In ESTER, Peter, HALMAN, Loek, MOOR, Ruud de (eds.), *The individualizing Society: Value change in Europe and North America.* Tilburg: Tilburg University Press, p. 163-181.

ESTER, P. , MANDEMAKER, T. (1994). '"Socialization of environmental policy objectives: tools for environmental marketing." In: Dutch Committee for Long-Term Environmental Policy, *The Environment: Towards a Sustainable Future.* Dordrecht: Kluwer Academic Publishers.

ESTER, P. and VINKEN, H.. (2001-forthcoming). "Sustainable Development and the Attitudes, Values and Behaviors of Dutch Citizens. In: SIMÕES, S., ESTER,. P., VINKEN, H., AOYAGI-USUI, M (ed.) *The Global Environmental Survey: An International Comparative Study Of Cultural Change And Environmental Issues Among Mass Publics and Elite*".

HOGAN, D.J. and VIEIRA, F. (1992). *Dilema Socioambientais e Desenvolvimento Sustentavel.* Campinas: Editora da Unicamp.

INGLEHART, R. (1997). *Modernization and Post-modernization: Cultural, Economic and Political Change in 43 Societies.* Princeton: Princeton University Press.

INGLEHART, R. (1995). Public Support for Environmental Protection: Objective Problems and Subjective Values in 43 Societies. PS: *Political Science and Politics* 15, 57-71.

LIPSEY, M.W. 1977. "Personal antecedents and consequences of ecologically responsible behaviour: a review." JSAS catalog of selected documents in psychology.

MCEYOY III, J. 1972. "The American concern with the environment." In: BURCH, W., , CHECK, N.H. , TAYLOR, L. (eds.), Social behaviour, natural resources, and the environment. New York: Harper & Row, p 214-236.

MARTINEZ-ALIERr, J., HERSBERG, E. (1992) Environmentalism and the Poor - the ecology of survival. Items. *Social Science Research Council*, vol. 46, nº 1. March.

MOLHER, P. and NORRIS, P. (1993). *Comparing Environmental Public Opinion' West and East Germany.* Paper presented at GOES Bar-celona Workshop, September.

NELISSEN, N.J.M., PEEREBOOM, P., PETERS, R., PETERS, V. (1987). *De Nederlanders en hun milieu: Een onderzoek naar het milieubesef en het milieugedrag van vroeger en nu.* Zeist: Kerckebosch,

OLSEN, M.E., LODWICK, D.G., DUMLAP, R.E., (1992), *Viewing the world ecologically* Boulder: Westview. p.47-81.

SIMÕES, S. (1995). *Comparing Public Opinion on the Environment in Latin America.* Paper presented to the XVI World Congress of the International Political Science Association, Berlin, August 21-25.

SIMÕES, S., STYCOS, J. (1996). *GOES WORKPLAN*, unpublished manuscript.

SIMÕES, S. and HOGAN, D. (1997). *Brazilian GOES Pretest Report*, unpublished manuscript.

SIMÕES, S. and Porto, J. H. (1998) *Nosso Rio, Nossa Gente : Percepção e Comportamento Ambiental da população da Bacia Do Rio Das Velhas.* Belo Horizonte: FEAM, 1998.

SIMÕES, S. (2001-forthcoming) "Description and Explanation of the Greening of the World: a methodological and theoretical challenge for survey methodology (as illustrated by research in Brazil and Mexico)". *Policy Studies Annual Review 2001, special issue "Knowledge, Power and Participation in Environmental Policy*" (eds. W. N. Dunn, M. Hisschemoller, R. Hoppe, and J. Ravetz - forthcoming)

SIMÕES, S. and HOGAN, D. (2001-forthcoming). "Beyond the Rain Forest: environmental attitudes and behavior among Brazilians" in SIMÕES, S., ESTER.P., VINKEN H, AOYAGI-USUI, M (ed.) *The Global Environmental Survey : An International Comparative Study Of Cultural Change And Environmental Issues Among Mass Publics and Elite".*

SIMÕES, S. And VIOLA, E. (2001-forthcoming). "The Brazilian Sustainability Challenge: Combining the Environment, Modernization and Inequality Reduction" in in SIMÕES, S., ESTER.P., VINKEN H., AOYAGI-USUI, M (ed.) *The Global Environmental Survey : An International Comparative Study Of Cultural Change And Environmental Issues Among Mass Publics and Elite".*
STERN, P. and T. DIETZ. (1994). The value basis of environmental concern. Journal of Social Issues, Vol. 50, No 3, 65-84

STERN, P. and G. GARDNER. (1996). *Environmental Problems and Human Behavior.* Boston: Allyn and Bacon.United States Information Agency. (1992). *Brazilians Pessimistic About Environment and Brazilians Choose Environmental Protection Over Economic Growth.* Research Memorandums.

VIOLA, E. (2000) "The Global Politics of Climate Change. How Strong are the Forces Supporting Sustainability? in *10th World Congress of Rural Sociology*, article # 305, CD-ROM, Rio de Janeiro, International Rural Sociology Association & Brazilian Council for Scientific and Technological Development.

WITHERSPOON, S., MARTIN, J. (1992). What do we mean by green?, In: JOWELL et.al., eds., *Bristish Social Attitudes - the 9th Resort (SCPR).* Aldershot, GB: Dartmouth Press.

WITHERSPOON, S. (1994). The greening of Britain: romance and rationality. In: JOWELL et.al., eds., *Bristish Social Attitudes - the 11th Resort (SCPR)*, Aldershot, GB: Dartmouth Press.

WITHERSPOON, S. and MOHLER, P. (1995). *Research into Environmental Attitudes and Perception* (The REAP project). European Consortium for Comparative Social Surveys

WORCESTER,R. (1993). *Societal Values and Attitudes to Human Dimensions of Global Environmental Change.* Paper presented at the International Conference On Social Values, Complutense University of Madrid, September 1993.

WORCESTER,R.(1995). Environmental Activism in Countries from Australia to Finland, West, East, North and South. Belo Horizonte GOES Workshop, Dezembro.

Global Food Risks: Enviromental and Heath Concerns in Brazil

Julia Silvia Guivant
Federal University of Santa Catarina

INTRODUCTION

One of the dimensions of global environmental change is related to the food system. This does not refer only to production and its environmental consequences (water and soil pollution, declining soil fertility, water scarcity, for example), but obviously to the effects on consumers' and producers' health. Environmental and health food risks are global because the modern food system involves international trade, with more integrated international markets, and also because their effect on health and the environment is not restricted to national boundaries or poverty pockets. Pesticides, for example, through the circle of poison, may affect populations that consume imported food harvested with illegal pesticides (Guivant, 1992, 1995, 2000a). Other characteristics of modern food risks are that they can be invisible and irreversible (as in the case of DDT, whose effects were discovered much after commercial use began).

The recent highly publicized debates on food-borne diseases, especially Bovine Spongiform Encephalopathy (BSE) and its links with new variants of the Creutzfeld-Jacob Disease (CJD) in humans and the onset of genetically modified (GM) organisms in the food system, have deepened public concern about how food is produced on farms and in factories, processed, transported, stored and traded. More importantly, they have launched a crisis related to central social issues, such as the roles of science, politics and business corporations in the decision processes for determining which risks societies should or wish to assume. So far, consequences of these debates include governmental and ministerial destabilization, discredit of scientists, threats to international trade relations, and demand for public debates about how to regulate uncertain, open-ended and indeterminate technological risks. Also, public interest groups are attempting to influence the pace and trajectory of the globalization process itself (Tait and Bruce, 2001). This is the scenario of a global risk conflict.

This crisis was anticipated by German sociologist Ulrich Beck, who over ten years ago transformed the field of risk analysis when he introduced the distinctive concept of "risk society" in his 1986 book (translated into English in 1992), to refer to the radicalization phase of the principles of modernity. "Risk society" is a term for contemporary or late modern societies which, contrary to the industrial and class society of early modernity, face environmental and human health risks that are not simply side effects of technological progress, but central and constituent components of these societies. Moreover, the risks are structurally different in terms of their sources because they threaten every life form on the planet. Humanity has always been faced with risks, but the current ones are specifically unique because they are consequences of a certain type of science and technology development, and they have open-ended consequences. These risks have invisible global effects - regardless of class or nationality distinctions - which can be fatally irreversible once they are identified. Examples of these risks are global warming, pollution of water sources, food contamination, AIDS, the ozone layer hole, desertification, eco-toxicity, and radioactivity, with short- and long-term effects on people, animals and plants.

Much social research on manufactured risks focuses on the situation of highly industrialized countries.[1] Particularly for Beck, they are risk societies because they are going through a process of reflexive modernization where there is a growing realization (reflexivity) of the dangers involved in modernity and a consequent questioning of the main structures of society through new forms of doing politics. Another type of reflexive modernization characterizes a previous stage of modernity, where risks are produced but not yet the subject of sustained public or personal debate or political conflict.

If global risk society theory can demonstrate how risks with serious consequences are central to understanding highly modern society, now, with the global dimension that food-borne risks have assumed, this theory can prove its heuristic value (Guivant, 1998). In this article I will argue for the need to develop a more complex perspective on the global dynamics of food-borne risks and how the Brazilian debates around manufactured global food risks, such as BSE and GMs, can contribute to this objective. A significant bibliographical production has emphasized how food production is a key issue of environmental security, one of the components of global environmental change (related to food supply and national sovereignty). Here I will consider how food safety is also central for the discussion of global environmental change. Its analysis, although still very marginal in Brazilian academic and policy debates, can contribute for a better understanding of the new characteristics of the global food system and also of the specificities of global challenges for more and less industrialized countries in terms of the food policy research agenda in safety standards. Brazil is already affected by the increased concerns in highly industrialized countries on food safety issues. Pinstrup-Andersen (2000) identified two major consequences of this process for less industrialized countries that can be applied to Brazil. First, exports of food commodities will be exposed to new and more demanding food safety standards, partly included in changes of the Codex Alimentarius and partly through unilateral demands by importers. Second, the influence on consumers in less industrialized countries of the changing attitudes and new legislation for food safety in highly industrialized countries.

In the first section I briefly introduced world risk society theory. In the following sections, I analyze the character of the coalitions organized around GMs and BSE in Brazil and how the heterogeneous social actors involved interpret those risks. To understand the peculiar coalitions and meanings of global food risks in Brazil it is necessary to consider them in a broad perspective that includes a reference to the progressive awareness of consumer rights and the growing food safety control that coexists with a significant amount of ignorance in relation to food-borne risks. In the last section, I will examine how little discussion has been raised about global food risks and their implications for the population's quality of life in the academic context of social sciences in Brazil. Moreover, we will discuss the relevance of this analysis for suggesting a research agenda.

POLITICS AND SUBPOLITICS IN A WORLD RISK SOCIETY

Given the global nature of risks, surpassing class and national borders, Beck argues that this new type of society can no longer be explained as a class society. If in the past hunger was hierarchical, now pollution is democratic. If before risks were seen as personal, now they are global: "food chains connect practically everyone on earth to everyone else" (Beck, l992: 36). It results in a boomerang effect, since risk society is a **global** risk society, and it is not easy to escape. Consumers who are more informed and who have higher incomes might be able to avoid eating vegetables with higher amounts of pesticides, for example, but complete protection from risks would only be possible if one did not eat, drink or breathe. With this argument, Beck does not deny that some people are more affected by the risks than others, and he recognizes that the distribution parallels class inequalities and positions on the social scale, but with a different distribution logic.

Scientific knowledge, according to Beck, has caused two serious problems related to risks. One is linked to how "side effects" of a chemical substance are determined. The second problem relates to the fact that risks emerge not only as "accidents," "side effects" or because of improper use of technology. They are submersed and masked in the "levels of acceptability" of contamination that are defined through laboratory research standards, and thus unreal when expressed in formulas that do not reflect the multiplicity of chemical bombardment we suffer daily. Science, by stipulating these acceptable levels of contamination, is issuing a type of *carte blanche* for polluting and poisoning the environment. Yet the criteria of acceptability for residues and exposure are not fixed or definite. Often, what was said to be harmless to human health is later found to be harmful, through new evidence resulting from long-term research.

Beck's main criticisms of technical methods of risk analysis can be summarized as follows (Mol e Spaargarem, l993): 1) limited method range, since not all substances can be evaluated for their risk potential, and neither can the effects of the combinations in our bodies and in the environment; 2) they do not consider cumulative long-term effects; 3) they project results for humans that were obtained with animals in a controversial manner and; 4) they ignore the social factors that may influence the peculiarities of individuals' sensitivities.

In regard to globalization, Beck (1997, 1999, 2000) has emphasized that the contribution of his theory of global risk society would be to demonstrate not only that there is a synchronism between class societies and risk society, but that both Western and non-Western societies can simultaneously face the same challenges of high modernity. He believes there is a plurality of modernities in the global risk society, in which non-Western and Western societies share the same challenges of second modernity, through different cultural perceptions.

However, it is surprising how Beck maintains an impoverished view of globalization, by considering the maximum difference within its synchronism to be the one between Western and non-Western societies, and ignoring the different types of Western and non-Western societies and the internal differences within countries (Guivant, 2000b,c). Poor countries are not found exclusively in non-Western societies. Neither are poor regions far from highly industrialized countries. By only referring to the latter countries, Beck disregards the complexities of possible combinations within the block of Western countries. An isolated reference to this problem is mentioned by Buttel (2000), but it is also absent from the various criticisms that Beck and his theories have received recently (see, for example, Adam, Beck and Van Loon, 2000; Spaargaren, Mol and Buttel, 2000).

Another characteristic of risk society is that it leads to a redefinition of the social actors and arena for conducting politics (Beck, 1994,1995). What once belonged to a private realm of responsibility and scientific creativity, has become a subject of debate for what Beck terms subpolitics (in the sense of political subsystems). Subpolitics distinguishes itself from politics (in the sense of the official political system) by involving actors who are outside the latter sphere (professional and occupational groups, technical intelligentsia of companies, research and management institutes, qualified workers, etc.) and who begin to participate in the public debate on various topics. Since there are limits for determining the standards of potential risks exclusively through scientific knowledge, the rules and bases for making such decisions should be reestablished: opening the dialogue and the decision process, recognizing that there is an inevitable openness of risk problems and solutions (Beck, 1994).

How are subpolitics implemented? How can risks be controlled? An alternative would be to set up negotiation forums with government authorities and corporations as well as unions, political representatives and other entities or individuals. These forums would not necessarily seek consensus, but would enable the parties to take precautionary and preventive measures, integrate doubts, define winners and losers, make the issue public, and thus finally, improve the pre-conditions for political action. These negotiating forums are not necessarily "consensus producing machines with guaranteed success," or do they eliminate conflicts or out-of-control industrial dangers. But they can contribute to the prevention of risks, guarantee a symmetry of sacrifices that can not be avoided and make it clearer as to who are the winners and the losers. Risk society's typical science, working behind closed doors, must be replaced by another science, which paradoxically would be more rational than the former. Beck is distant from the discourse of post-modernity or from recognizing any possibility of defending irrationality.

Research carried out recently in the EU in response to the challenges presented by BSE and GMs is, in part, greatly synchronized with Beck's proposed de-monopolization of the decision-making process. But this research has advanced significantly more than Beck in the formulation of more viable, yet not always easily implemented, strategies (Grove-White et al, 1997, 2000; Green Alliance, 2000; Stirling and Grove-White, 1999; Adam, 1998). The research is partially a product of the criticism of the way expert systems dealt with BSE and GM risks and so it has developed from a consensus about the need to substitute unidirectional information systems with more transparent and open ones, so that consumers receive information that helps them make choices and decisions. The biggest challenge pointed out in the research is the lack of recognition on behalf of information providers about how to deal with risks about which there are divergent opinions, ignorance or uncertainty among experts. The tendency to deny such situations, instead of calming the public, can foment its skepticism and lack of confidence in the expert systems, in addition to stimulating suspicions about the opacity of the techniques and the potentially irreversible and unknown side effects.

Another consensual recommendation from the research cited above is that a participatory decision-making system is required in order to build legitimacy into the political decisions on food-borne risks. This can imply difficult challenges for politicians and scientific experts, as well as the need to accept imperfect decision-making and open learning processes: "there are no easy answers in the face of profound uncertainty" (ESRC Global Environmental Change Program, 1999).

In the following sections we will argue how the theory of world risk society can contribute to the analysis of the Brazilian debate on GMs/BSE and how this analysis can help to overcome the limits of a largely European focus.

UNCONVENTIONAL DISCURSIVE COALITIONS IN RESPONSE TO GMS

GM crops and food involve two types of uncertainty: how will they affect the natural life around them? And what will food from such plants do to humans when they become part of the standard diet? In regard to these two questions there are not any simple and definitive answers: "there could be hundreds of GM foods, all from entirely different gene adjustments. Even if one of them is proved safe – after a lifetime of human consumption – what does that say about the others? And if a consumer does die after many years of consumption, how will anyone be sure it was the GM food, and not the crisps, the hot dogs, the beer or the cigarettes? Can resistant genes escape into wild plants? Probably. Will it happen on a massive scale? So far, the answer is probably not. Will the world eventually be over-run by superweeds? Who knows? The alternative scenario painted by environmentalists is that the effect of herbicide resistant crops will be to kill off all the weeds, and with them the birds and insects that depend on them, turning farmland into sterile monocultures (Radford, 1999). The answers to these questions are part of the struggle between coalitions – on the one hand, there are those who consider the uncertainties to be unfounded and irrelevant, and on the other those who prefer to assume an attitude of precaution.

The debate about GMs in Brazil began receiving media attention in September, 1998, when a report by the National Bio-Safety Techniques Commission (CTNBio) was challenged. This agency was created by the State within the Ministry of Science and Technology to regulate and study companies' requests to test and plant crops on Brazilian soil. It was initially comprised of scientists, representatives from interested companies and consumer organization representatives – who later withdrew from the commission due to disagreements over its operation. The disputed report was favorable to the first request (presented by Monsanto) for large-scale planting of genetically modified soybeans intended for commercialization, pending later approval by the Ministries of Health and the Environment. Reaction against this report was basically led by the Institute for the Defense of Consumers' Rights (IDEC) and by Greenpeace (as part of its international campaign), who initiated a public civil action suit against the government and Monsanto. The judge of the 6th District Court of Federal Justice in Brasilia sided with the plaintiffs and, with this decision, the government will have to elaborate safety and labeling norms so that in the future it can evaluate, and eventually authorize, the cultivation and commercialization of these foods (Leite, 2000; Gorgen, 2000).

The movement against the liberation of GMs was joined by other NGOs and political parties, such as the Workers Party (*Partido dos Trabalhadores* - PT), the Landless Workers Movement (*Movimento dos Trabalhadores Sem Terra* - MST), the Brazilian Society for the Progress of Science (SBPC), the Federal Prosecutor's Office, the Brazilian Institute of the Environment and Renewable Resources (IBAMA), and state Programs for the Defense of Consumer Rights (PROCON). These social groups and state agencies began to voice their concerns through the media, promoting debates, writing manifestos, and taking similar actions to publicize the lack of information about the risks involved and effectively press authorities to be cautious before approving any release of GMs (http://www.consumidorsa.org.br).

A new coalition of heterogeneous social actors is creating a space of supolitics against GMs. Although subpolitics refers to politics outside and beyond the representative institutions of the political system of nation-states, it can put unconventional political actors and discourses together in *ad hoc* ´coalitions of opposites´

with more conventional political actors (political parties, governments, unions, etc). This mix is also considered by Hajer (1995) in his characterization of the unconventional discourse coalitions in environmental issues: they are unconventional because they are constituted by a heterogeneity of social actors, such as scientists, politicians, activists, or consumer organizations, who do not necessarily share the carefully planned action strategies. However, as Hajer explains (1995: 13), "although these actors might share a specific set of story-lines, they might nevertheless interpret the meaning of these story-lines rather differently and might each have their own particular interests." The same risks assume different meanings for different social actors, as Douglas and Wildavsky (1982) had already pointed out. Unconventional discursive coalitions assume their own characteristics in Brazil, mainly in sectors that oppose the liberation of GM crops.

Within Brazil's unconventional discursive coalition against GMs we find "conventional" social actors such as PT and MST, who incorporate the GM theme within their classicist discourse against globalization, imperialism, multinational corporations, the USA, the International Monetary Fund, etc. The charges against GM crops are justified through economic and political arguments: who benefits and who suffers? How can we defend the survival of small rural producers and squatters against globalization and social exclusion markets? (Federal Senate/ PT, 1999) The battle against GMs is presented by the MST leaders as an element within the fight against the government's agricultural policies, against the agricultural model and in the last instance, against capitalist society (Stedile's declarations, *O Estado de S. Paulo*, June 1, 2000). But within the "conventional" discourse new issues are also integrated, mainly the environmental and health consequences of GMs. Environmental issues are more and more present in recent MST documents, although with marginal concern on issues like pesticide contamination.

Among these actors' actions are, in July of 2000 in Recife (Pernambuco), the invasion of a ship loaded with GM corn, several incidents of destruction of clandestine and trial fields of genetically modified crops of soybeans and manifestations in front of supermarkets and McDonald´s, considered to be representatives of "imperialism" and the power of multinational corporations.[2] It is argued that the use of GM seeds encourages dependence by farmers on a single seed supplier and may involve the purchase of both the seed and herbicide from one supplier. The farmer can then be at the mercy of the seed company who may vary prices of both seed and herbicide at will.

The group of actors who assume a less conventional discourse includes international NGOs such as Greenpeace and organizations that represent consumers, such as PROCONs, IDEC and federal public prosecutors. IDEC and Greenpeace are part of the Brazil Free From Genetically Modified Products Campaign, which groups various NGOs and are part of an international coalition including powerful actors of a globalization from below, that pressure the activity of world corporations and national governments (Beck, 1999:38). Their intervention is centered on legal actions, with the objective of redefining CTNbio duties and decisions, advocating labeling and employment of the precautionary principle.[3] Tests are also performed on food products to detect the presence of GM elements (*Revista Consumidor S.A*, 47, February 2000) and consumer awareness campaigns are carried out. Arguments are focused primarily on health and environmental risks and on defense of consumer rights, a category significantly absent from the discourse of PT and MST. Among those who are directly allied to this group are governmental agencies such as IBAMA, and scientists who focus mainly on the level of uncertainty surrounding GM product risks and who also recommend applying the precautionary principle. Several manifestations of the Brazilian Society for the Progress of Science are oriented in this direction (Pavan, 1999).

The criticisms raised against GMs by these unconventional groups, parallel to the criticisms presented internationally by equivalent groups, point out the limits affecting the trustworthiness of conclusions obtained through scientific methods that tend to deny uncertainty and indeterminacy.

The model of society presumed in these groups, different from the model assumed by MST and PT, is about organizing actions that guarantee the strengthening of civil society, with special attention to the rights of the population as consumer citizens. It does not propose a model of society other than the capitalistic one, but something closer to the idea of ecological modernization (Mol and Spaargaren, 1993). Although IDEC's discourse clearly recognizes the precarious situation of consumer rights, when the Institute argues in favor of GM labeling, it may be putting this idea aside. The efficiency of labeling is important for a minority of persons with food allergies, for example, but in general it should be evaluated in the specific context of Brazilian consumers, who are fairly inactive in terms of reading food labels.[4]

This part of the coalition does not intend to assume an irrational or anti-scientific position regarding scientific knowledge. This becomes clear in its defense of the precautionary principle, which presumes, to a certain extent paradoxically, that in the future we might possibly be able to obtain precise and definitive scientific knowledge regarding risks (Latour, 2000). However, food risks related to BSE and GMs might require the challenge of generating strategies to deal with technologies and productive processes that involve an uncertainty that is constituent and not necessarily surmountable. Yet this issue is not significantly considered. Science in general is questioned as to its limits in evaluating GM environmental and health risks and the government is criticized because of its narrow approach to the decision-making process, responding directly to the needs of the biotechnology companies. There is no extensive criticism of how expert systems in Brazil deal and advise governmental agencies on how to deal with other food-borne risks.

In the coalition in favor of GMs we find scientists – who defend CTNbio criteria and decisions - and representatives of biotechnology companies such as Monsanto. They present a conventional scientific argument, defining the opposing sectors as irrational and uninformed catastrophists who are against progress. They even reverse the "imperialist" charge they receive from others back onto GM critics, arguing that the latter group is in fact the imperialist one, for impeding the distribution of seeds which could reduce hunger in less industrialized countries. The CTNbio scientists also assume an environmental argument, just as scientists from other countries, by stating that the use of GM crops significantly reduces the use of pesticides (http://www.ctnbio.gov.br) and can be the key to richer wildlife and efficient food production.[5]

This coalition's discourse does not hedge towards the view that standard decision-making tools, which are based on quantitative procedures and objectives, fail to capture the level of uncertainty and indetermination in relation to the effects caused by the new agricultural technologies, and neither does it recognize that environmental and health problems and their solutions are often complex and value-laden. That is, there is a resistance to recognizing the inevitability that risk analyses are permeated with "wider tacit political and institutional contingencies and commitments" (Grove-White, 1999), in this manner maintaining the presumption of a model of society where the role and contribution of science should not be publicly questioned.

Among rural producers there is a certain division between those who recognize themselves as beneficiaries of the GM crop prohibition and those who do not see the benefits and who, despite the sentence of up to three years in prison for infractors, prefer to import seeds clandestinely from Argentina, where

GM use is approved. Although there is no precise data available, some estimates suggest that in states such as Rio Grande do Sul, despite its policy of tight control over GM use, 30% of the area planted consists of GM soybean crops (*Veja*, March 28, 2001; *Consumidor S.A*, N. 47, February, 2000). This use would be justified as an attempt to reduce pesticide expenses and to increase profits.[6]

Farmers who continue harvesting conventional crops are discovering a segmented market in their favor. The European food industry, supermarkets and processing companies, in response to consumer demand for products without GM ingredients, are taking concrete actions that significantly affect tendencies in the international commodities market. For example, in January 2001, Tesco and Asda, two British supermarkets that control 42% of the grocery market and import more than one million tons a year of GM animal feed, announced that they would switch their imports from the United States to Brazil, and that they would no longer sell the meat or milk of any animal fed with genetically modified soy or corn. Brazilian soy exports to the EU have increased from 2.99 million tons in 1996 to 6.87 million in 1999.[7] In turn, this change has caused United States soy exports to Europe to fall from 9.85 million to 6.75 million tons between 1995 and 1999, following concern about GM crops (*The Guardian*, January 27, 2001).

Conventional crop producers are not the only ones who benefit; producers who cater to the organic market are also expanding their business. For example, the French grains dealer Cabinet Boyer is setting up a facility in Brasília in May of 2001. The goal is to export 45 thousand tons of pesticide-free grains to the EU, Japan and the United States (*Gazeta Mercantil*, March 29, 2001).

The sectors in favor and against GMs encompass lay people - including rural production sectors – as well as experts, but they apparently do not significantly mobilize consumers. Each one of these heterogeneous alliances articulates itself with others on an international scale, but without necessarily involving identification of themes other than GMs or presenting the same interpretation of risks. Concern for the environment and consumer health might be more central for European and United States anti-GM groups than for those in Brazil, but it is expressed by part of the unconventional discursive coalition. This is due less to the confidence crisis in the expert systems - as would be happening among European consumers (Global Environmental Change Programme, 1999; Adam, 1998) – and more to an advance of the demands for consumer rights in defining scientific/technological and economic policies.

On one hand, NGOs in the coalition do not raise larger questions about the expert system in control of global food risks. On the other, we suggest that the expert systems in Brazil have a limited role in consumers' decisions, who are perhaps more guided by trust in the food companies and supermarkets. Next, we will consider in more detail the recent BSE crisis in Brazil.

THE DISCURSIVE COALITION IN RESPONSE TO MAD COW DISEASE

For twenty days in February 2001, Canada and then NAFTA embargoed Brazilian meat because of a suspicion of BSE contamination. Cattle raisers in southern Brazil had bought animals from England, Germany and France (which used British feed for their cows) between 1989 and 1990, after BSE infection in British cattle had been confirmed.

BSE started to noticeably affect British cattle in the mid-1980s (due to feed made from the carcasses of contaminated sheep), and its transmission to humans in the form of a variety of the Creutzfeld-Jacob disease (CJD) was only recognized around ten years later. In the last few years, several EU countries that considered themselves to be BSE-free have identified cases that have generated consumer mistrust of the expert and political systems, causing serious economic damage. After almost a decade of denial, BSE is now being considered a serious risk to human health, since it is irreversible and lethal. However, there is a large amount of uncertainty about the mechanisms through which the disease passes from cows to humans: whether it can be passed along through vaccines containing animal by-products; whether it is contagious among humans; how much time it takes to manifest itself; which are the risk groups, etc. The current debates surrounding BSE affect governments and sanitary inspection sectors that had not considered taking adequate measures to protect the population, especially in countries that considered themselves to be disease-free but that soon suffered bouts of contamination. For these reasons, BSE requires rigorous cattle monitoring.

With NAFTA's claims, the quality of Brazil's system of animal sanitary control was under scrutiny, as the country could not immediately identify the destination of the cattle imported from England. With the embargo, Brazil's meat market, which has one of the biggest herds in the world – 160 million heads – was at stake. Brazilian GM-free grain producers, but mainly meat producers, would gain from the country's elimination of BSE risk.

NAFTA's position led to an alliance in Brazil among sectors that were in opposition regarding GM crops and on different sides of national political questions: PT, environmentalists, MST, the Brazilian Association of Meat Exporting Industries, diplomats, businessmen, the government, cattle raisers and packing house owners.[8] In this case, the discursive coalition was based on nationalistic arguments that disagreed with NAFTA's possible foundations in regards to the Brazilian governmental control system over the imported cattle's destination.[9]

The most consensual interpretation seemed to be that Canada had decided to embargo Brazilian meat for political and economic reasons, related to the commercial crisis at the WTO between the two countries' aviation companies, Bombardier and Embraer. Despite the fact that governmental authorities denied that the cause was the delay in sending information about the imported cattle, during the crisis more rigorous rules were defined for animal sanitary control, including registration and tracking programs for imported animals.[10]

The nationalistic reaction consisted of various acts: threats of aggression towards the NAFTA technical mission (which arrived in Brazil to examine cow herds and inspect farms, laboratories, feed factories and packing houses), notes in the newspapers about the Brazilian meat the delegates ate, demonstrations opposing Canada's policy, barbecues held in front of the Canadian consulate in São Paulo, a declaration from the local government representatives of São José dos Campos – where Embraer's factory is located - stating that Canada's prime minister is *persona non grata* in the city, bar boycotts of Canadian beverages, delivery of a cow to the embassy in Brasilia, the refusal of port workers in Santos to unload Canadian ships, stereotypical Canadian jokes published in the media, etc.

This conjunctional coalition presents the same criticism of globalization as a phenomenon contrary to national sovereignty as is assumed in the discourse of part of the coalition against GMs. However, there is one important difference. In relation to the BSE crisis there was a large degree of unification against more industrialized countries, and the division between government and political oppositions was attenuated.

Who was left out of the coalition? IDEC, Greenpeace and other NGOs that work together in the GM alliance were not very present in the debate. If we look in the newspapers of that period, only a few critical commentators voiced their opinion on the Brazilian reaction. Pedro de Camargo Neto, president of the Cattle-raising Development Fund (Fundepec), interpreted the crisis to be a warning against the lack of Brazilian concern for internal consumers and the belated attention to the disease's risks (*O Estado de S. Paulo*, February 21, 2001). Larger questions were raised by Pedro de Felícia, a professor in the department of technology at Unicamp (Universidade de Campinas) and president of the Brazilian Association of Meat Sciences, and researcher Maristela Pituco, from the Biological Institute of São Paulo (*Folha de S. Paulo*, February 20, 2001). Both of them questioned the inversion of values inherent in the debate, which had a heavy emphasis on the economic aspect of the government's and society's positions, while ignoring the situation of internal consumers, threatened with diseases that attack cattle (rabies, leptospirosis, among others) and compromise meat consumed in the country. One of the risk management proposals that researchers suggest is the creation of a Food Ministry which would substitute the Agriculture Ministry and could enter into agreement with states to regulate demands such as animal prophylaxis and food inspection.

But this criticism of the internal management of food-borne risks was not pronounced publicly by the main social actors, as if the GM and BSE debates had no connection. It seems that the GM debate generated a reduced space for the discussion of risks of serious consequences in Brazil. So, consumer rights are gently and slowly being legitimized while the nationalistic discourse against commercial consequences of globalization is still strongly defended. Food risks were "translated" to the Brazilian situation through conventional political codes, as if any suspicion of the food safety system were impossible.[11]

Through this discussion of the Brazilian reaction to BSE accusations I tried to address how the GM debate has involved little questioning of the food safety system. New technological risks are perhaps strengthening conventional political categories (North against South, left against right, nationalism against globalization, etc.), while consumer organizations are becoming important social actors in the decision process, as part of a struggle in the process of defining consumers as citizens.

WHERE ARE CONSUMERS AND ENVIRONMENTAL GLOBAL RISKS?

In Brazil few food panics have been registered (Guivant, 1995, 2000). But this does not imply, obviously, that risks do not exist because the systems that attempt to guarantee food safety are more efficient, but because there are no technical resources or sufficient workers to analyze and detect these risks (*Revista Amanhã*, year XV, 158, September 2000). According to the daily newspaper *Folha de S. Paulo*, the Ministry of Agriculture recognizes the lack of any kind of inspection of 42% of beef, 19% of pork, 25% of chicken and 46% of dairy products produced in the country for immediate consumption or for production of food derivatives. Very little is known about the risks of pesticide residues in food (Guivant, 1995; 2000). The Federal Inspection Service (SIF) is the agency technically equipped to carry out inspections of foods of animal origin, but its inspection stamp is no guarantee of quality control, either. The inefficiency of inspection opens the possibilities for adulterated products, especially in industrially processed foods. From this year on, even this inefficient inspection will cease to exist, because the government has plans to create an agency to supply inspection services only for slaughterhouses, which pay for the service. Priority will be given to export products. For this reason, the purchase and sale of sick cattle, contaminated dairy products, falsification of sanitary inspection stamps, medieval looking cattle pens hidden on the outskirts of large cities (*Folha de S. Paulo*, 30/08/98) are but a few phenomena that escape inspection.

The food and environmental risks are part of a general picture that Port (2000) and Freitas (2000) termed "socio-political amplification of risks," that is, a generalized aggravation of risks to the population due to, among other factors, the lack of political, economic, technical-scientific and institutional infrastructure related to prevention and control of technological industrial risks. Faced with the generalized precariousness in food control and inspection, the consuming public is probably far from having an expectation of zero risk control, as the European consumer might have had in the pre-food scandal period. The attitude is perhaps one of resignation or indifference in view of the possible risks, together with the lack of knowledge of their dimension. Within this dimension we could discuss a reflexivity in the sense of reflection, with the risks of modernity being generated, but without becoming a public matter or the center of political conflicts.[12] It would still be valid to speak of "industrial society," which simultaneously produces and legitimizes the risks as residual (Beck, 1994:5).

Yet the vulnerability of the population regarding this precarious protection of its rights is slowly changing, and one of the consequences of the coalition initiatives for control over GMs may be an increase of the space for consumer citizenship. According to Sorj (2000: 53), consumer protection emerged in association with a combination of factors, among which he highlights the consolidation of a middle class with globalized consumption standards, establishment of a mass consumption industry, strengthening of civil society and of citizenship rights. The advanced Consumer Rights Code, approved ten years ago mainly after pressure from organized groups such as IDEC, has become a decisive tool for consumer education.

The food-borne risks that have stimulated a more reflexive attitude among consumers in the last decade are those related to diets/health/body, increasing consumers' search for safety niches and refuge in control systems of the private sector (supermarkets, large food sector companies, and the growing offer of organic products). The body is being converted into a phenomenon of options and choices, in the sphere of biological reproduction, genetic engineering, and medical intervention. These transformations on the individual level are linked to global transformations, in a complex process of "nature's transmutation into a field of human action" (Giddens, 1991). This possibility and necessity of choice about our self-project (on the physical and psychic levels) is accompanied by a general disbelief in expert knowledge, which is always undergoing revisions and being debated. In order to make their decision, consumers must navigate through a sea of information that emerges from the means of communication, science, and local knowledge.

The manner in which expert systems take specific actions towards food-borne risks (pesticide use control, farmers' poisoning and contamination of food, water sources and soils) is apparently not a concern. This is quite different from Beck's characterization of the debates on food risks caused by pesticides and GMs as creating a trust crisis in criteria, rules, institutions and scientific production involved in guaranteeing safety of the foods consumed. Brazil has the problems of a scarcity society – in which the distribution of wealth between social classes is highly unequal – in conjunction with the problems of risk society (more than the ones affecting the food we consume), which does not rely on the same kind of reflexivity Beck identifies in more industrialized societies. This peculiar social hybridization affects not only Brazil, but also plays a role in the complex and multilinear dynamics of the globalization of risks. The type of reflexivity that may predominate in the middle and upper classes is not one that necessarily denies or ignores risks, but one that perceives a certain improvement in food risk control since the Consumer Code was approved. Without expectations of zero risk control, the current situation can be evaluated as positive, together with a feeling of impotence in relation to a still gray area of unknown risks.

TOWARDS A RESEARCH AGENDA

The issue of food-borne risks to internal consumers has not received special attention within rural sociology, which has tended to consider consumer rights and the question of quality of life to be secondary or non-problems. In some way, this is a reproduction of a general division of labor that seems to have dominated internationally the study of food. Tovey (1997) calls our attention to how the study of the organization of agriculture and its variable local forms have been the main concern of rural sociologists, as a separate area from the study of the sociology of food and its relations with consumption (eating, diet and culture) and not with production. In Brazil we can observe how social research on agricultural sustainability has assumed a predominantly agro-environmental productivist view (production methods and techniques with less or no use of pesticides, for example) and a socio-economic view (protection of family agriculture against agricultural modernization). This approach has contributed to criticism of the social and economic consequences of the Brazilian agricultural modernization process (for example, Ehlers, 1996; Almeida and Navarro, 1997), while not considering with the same emphasis health and quality of life questions for both the rural population and the population that consumes food/water with pesticide residuals. It is possible to identify, on an academic level, a reproduction of the principles and presuppositions of the alternative agricultural movements in Brazil, significantly influenced by early versions of Miguel Altieri's approach to agro-ecology (Guivant, 1997).[13]

There are recent case studies addressing health and environmental issues related to food production/consumption. But urban environmental policies still receive more interdisciplinary analysis than rural environmental policies or enforcement of those already existing. In part this is due to the objective limits in identifying and controlling agricultural pollution and also to the strong orientation in the academic community to avoid blaming family farming for environmental and health risks.

In relation to the global dynamics of the food system, integrating production and consumption, there is still limited debate (among the exceptions cab be mentioned, for example, Wilkinson, 2000, and Cavalcanti, 1999). There are few studies about food safety (Sell, 1997); the preferred issue has been food security, especially interpreted as a question of national sovereignty in the context of globalization[14] (for example, Menezes, 1998, Galeazzi, 1996).[15] However, recently researchers who had studied primarily food security have also been incorporating the issue of food safety (Maluf, 2000), with special attention to the question of consumer citizens.

For a tentative research agenda for studying these risks in Brazil, we can identify the following basic points about how diverse food-borne risks are managed and controlled. Some of the issues included would be: the arguments raised by each of the sectors active in debates on these risks, attempting to identify and trace the history, stages and characteristics of the specific debate and the causes that can be attributed to such characteristics. Questions related to these issues are: How are scientific, political and economic arguments articulated with arguments related to the population's quality of life and health? How does each sector represent the interest and arguments of the others?

Also crucial is the analysis of the challenges posed by food safety to policy research. Pinstrup-Andersen (2000) enumerates questions that can apply to the Brazilian debates: what policies and institutions will be needed to promote the required changes in food production, processing, and distribution to meet the new standards, while still being competitive in the export market? Will these policies and institutions be appropriate for the domestic food supply or should countries like Brazil have different standards? What are the implications of disagreements on food safety standards and environmental concerns between more and less industrialized countries and between poor and affluent population groups within Brazil, when different sectors have different acceptable risk levels? How will global changes in food safety criteria affect environmental and health risks in Brazil? What will the relations between food safety standards and food security be?

CONCLUSION

In this article I attempt to advance analysis of the specific ways in which global food-borne risks, such as BSE and GMs, are interpreted in Brazil. I argue that a heterogeneity of social actors with different "translations" of risk, generated from different models of society, is involved in the discourse coalitions constituted around these risks. Within the same alliance against GMs, for example, there are conventional political discourses, that give priority to class and national identities, coexisting with new discourses, oriented to the protection of consumers and the environment. In its turn, this national coalition is involved in a broader global alliance, where individual and collective social actors, both lay and experts, may only share an antagonism to GMs without a common interpretation of their potential risks or the same appreciation of the range of issues in which GMs are incorporated as significant sources of risks.

In addition there was no significant questioning of the expert/governmental systems' role in food safety. The GM debate has been restricted to this specific issue, with no "contamination", or overflow, to other sources of risk in the food chain. Pesticides are still out of the scenario. The debate is centered mainly on how to control the marketing of GMs, and since it does not focus on alternatives, it ends up strengthening conventional crops, although organic production may also begin to be benefited by the limitation on GM markets. The debate, therefore, is more in terms of "opposition to" than "in favor of" alternative models. In the case of the BSE crisis, it was argued that instead of provoking a critical reflection on Brazil's food system, it generated a nationalistic response. This response reinforced diffuse beliefs in the quality of the food system instead of opening up a discussion on the vulnerability of consumer health and the sustainability of the rural environment through insufficient controls and lack of implementation of existing legislation.

Another characteristic of Brazilians' "translation" of global food risks is the marginal involvement of consumers in these debates. This can be attributed, via a complex causal articulation, to the absence of previous zero level control expectations with regard to risks. Brazil is a very dynamic society, characterized by improvements in the implementation of food safety procedures, the social construction of consumers as citizens, growing levels of consumption, the globalization of markets coexisting with a still very inefficient system for enforcing food safety controls. Instead of looking at food safety as a half empty glass, the middle and upper classes are seeing rather a half-full glass.

Analysis of the peculiar "translation" of these recent global food panics/risks in Brazil reveals the complexities of the discourse coalitions which have emerged, and point the way to the need for a more differentiated approach to the global dynamics of risk society and also for the formulation of a research agenda which will permit a more complex examination of Brazil's specific characteristics facing global environmental and health risks and also contribute to a more open discussion of how to deal with open-ended risks.

NOTES

__1.__ The wide influence of Beck's ideas, not only in social sciences but in other fields and in the political sphere as well, can be gauged from the analyses of scholars from various fields: Franklin (l998), Adam et al (2000) and Spaargaren et al (2000).

__2.__ McDonald's has already acted to remove GM in animal products.

__3.__ The fact that product labels do not indicate the presence of GM products violates article 66 of the Consumer Defense Code, which stipulates a punishment and fine for whoever omits relevant information regarding the nature, characteristics, quality and quantity of a product.

__4.__ According to Zancan (1999), labeling foods that are obtained from GM organisms is fundamental for the availability of epidemiological data on allergic reactions and other food intolerance reactions, about which the molecular mechanisms are unknown. The available US data estimate food allergy incidence in 1.5% of the adult population and 5% of children under three years of age.

__5.__ This is also a very controversial issue. Crops which have been Genetically Modified to resist herbicides can encourage the use of larger quantities of herbicide, with the effect that both weeds and beneficial plants are killed indiscriminately. These herbicides are harmful to both the environment and to humans (29 March 2000, LA Times; 21 September 1999 Daily Mail UK; 8 July 1999,Times UK). GM plants may crossbreed with wild species to produce "super-weeds", which cannot be eliminated using standard herbicides.

__6.__ The benefits of GM seeds for small rural producers is still a subject of international debate (Wilkinson and Castelli, 2000; Marsden, 1999). Divergencies in production strategies are also found in the United States, with sectors reviewing their past decisions in light of the resistance of European markets.

__7.__ Soy is in about 60% of all processed foods, including biscuits, soft drinks, ready-made meals, bread, chocolate and baby food while corn is found in baby food, crisps, confectionery and a variety of snacks (The Guardian, March, 20, 2000).

__8.__ The declarations of PT president José Dirceu in relation to the meat embargo illustrate the assumed nationalist position: "It had been many years since something of such importance, related to international politics, had happened in the country. In a few days, Canada accomplished what we could not in years in terms of the awareness of millions of Brazilians. It had never been so clear what globalization is and what its limits are. And mainly, Brazil's need to change its course and project itself as a sovereign nation." (Folha de S. Paulo, March 14, 2001). Environmentalist and Federal Representative Fernando Gabeira believed the crisis with Canada was a lesson to Brazilians on globalization and colonialism: "...on this question with Canada, we were condemned by our qualities... It is as if history arrives a year late, for this episode with mad cow disease is a starting point from which to review 500 years, and in a certain way, to re-situate ourselves in the world. Who knows? Maybe one day we will thank Canada for waking us up from an old colonial dream." (Folha de S. Paulo, February 19, 2001).

9. *In some European countries it is also possible to observe the same nationalistic reaction against the implications of mad cow disease for meat producers. For example, the case of French agricultural producers who were against the import of Argentine meat in the beginning of 2001. However, this reaction has not led to an alliance between opposing camps in the GM crops question. (see Le Monde, January 29, 2001)*

10. *Since 1998, Canadian sanitation authorities have requested that Brazil present documents concerning the imported cattle, but have received no response.*

11. *We can draw a parallel between the nationalistic discourse in relation to BSE in Brazil and in European countries. When the British government started to receive criticism for how it was dealing with the disease, the question became a nationalistic test of confidence in British beef. The same took place in other European countries, such as France and Germany, that denied any possibility of infection in their cattle. Any criticism was considered to be anti nationalist. However, while in European countries the nationalistic discourse was stronger before the GM crisis and the emergence of evidence of local infection and human contamination, in Brazil this discourse became stronger after the GM debate.*

12. *There is not enough data on Brazilian consumers' perception of risk. Internationally, the opposition to GM foods and crops is growing and now includes North America. The latest industry-funded polls show 59% of Canadians flatly opposed to GM foods, while 51% in the US say the same thing - with the overwhelming majority in both countries expressing mounting concern over the lack of safety-testing and labeling. In Europe up to 70% adamantly oppose GMs, while 82% of Japanese are opposed. In response, North American farmers are already starting to cut back on major GM crops - corn, soybeans, and canola. The latest July USDA estimates show a 20% drop in GM corn acreage and a 6% drop in soybeans, while Canada analysts predict a 10% reduction in GM canola (Biodemocracy News, n.28, www.purefood.com - site of the Organic Producers Association, from the USA).*

13. *Exposing this perspective, that excludes the issues mentioned above, Almeida (1998: 69) enumerates the principles of sustainable agriculture: "1) alternative agriculture determines a stricter and more balanced relationship between the natural environment and the one created by man; 2) it improves social, economic, ecological and cultural diversity; 3) it implies the creation and management of production systems that seek a greater personal involvement and more direct production, promoting greater autonomy in the productive realm; and 4) it leads to the building of a future in which individuals are free and can comprise a truly autonomous and democratic society."*

14. *Graziano da Silva (2000) illustrates this position by asking: "In the era of globalization, can a peripheral country such as ours implement a national food safety policy? Can we exercise our sovereignty by choosing our population's diet? Is it possible, in an open economy, to implement a policy of regulatory stocks which are supplied preferably by local production? Or should we simply import our food every time internal prices threaten to increase? In summary, can we eat what we want or must we follow the food standards imposed upon us in the* *name of globalization and modernity? And who sets these standards? Is it the consumers of developed countries or the large transnational companies of the agricultural food system?"*

15. *According to the White Paper on Food Safety (2000), of the European Union, food safety includes primary food production (agricultural and veterinary aspects), industrial processes, storage, distribution and retailing, remitting to both risk and nutritional issues related to animal feed, animal health and welfare, hygiene, contaminants and residues, novel food, additives, flavorings, packaging and irradiation.*

BIBLIOGRAPHY

ADAM, B. (1998). *Timescapes of Modernity: the environment and invisible hazards.* London : Routledge, 1998.

ADAM B., BECK, U. e VAN LOON, J.(eds), (2000). *The risk society and beyond. Critical issues for social theory.* Londres : Sage, 2000.

ALMEIDA, J. (1998). *A construção social de uma nova agricultura.* Porto Alegre: Editora da UFRGS, 1998.

AI MEIDA, J., NAVARRO, Z. (orgs.), (1997), *Reconstruindo a agricultura. Idéias e ideais na perspectiva do desenvolvimento rural sustentável.* Porto Alegre: Editora da UFRGS, 1997.

BEARDSWORTH, A, KEIL, T. (1997). *Sociology on the menu. An invitation to the study of food and society.* London: Routledge, 1997.

BECK, U. (1992). *Risk society. Towards a new modernity.* London: Sage Publications, 1992.

BECK, U. (1994). The reinvention of politics: towards a theory of reflexive modernization. In: BECK, U., Giddens, A. e Lash, S. *Reflexive Modernization. Politics, tradition and aesthetics in the modern social order.* Cambridge: Polity Press, 1994.

BECK, U. (1995). *Ecological politics in an age of risk.* Cambridge: Polity Press, 1995.

BECK, U. (1997). *The reinvention of politics. Rethinking modernity in the global social order.* Cambridge: Polity Press, 1997.

BECK, U. (1999). *World risk society.* Cambridge: Polity Press, 1999.

BECK, U. (2000). *The brave new world of work.* Cambridge: Polity Press, 2000.

BUTTEL, F. (2000). Classical theory and contemporary environmental sociology: some reflections on the antecedents and prospects for reflexive modernization theories in the study of environment and society. In: SPAARGAREN, G., MOL, A e BUTTEL, F. (orgs.), *Environment and global modernity.* Londres: Sage/ISA, 2000.

CAVALCANTI, J. (org.), *Globalização, trabalho e meio ambiente. Mudanças socioeconômicas em regiões frutícolas para a exportação.* Recife: Editora Universitária/UFPE.

DOUGLAS, M. e WILDASKY, A (1982). *Risk and culture. An essay on the selection of technical and environmental dangers.* Berkeley, CA: University of California Press, 1982.

EHLERS, E. (1996), *Agricultura sustentável.* São Paulo: Editora Livros da Terra, 1996.

ESRC Global Environmental Change Programme (1999) The politics of GM food: Risk, science & public trust. Special briefing n.5, Oct. 1999.

EUROPEAN UNION (2000). White paper on food safety. Disponível em: http://europa.eu.int/comm

FRANKLIN, J. (ed.), (1998), *The politics of risk society.* Cambridge: Polity Press, 1998.

FREITAS, C.M. (2000), Acidentes químicos ampliados, vulnerabilidade social e planejamento de emergências. In: Herculano, S., Porto, M e Freitas, C.(Org.), *Qualidade de vida e riscos ambientais.* Niteroi: Editora da UFF, 2000.

GALEAZZI, M. A. (org.), (1996), *Segurança alimentar e cidadania.* São Paulo: Mercado de Letras, 1996.

GIDDENS, A (1991), Modernity and self-identity. Self and society in the Late Modern Age. Stanford: Stanford University Press, 1991.

GORGEN, S. (org.) (2000), *Riscos dos transgênicos.* Petrópolis: editora Vozes, 2000.

GRAZIANO DA SILVA, J. (2000), Resenha Disponível em: www. cria.org.Brasil /gip/gipaf/ itens/publ /livros/belik-maluf.html.

GREEN ALLIANCE (2000), *Steps into Uncertainty: Handling risks and uncertainty in environmental policy-making.* London: Green Alliance, 2000.

GROVE-WHITE, R. (1999), Afterword: on sound science, the environment, and political authority. *Environmental Values*, n.8, 1999.

GROVE-WHITE, R., MACNAGHTEN, P. e WYNNE, B. (1997), *Uncertain world: Genetically Modified Organism, food and public attitudes in Britain.* Centre for the Study of Environmental Change, Lancaster University, 1997. (research report)

GROVE-WHITE, R., MACNAGHTEN, P. e WYNNE, B. (2000), *Wising up. The public and new technologies.* Centre for the Study of Environmental Change, Lancaster University, 2000. (research report)

GUIVANT, J. (1992), "*O uso de agrotóxicos e os problemas de sua legitimação. Um estudo de sociologia ambiental no município de Santo Amaro da Imperatriz, Santa Catarina.*", 1992. Tese (PhD) , Unicamp, Campinas.

GUIVANT, J.(1995) Percepção dos olericultores da Grande Florianópolis (SC) sobre os riscos decorrentes do uso de agrotóxicos. *Revista Brasileira de Saúde Ocupacional.* Fundacentro. São Paulo, v.22, 1995.

GUIVANT, J.(1997). "Heterogeneidade de conhecimentos no desenvolvimento rural sustentável", *Cadernos de Ciência e Tecnologia.* EMBRAPA. V.14, n.3, set/dez 1997.

GUIVANT, J. (1998).., A trajetória das análises de risco: da periferia ao centro da teoria social, in *Revista Brasileira de Informação Bibliográfica em Ciências Sociais.* ANPOCS, n.46 1998.

GUIVANT, J.(2000a). Reflexividade na sociedade de risco: os agrotóxicos nos alimentos. In: HERCULANO, S., PORTO, M e FREITAS, C. (Org.), *Qualidade de vida e riscos ambientais.* Niterói: Editora da UFF, 2000.

GUIVANT, J. (2000b). Cross alliances: lay and expert knowledge facing food risks. *World Congress of the International Rural Sociology Association*, Junho 2000, Rio de Janeiro, Brasil. (paper presented). CD rom.

GUIVANT, J. (2000c). The theory of world risk society: between diagnosis and prophecy. *World Congress of the International Rural Sociology Association*, Symposium ISA RC24 (Environmental Sociology) Junho 2000, Rio de Janeiro, Brasil. (paper presented). (To be published in T. Roberts, E. Viola, and F. H. Buttel, (eds.) *Sustainability and Unsustainability on the Road from Rio*).

GUIVANT, J., MIRANDA, C. (1999), As duas caras de Jano: agroindústrias e agricultura familiar diante da questão ambiental. *Cadernos de Ciência e Tecnologia*, EMBRAPA. v. 16, n.3, set/dez 1999.

HAJER, M. (1995), *The politics of environmental discourse. Ecological modernization and the policy process*. Oxford: Claredon Press, 1995 .

LATOUR, B. (2000), *Prenons garde au principe de précaution.* Le Monde, January 3, 2000.

LEITE, M. (2000), *Os alimentos transgênicos.* São Paulo: Editora da Folha, 2000.

LUOTON, D. (1999), *Risk*. London: Routledge, 1999.

MALUF, R. (2000), *Consumo de alimentos no Brasil: traços gerais e ações públicas locais de segurança alimentar.* Polis paper n.6. Disponível em: www.polis.org.br

MARSDEN, T. (2000). *Agri-business as Usual: genetically modified foods in perspective.* Document of the ESRC Global Environmental Change Programme, 2000.

MENEZES, F. (1998). Sustentabilidade alimentar: uma nova bandeira? In: DUARTE, A, BRANDENBURG, ALFIO. (orgs.), *Para pensar outra agricultura.* Curitiba: Editora da UFPR, 1998 .

MOL, A e SPAARGAREN, G.(1993). Environment, Modernity and the Risk-Society: the Apocalyptic Horizon of Environmental Reform. *International Sociology*, v.8, n 4, p. 431-459, 1993 .

PINSTRUP-ANDERSEN, P.(2000). Food policy research for developing countries: emerging issues and unfinished business. Food Policy, v.25, n. 2, April 2000.

PORTO, M. (2000), Considerações sobre a dinâmica de regulação dos riscos industriais e a vulnerabilidade da sociedade brasileira. In: HERCULANO, S., PORTO, M. e FREITAS, C. (Orgs.), *Qualidade de vida e riscos ambientais.* Niterói: Editora da UFF, 2000 .

RADFORD, (1999). A bluffer's guide: why genetic modification matters. *The Guardian*, February 16 1999.

SELL, S. (1997), *Controle Institucional de Riscos Alimentares: Uma análise a partir de Florianópolis.* 1997. Dissertação (Mestrado em Sociologia Política), Universidade Federal de Santa Catarina , Florianópolis.

SEMINÁRIO INTERNACIONAL SOBRE BIODIVERSIDADE E TRANSGÊNICOS, 1999, *Brasília. Anais.* Brasília: Senado Federal, 1999.

SPAARGAREN, G., MOL, J. e BUTTEL, F. (eds.), (2000), *Environment and global modernity.* London: Sage/ISA, 2000.

SORJ, B. (2000), *A nova sociedade brasileira.* Rio de Janeiro: Jorge Zahar Editor, 2000.

STIRLING, A., R. GROVE-WHITE (1999). Submission to the House of Commons Science and Technology Select Committee inquiry on The Scientific Advisory System for Genetically Modified Foods.

TOVEY, H. (1997). Food, environmentalism and rural sociology: on the organic farming movement in Ireland. *Sociologia Ruralis*, v.37, n.1, 1997.

WILKINSON, J. (2000). Da ditadura da oferta à democracia da demanda, *Teoria e Sociedade*, n.6, Outubro 1997.

WILKINSON, J. (coord.) e CASTELLI, P. (2000). *A transnacionalização da Indústria de Sementes no Brasil. Biotecnologias, patentes e biodiversidade.* Rio de Janeiro: Action Aid Brasil, 2000.

ZANCAN, G. (1999). Alimentos transgênicos: riscos à saúde. CONGRESSO DE HIGIENISTAS DE ALIMENTOS, 5, 1999, Foz de Iguaçu, Brasil. (Lecture). Disponível em: www.idec.org.br.

Consumption Patterns and Environmental Impacts in a Global Socioecosystemic Perspective[1]

DONALD SAWYER
Institute for Society, Population and Nature - ISPN

INTRODUCTION

The environmental impacts of consumption patterns have been one of the most elusive and neglected issues in debates on sustainable development and the human dimensions of global change in the last 30 years. Consumption patterns were simply not on the environmental agenda between the Stockholm Conference in 1972 and the Rio de Janeiro Conference in 1992, except as a subtopic of population pressures (Tolba and El-Kholy 1992:473-503). Building on ideas disseminated in Our Common Future (World Commission 1987), after broad consultations, consumption patterns were introduced explicitly in the agenda at the United Nations Conference on Environment and Development held in Rio de Janeiro in 1992 (United Nations 1992).

Even so, more attention has been paid to biodiversity conservation (the "green agenda"), pollution, sanitation and emissions (the "brown agenda") and water resources (the "blue agenda"), for example, than to consumption patterns (cf. World Bank 1995). The components of the three main agendas are generally tangible and easily classified as positive or negative, while consumption is a complex, diffuse and sensitive issue. Documents and debates on consumption patterns, to the extent that they surpass silence, are replete with confusion, misconceptions, biases, fallacies and empty rhetoric. This paper seeks to clarify the issues and identify possible research priorities and actions to be taken with regard to environmental impacts of consumption patterns at the international, national and local levels.

The literature and the press have paid much more attention to consumption in developed countries in the North than in developing countries in the South. Indeed, it is often presumed that environmental degradation is consumption-driven in the North while being poverty-driven in the South (Sawyer 1993:21). The rationale is, for example, that 80 percent of the world's energy is consumed by only 20 percent of its population, in the North. The implicit conclusion is that the North bears responsibility for environmental problems due to consumption. The apparently logical but fallacious conclusion is that poverty reduction in the South would have net positive environmental impacts. The North-South dichotomy was one of the ways around the limits to growth approach formulated by the Club of Rome (Meadows 1972). The limits were was unacceptable to countries which insisted on their "right to development" and the need for economic growth, which is a prerequisite for equity among present and/or future generations.

This paper seeks to achieve greater North-South balance as well as stressing links among consumption, population, environment and development, in a "socioecosystemic" approach, joining ecosystems and social systems. Such interlinkages can also be considered as multifunctionality. The thrust of the argument is that consumption patterns are the most basic root cause of unsustainable development and should be taken into account in scientific research agendas and official and non-governmental policy agendas. It is argued that countries in the South suffer impacts of consumption that occurs in the North as well as the South. At the same time, Southern countries could reach sustainable consumption before developed countries and should actively pursue this goal as part of their negotiating strategy with the North.

Initially, the various environmental impacts of consumption and their underlying causes are considered in a global perspective, as are global and local causes. Then, the results of international conferences and agreements are examined, as are trends in public policies at the national and local levels. Attention is drawn to the importance of demographic and social composition and trends. Finally, conclusions are proposed and recommendations formulated.

GLOBAL AND LOCAL ENVIRONMENTAL IMPACTS

The Global Environment Facility (GEF), the financial mechanism for international cooperation in environment, defined global environmental problems as biodiversity loss, climate change, ozone depletion and contamination of international waters (GEF 1996). It reluctantly included desertification and land degradation, but only as related to the other four "focal areas". The first basic problem with the GEF approach is that it involves compartmentalization of the responses to ecological processes that are intimately related, like biodiversity conservation and carbon sequestration, not to mention water resource management. Such fragmentation is starting to be overcome by an approach called "integrated ecosystem management," which so far is only a separate operational program rather than a general principle.

A second problem is that, in the GEF definition, "global" refers only to impacts that are felt far beyond the local level at which the immediate causes occur. This limited scope for international cooperation excludes common environmental problems such as water resources, sanitation, pollution, waste disposal, vector control, contamination of food and sustainable livelihoods, which are considered to be national or local responsibilities ("baseline costs") that are not eligible for international support, which is restricted to "incremental costs." In practice, this means that the North is willing to provide support to the South for reduction of the environmental impacts which the developed countries will eventually feel themselves. The principal impacts the North would like to avoid are loss of possible cures for diseases, rising sea levels, infrared radiation and polluted oceans. Consequently, the "common but differentiated responsibilities" mentioned in Principle 7 of the Rio Declaration on Environment and Development (United Nations 1992:9) are seen in a self-serving way. The "global" goods worth protecting are those that interest the North.

A third problem with the existing global framework is that, in addition to separating global from local problems, it also separates environment from development or human concerns. Such separation violates Principle 4 of the Rio Declaration, which establishes that "environmental protection shall constitute an integral part of the development process and cannot be considered in isolation from it" (United Nations 1992:9). By separating partial from systemic, global from local and human from environmental dimensions, this approach is anything but integrated and its efficacy is limited.

In order to define appropriate priorities for ecosocial research and action at all levels, a broader and more integrated framework is needed. Global impacts could be defined as those which are most widespread, not just far-reaching. It would be fitting to include all environmental impacts that are due to global causes, especially the impacts that have root causes in the North. The inseparable human dimensions could also be made more explicit.

In this broader perspective, incorporating international, intersectoral and human dimensions, the environmental impacts which seem to pose the greatest threats to global sustainability can be organized according to the following categories, with respective examples:

1. Water: sufficient quantity, quality, security and equitable distribution of water resources for human consumption, production of crops and livestock and generation of hydropower;

2. Food: depletion of soil fertility, deforestation, soil erosion, water pollution, salinization, depletion of fish stocks, food scarcity and insecurity, toxins in food chain, spread of plant and animal pests and diseases and risks of genetically modified organisms;

3. Energy: depletion of fossil fuels, impacts of dam construction, nuclear risks and scarcity of fuelwood;

4. Biodiversity: ecosystem disturbances, habitat destruction, species extinction, invasive species and genetic erosion;

5. Pollution: sewage, garbage disposal, air pollution, accidents such as oil spills, toxic materials, hazardous wastes;

6. Emissions: greenhouse gases that contribute to global warming and climate instability, primarily from generating electricity, automobiles and deforestation, as well as ozone depletion;

7. Congestion: cramped space for housing and leisure, traffic congestion, noise pollution and increased potential for spread of disease.

These impacts have to do with interactions between social systems and ecosystems, between society and nature. The first four categories of impacts are related primarily to depletion of natural resources, involving withdrawals from nature, while the others are externalities related to pollution, i.e. returning materials or energy back to nature or society. The various impacts are clearly inter-related. Ozone depletion seems to be the only problem that can be dealt with in isolation from the others, without involving society (cf. Banco Mundial 2000).

In this broader context, it can be seen that the GEF priorities are for the middle and long-term, while most of the other impacts are already being felt intensely. The GEF priorities are fundamental for future generations, while present generations are suffering from the other impacts, especially in the South.

GLOBAL AND LOCAL CAUSES

The causes of the various environmental impacts operate at various levels. The proximate causes are fairly well known: industrial production, agriculture, energy generation, use of automobiles, inappropriate waste disposal, deforestation, etc. An intermediate level of causation has been summed up as population, affluence and technology or "PAT," factors which can operate independently (Erlich and Erlich 1970, Harrison 1992, Hogan 2000). The argument presented here is that the deepest root cause of all of the above-mentioned global impacts is increased consumption. Population growth and inappropriate technology are relevant, but contribute to the impacts of increased consumption rather than being truly independent.

In this perspective, few causes of global environmental impacts can be considered to be entirely local. With globalization, consumption cannot be viewed only within isolated local or even national contexts, but must take into account international markets, flows and influences. The global causes of increased consumption can be characterized as population growth, industrialization, urbanization, frontier expansion, material consumption, circulation and energy, all of which are integral parts of globalization:

- Population growth in the South is due primarily to reduction of mortality rates through incorporation of modern means of sanitation and medical technology from the North, there being a lag in fertility decline (demographic transition).

- Industrialization is led by international markets, while industries that are intensive in natural resources (minerals, pulp, raw materials) and energy or that generate pollution tend to move to countries where environmental and social controls are not as strong as in the developed countries (cf. Torres 1993).

- Urbanization is due to industrialization in the North and South and to dissemination of modern consumption aspirations from North to South.

- Frontier expansion is due in part to population growth and technical progress, which cause migration, but also to the need to export primary products (timber, minerals, crops and meat), pressures for which are intensified by structural adjustment policies.

- Material consumption, emulating models of developed countries through a demonstration effect, is part and parcel of market expansion; use of natural resources in material form is reduced in relative but not absolute terms by services and information.

- Circulation of people and goods among various places of residence, work, production, leisure and tourism is accelerated by transportation technology, causing its own damage and amplifying the spatial distribution of environmental impacts.

- Energy from fossil fuels, dams and/or nuclear plants is necessary for all of the above.

None of these causes of increased consumption can be considered entirely local, even in the most remote parts of the planet. In sum, both global and local impacts have global causes.

INTERNATIONAL CONFERENCES AND AGREEMENTS

The question of policy responses to consumption patterns can be seen in light of the international conferences of the United Nations and associated agreements during the 1990s, which were cumulative and systemic (Viola 1996) and constituted a truly global initiative. The most pertinent are the United Nations Conference on Environment and Development (UNCED), also known as the Earth Summit, held in Rio de Janeiro in 1992, and the International Conference on Population and Development, held in Cairo in 1994.

Principle 8 of the Rio Declaration on Environment and Development states that "To achieve sustainable development and a higher quality of life for all people, States should reduce and eliminate unsustainable patterns of production and consumption and promote appropriate demographic policies" (United Nations 1992:9). The interrelated issues were on the official agenda, albeit in vague terms.

The Earth Charter, proposed by non-governmental organizations at the Global Forum, also mentioned consumption. On the other hand, the NGOs recognized a possible contradiction between poverty alleviation and environmental protection when they stated that "Efforts to eradicate poverty do not constitute a mandate for environmental degradation" (Carta da Terra 1992).

Chapter 4 of Agenda 21 (United Nations 1992:37-42) deals specifically with changing consumption patterns, but also makes reference to chapters on energy, transportation, wastes, economic instruments and the transfer of technology, as well as demographic dynamics and sustainability. It is significant that no reference is made to the chapters on poverty or trade. While environment-development linkages and interrelations among environmental impacts are recognized, North-South differences are emphasized as separate realities, to the detriment of a truly global approach. The international dimensions are restricted to technological and other assistance, while issues such as international trade and North-South diffusion of unsustainable consumption patterns are practically ignored. The recommended activities for national policies and strategies are: a) encouraging greater efficiency in the use of energy and resources; b) minimizing the generation of wastes; c) assisting individuals and households to make environmentally sound purchasing decisions; d) exercising leadership through government purchasing; e) moving towards environmentally sound pricing; and f) reinforcing values that support sustainable consumption.

Two years later, sustainable consumption and demographic policies were also dealt with at the International Conference on Population and Development. At the insistence of G-77 countries, led by India, references to "overconsumption" and "unsustainable patterns of production and consumption" were included in the Cairo Program of Action. However, "sustained economic growth" was juxtaposed with sustainable development, with no recognition of incompatibility among sustained economic growth, population growth and environmental sustainability. The issues were addressed in parallel fashion, without regard for inconsistencies.

On the specific issue of consumption, a United Nations Inter-Regional Expert Group Meeting on Consumer Protection and Sustainable Consumption was held in São Paulo in 1998. The main concern was revision of the United Nations Guidelines for Consumer Protection, which could not be renegotiated, not environmental impact. Nonetheless, some links between consumption and environment were recognized.

In spite of advances in recognition of consumption-environment linkages at the international level, the North-South fallacy persists. While it is often recognized that wealthy elites in the South have patterns of consumption similar to or even more wasteful those of the North, it remains implicit that since poverty causes environmental stress and degradation, poverty alleviation is good for the environment. No consideration is given to the delicate subject of environmental impacts of increased consumption resulting from poverty alleviation, except at a very abstract level, or with specific reference to China.

A "sustainable livelihoods" approach has been advanced by international agencies such as the United Kingdom's Department for International Development (DFID) (Carney 1998) and the Small Grants Program of GEF and the United Nations Development Program (UNDP) (UNDP 1997). It holds promise for reconciling poverty alleviation and environment, but has not been developed consistently in environmental terms.

PUBLIC POLICIES AT THE NATIONAL AND LOCAL LEVELS

Although remarkable progress was made at the global level in terms of international agreements and official documents concerning the "green", "brown" and "blue" agendas, with numerous recommendations for specific actions, there are few global guidelines as to action regarding consumption patterns. Global guidelines do not translate automatically into action at the national and local levels. In spite of all the limitations, there has been much more progress in biodiversity conservation (saving threatened species and protected areas), water resources, climate change and especially ozone depletion than in guidelines or action concerning consumption patterns.

At the national and local levels, few if any public policies have been aimed at changing consumption patterns in the broad sense. Very few of the recommendations for action of Chapter 4 of Agenda 21 have been put into practice. On the other hand, various policies have targeted specific consumer behavior related to efficient use of natural resources such as energy conservation and saving water, areas in which the public sector is directly involved in production, distribution or regulation.

There are numerous other possibilities for public policies to promote sustainability in the broad sense (e.g. Guimarães and Maia 1997). The public policies that can influence sustainable consumption most directly include those related to:

- Public transportation and mass transit;
- Restrictions on use of automobiles, for example in city centers and/or during certain days of the week;
- Reduced-impact transportation such as walking, carpools, bicycles and animal traction;
- Increased durability of consumer goods;
- Increased energy efficiency of appliances and lighting;
- Biodegradable products;
- Recycling of paper, glass, iron, aluminum and plastics;
- More settlement in climates that require less heating or air conditioning;
- Decreasing frequency or distance of travel from home to work;
- Organic agriculture;
- Food with more plant and less animal protein;
- Processed foods that take advantage of less valuable species or parts;
- New materials;
- Use of electronic media instead of paper.

It can be seen that governmental action is more likely to occur when: a) there are pressing needs, like crises of supply; b) government agencies are responsible; and c) important economic and political interests are not threatened.

Public opinion regarding consumption patterns is not strong and is rarely a motivation for governments to attempt to change them. There can be inconsistencies among knowledge and attitudes and well as inconsistency between these and behavior. Little is known about how combinations of command and control and economic incentives influence consumer behavior.

More general public policies such as environmental education contribute to public awareness of the impacts of consumption of individuals and families. One of the important mechanisms for translating this awareness and concern into effective action by consumers and producers would be certification of goods and services that are environmentally friendly. This can be done through mechanisms such as the International Standards Organization (ISO), the Forest Stewardship Council (FSC), organic certification and more limited schemes such as certification of origin. Certification is a market mechanism, but its use can be supported by governments.

Public policies with regard to environmental control, scientific and technological development and official credit and incentives would also be important for changing consumption patterns. Very little has been done about Agenda 21's recommendations regarding environmentally sound pricing, which has more to do with markets than public policy. Use of government purchasing power does not seem to have been adopted anywhere.

Population policies would also be relevant for mitigation of the environmental impacts of consumption. While in the case of fertility declines there is a coincidence between people's aspirations and global needs, this is not true in the case of consumption, where aspirations are generally to increase consumption. It may be "easier" to promote fertility decline in the South than to change consumption patterns in the North. Indeed, many people in the South limit their fertility in order to be able to consume more (Carvalho, Paiva and Sawyer 1981). However, there are other good reasons for promoting access to reproductive health and family planning, as established in the Cairo Program of Action. There are also many reasons to promote more sustainable settlement and mobility patterns.

In sum, public policies regarding consumption patterns will probably be more effective to the extent that they take into account the multiple functions, impacts, costs and benefits of different alternatives.

SOCIAL AND DEMOGRAPHIC TRENDS

Consumption patterns depend more on markets and social and demographic trends than on international agreements or national public policies. Overconsumption may only decrease in the North as incomes decline because of increasing distance between the well-paid managers and the low-paid service workers. In the South, on the other hand, there are stronger income constraints on overconsumption and poverty is widespread. However, in spite of the distance between the rich and the poor, there is a very significant middle class, and the poor increasingly have access to consumer goods, water, electricity and transportation.

Many of the environmental impacts are made more severe by overconsumption of the wealthiest sectors of the population, which consume energy and materials and cause pollution and emissions directly or indirectly, through the production of the goods and services they consume. On the other hand, most of the impacts are aggravated by even modest levels of consumption of the majority. Thus, global environmental impacts are not just a matter of wasteful consumption in developed countries, but even of meeting basic human needs in all countries.

In order to understand the dynamics of changing consumption, it would be appropriate to view consumption in terms of different "classes," distinguishing the middle classes from the wealthy and the working classes from the poorest of the poor. Using a division common in marketing, classes A, B, C, D and E can be distinguished according to criteria such as the following household characteristics:

A. Family income above $20,000 per year, late model cars, more than one car per family, second home, microwave oven, freezer, mobile telephone, house or apartment with more than one bathroom.

B. Family income between $8,000 and $20,000 per year, older car, color television, refrigerator, freezer, telephor.e.

C. Family income between $3,000 and $8,000 per year, old or used car, various electrical appliances, plumbing.

D. Family income between $2,000 and $3,000 per year, some eletrical appliances, some plumbing.

E. Family income below $1,000 per year, no appliances, no plumbing.

Obviously, consumption depends on intra-national class composition. Since class A is small in developing countries, on the order of 5%, its overconsumption has less environmental impact that more moderate consumption of classes B and C, which usually constitute at least half of the population. Classes D and E are more common in rural areas. Their consumption can be made more efficient but should not be reduced. Policies could stimulate convergence of income from both extremes toward class B. Upward social mobility from classes D and E increases environmental impact, as does income redistribution, once the effects of poverty-driven degradation are discounted.

There is some correlation between class and environmental consciousness, which is remarkably widespread in terms of attitudes but is not necessarily accompanied by behavior. The middle groups generally have greater consciousness, but their choices are limited by their means. Changes in practices depend to a large extent on costs.

In the developed countries, and in some cases in classes A and B in developing countries, there have been recent trends toward healthy living, involving, among other things, physical exercise, less tobacco and healthy food. Healthy food, with desire for less beef and cholesterol and more fiber and poultry, often produced locally, is closely related to environmental impact of consumption. While a minority seeks natural food, there has been rapid expansion of consumption of processed foods such as breakfast cereal and processed meat. Some of these spontaneous consumption trends can be positive in terms of environmental impact.

"Green" consumer markets can be important for promoting sustainable production patterns. They are now more than small niches in segments of the North and classes A and B in the South. Even when they are small, they can be important for promoting sustainable use of biodiversity, especially when they involve social solidarity.

Non-consumptive use of nature can provide some of the funds necessary for conservation through revenues generated by visits to national parks and wilderness areas as well as ecological or sustainable tourism. These activities are important for public awareness, with positive influence on behavior in other spheres.

In the South, there have been strong recent trends toward urbanization, fertility decline, environmental consciousness and integration into various markets. The fact that the trends have been more or less spontaneous and occurred by diffusion, rather than being the result of explicit policies with these objectives, offers promise and challenges. Many policies have unintended effects. The promise is that sustainable consumption patterns may diffuse spontaneously.

The danger is that not even policies can help. Hopefully, a multifunctional, ecosocial and global perspective can help reduce unintended consequences that threaten global sustainability.

CONCLUSIONS

The main conclusions that can be drawn from the foregoing analysis are as follows:

1. Consumption patterns are at the core of sustainability but their environmental impacts, particularly in the South, have not received the attention they deserve, in academia, government, international organizations, mass media or society.

2. Global guidelines established in international conferences and multilateral agreements regarding environment are often based on fallacious reasoning and have generally skirted the issue of consumption patterns.

3. There is little scientific knowledge of the various environmental impacts of consumption patterns or of increases in consumption levels, especially in the South.

4. Little attention has been paid to multifunctionality, including social functions, and links among various environmental impacts of consumption patterns and their global dimensions.

5. Because of economic and political interests, it is improbable that meaningful multilateral agreements on consumption patterns can be reached in the foreseeable future, although generic statements of principle are possible.

6. At the national and local levels, there are specific public policies that could be expanded or adapted to discourage unsustainable consumption patterns, especially if multiple functions are considered.

7. Sustainable consumption patterns and practices could be defined to use as yardsticks to discourage unsustainable patterns and practices and reduce excess consumption.

8. Developing countries have the opportunity to "leapfrog" to sustainability before reaching irreversible unsustainablity of their consumption patterns.

RECOMMENDATIONS

On the basis of the foregoing analysis, the following research priorities and actions can be recommended as examples of ways to promote sustainable consumption, decrease global impacts and reduce the threat of global collapse:

1. Adopt a socioecosystemic approach to consumption patterns, considering multifunctionality, in addition to sectoral environmental approaches and fragmented actions with specific targets such as saving threatened species or establishing protected areas.

2. Seek better understanding of national and global causes and impacts of consumption patterns in a North-South perspective and identify possible pro-active interventions.

3. Promote studies on local and global environmental impacts of different consumption patterns ("footprints") and of poverty reduction and income redistribution and seek ways to minimize impacts.

4. Establish standards for consumption of energy, water and land and use economic instruments such as progressive taxation for overconsumption as defined by these standards, not just linear taxes on income or consumption.

5. Promote mass public transportation and non-motorized transportation as well as removing incentives for and taxing individual automobiles.

6. Establish progressive rate structures for consumption of energy and water, charging higher rates per unit (kilowatt/hour or cubic meter) when total consumption is higher.

7. Establish progressive taxation for use of space in settled areas and provide disincentives for frontier expansion.

8. Promote population distribution and settlement patterns which minimize impacts of consumption and favor production and consumption of local products.

9. Promote slower population growth as essential to achieving greater sustainability of consumption patterns, with emphasis on reproductive health and reproductive rights;

10. Seek common positions for the South to negotiate with the North in specific arenas, especially the Rio+10 Conference in Johannesburg in 2002.

NOTES

1. *This paper is based on research supported over the years by a grant from the John D. and Catherine T. MacArthur Foundation to the Institute for Society, Population and Nature (ISPN). The author is currently National Coordinator of the Small Grants Program of the Global Environment Facility (GEF) and a consultant for the United Nations Educational, Scientific and Cultural Organization (UNDESCO), assigned to the Ministry of Science and Technology. The points of view expressed here do not represent those of the institutions involved.*

BIBLIOGRAPHY

BANCO MUNDIALl. 2000. *No limiar do século XXI:* relatório sobre o desenvolvimento mundial 1999/2000. Washington.

CARNEY, Diana (Ed.). 1998. *Sustainable rural livelihoods:* what contribution can we make? London: Department for International Development.

CARTA da Terra. 1992. Rio de Janeiro: Global Forum.

CARVALHO, José Alberto Magno de; PAIVA, Paulo; SAWYER, Donald. 1981. *A recente queda da fecundidade no Brasil:* evidências e interpretação. Belo Horizonte: CEDEPLAR. (Monografia, 12).

ERLICH, Paul R; ERLICH, Anne H. 1970. *Population, resources, environment:* issues in human ecology. San Francisco: W.W. Freeman.

GEF. 1996. *Operational strategy of the Global Environment Facility.* Washington: Global Environment Facility.

HARRISON, Paul. 1992. *The third revolution:* environment, population and a sustainable world. London: Tauris, Penguin, WWF.

HOGAN, Daniel. 2000. A relação entre população e ambiente: desafios para a demografia. In: Torres, Haroldo; Costa, Heloisa (Orgs.). *População e meio ambiente:* debates e desafios. São Paulo: SENAC. p.21-52.

LEONARD, H. Jeffrey et al. 1989. *Environment and the poor:* development strategies for a common agenda. New Brunswick: Transaction Books.

MAIA, Kátia D.; GUIMARÃES, Roberto P. 1997. Padrões de produção e padrões de consumo: dimensões e critérios de formulação de políticas públicas para o desenvolvimento sustentável. In: LEROY, Jean-Pierre et al. (Orgs.). *Brasil Século XXI:* os caminhos da sustentabilidade cinco anos depois da Rio-92. p.385-97.

MEADOWS, Donella H. et al. 1972. *Limits to growth.* New York: Universe.

SAWYER, Donald. 1993. *População, meio ambiente e desenvolvimento no Brasil.* Presented at the seminar on "Políticas Públicas, Agentes Sociais e Desenvolvimento Sustentável," Belo Horizonte, June 14-15.

TOLBA, Mostafa; EL-KHOLY, OSAMA A. (Eds.) 1992. *The world environment,* 1972-1992: two decades of a challenge. London: UNEP, Chapman & Hall.

TORRES, Haroldo. 1993. Indústrias sujas e intensivas em recursos naturais: importância crescente no cenário industrial brasileiro. In: Martine, George (Org.). *População, meio ambiente e desenvolvimento:* verdades e contradições. Campinas: Editora da UNICAMP. p.43-67.

UNDP. 1997. *The Global Environment Facility Small Grants Programme:* protecting the global environment through community action. New York: GEF Small Grants Programme, UNDP.

UNITED NATIONS. 1992. *Report of the United Nations Conference on Environment and Development.* New York. (A/CONF.151/26).

VIOLA, Eduardo. 1996. A multidimensionalidade da globalização, as novas forças sociais transnacionais e seu impacto na política ambiental no Brasil, 1989-1995. In: FERREIRA, Leila da Costa; VIOLA, Eduardo (Orgs.). *Incertezas de sustentabilidade na globalização.* Campinas: Editora da UNICAMP. p.15-65.

WORD BANK. 1995. *Mainstreaming the environment:* the World Bank Group and the environment since the Rio Earth Summit. Washington.

WORD COMMISSION ON ENVIRONMENT AND DEVELOPMENT. 1987. *Our common future.* London: Oxford University Press.

TERRITORY

Human Dimensions of Global Environmental Change
Brazilian Perspectives

Amazonian Frontiers at the Beginning of the 21st Century

Bertha K. Becker
Dept. of Geography/LAGET
Federal University of Rio de Janeiro (UFRJ)

INTRODUCTION

Trans-frontier financial and informational flows and networks, institutions dedicated to global governance and globalization of environmental topics are characteristics of the globalization process, under which a new territorial division of labor and a new geopolitics are being designed.

Nevertheless, the virtuality of fluxes and networks does not imply the dissolution of geographic space and of the strategic value of wealth *in situ*. Currently, in the symbolic-cultural representation, the value of nature is conditioned by the centrality of life and Earth's sustainability in the modern world, in which the Amazon has become the uppermost symbol. Simultaneously, nature is reevaluated and considered important, conditioned by new technologies. This is the case, overall, of nature as a source of information for biotechnology, supported in the decoding, reading and instrumentation of biodiversity. It is the case of the effect of climatic alterations on global warming. But it is also the case of the theoretic possibility, not yet solved, of the use of hydrogen isotopes of water as a source of energy production. In other words, nature is valued as capital for present and future realization, and as a source of power for contemporary science.

But if financial flows are global, natural stocks are geographically located. The appropriation of decisions concerning the use of territories and environment as value reserves, or in other words, without an immediate productive use, constitutes a means of control of natural capital for the future. This establishes a new component in the disputes among technology-detaining nations: the competition for control of natural stocks, mainly those located in peripheral states and areas without judicial regulation.

Concerning geographic, territorial space, the strategic valorization of the Amazon region is a consequence of the new significance that it has assumed as a double asset: the land area itself and the huge natural capital contained in the circumscribed area. Three "Eldorados" can presently be recognized: the ocean floor that is still unregulated; Antarctica, partitioned by the major powers; and the Amazon region, the only one whose major portion belongs to only one national state.

The use of Amazonian nature cannot be reduced, therefore, to a global problem. It is true that important transformations that effect global environmental changes are processed there. Nevertheless, the apprehension of these facts cannot be separated from the interests that guide diverse interpretations and the actions derived from these. Interests that are not homogeneous and that are, in fact, conflictive at different geographic levels. It is a matter related to various coexisting frontiers in the same region. Frontiers understood as spaces not fully structured; potential generator of new realities, whose specificity is its historical virtuality (Becker, 1990).

At the global level, the Amazon is a frontier for science and technology and is perceived as an area that must be preserved for the survival of the planet. On this perception, coexist legitimate environmental, economic and geopolitical interests, respectively expressed in a process of mercantilization of nature and of the appropriation of the State's authority over territorial use.

At the State level, where diverse interests also coexist, dominant interests and perceptions attribute to the Amazon region the condition of resource frontier, or in other words, an area of population and economic expansion, which will guaranty Brazilian sovereignty over this immense territory. This does not mean that environmental policies do not coexist with development policies.

For Brazilian society, the frontier is the space for projecting the future. At the regional and local level, these perceptions and their resulting actions, as well as local demands, are expressed in rapid transformation of territorial dynamics and in a new Amazonian geography. The conflictive coexistence of projects that involve social and environmental burdens with competing projects – different space-time relations, distinct development strategies, integration processes that reaffirm sovereignty – represents a challenge that attributes to the Amazon region the quality of an experimental frontier with a new standard of development.

The proposal of this paper is the analysis of these diverse frontiers, considering three levels, which correspond to the frontiers at the global, national and regional/local levels, followed by lessons learned from this analysis.

GLOBALIZATION AND THE MERCANTILIZATION OF NATURE

The complexity of the Human Dimensions of Global Environmental Change implies recognizing diverse interests in the perception of questions. There are various interests at the global level: the scientific interest in the research of man's role in changes of the environment, as well as the impact of these changes on society, and the geo-political interests of controlling natural capital, associated to economic interests of gaining and/or protecting markets formed by snew elements of nature.

In this paper, we call attention to these last interests, which have been neglected in research and which must be recognized to alert society about their perverse effects. We are talking about the mercantilization of new elements of nature, in a process of their transformation into fictitious goods and objects, in real markets.

The ideas of Karl Polanyi (1944) about market economy, fictitious goods and their social and environmental impact are worth remembering. They pointed to the transformation of land into merchandise. We can now also relate them to air, life and water.

The commercialization of labor, land and money, inexistent in Mercantilism, has become a precondition to the market economy that emerged in the 19th century with industrialization, subordinating society to its requirements. But, as it happens, labor, land and money are **not** goods, objects produced for sale in the market. Labor is just another name for human activity, that accompanies life itself; it is not produced for sale, and cannot be stored. Land is just another name for nature, which is not produced by men. And money is only a symbol of purchasing capacity and, as a rule, is not produced, although it acquires life through the mechanisms of banks and financiers.

In any event, it was with the aid of this *fiction* that real labor, land and money markets were organized. The fiction that they are produced for sale has become the organizing principle of society, changing its very way of organization. But allowing market mechanisms to become the sole guide to the destinies of human and natural would result in the clash of society and environment. Therefore it is necessary to protect it by means of actions and policies integrated in powerful institutions directed toward the protection of labor, land and money in the market scenario. These processes have slowly generated the conscience that limits to human capacities are not a product of market laws, but of society itself. The emergent reality behind the market economy has been revealed: society (Polanyi, 1944).

Today, the goods' sphere is widened, including that of fictitious goods, that create real markets through their institutionalization. This can be seen in the attempts to implement mechanisms of global governance over the planetary environment, by means of establishing global environmental regimes; systems of "norms and rules specified by a multilateral legal instrument between states to regulate national actions related to specific topics" (Porter and Brown, 1991). It is worth remembering that it is society's role to establish the limits to the action of the market over nature.

Among the important environmental topics currently subjected to global regulations are the Climate Change Convention and the Convention on Biological Diversity, both of which tend to transform air and life into fictitious goods. These Conventions will be analyzed here.

Global Air Market

Industry and the central economies have created a new Use and a new market for air: the carbon captured and held by vegetation. Investments in carbon sequestration and the commercialization of carbon credits at the global level is the commercial mechanism suggested to industry, to balance their emissions, thus creating a billionaire business. Some estimate world investments in carbon sequestration in the years to come at US$ 100 billion; more moderate estimates expect investments around US$ 10 to 20 billion.

Concerning the Climate Change Convention, the countries that ratified the Kyoto Protocol in 1997 committed themselves to create mechanisms for reducing emissions of gases responsible for the greenhouse effect. Among these is carbon dioxide (CO_2), representing 55% of these gases, mainly emitted in the combustion of coal and petroleum derivatives.

Of the 84 signers, however, only 29 have so far ratified the Kyoto Protocol. Economic and geopolitical issues, as well as scientific questions, have slowed the evolution of the regulation of the air market.

The first obstacle is from the North-South conflict. The industrialized countries, holding 20% of world population, are responsible for approximately 71% of global CO_2 emission, while the peripheral countries, holding 80% of world population, are responsible for 18% of global emissions. The seven largest CO_2 emitters are: USA, Russia, China, Japan, India, Germany, and England. Brazil occupies the 17th position, mainly due to its energy matrix based on clean and renewable sources, where hydroelectric and biomass power predominate. The major global concern is that if developing countries follow the steps of industrialized nations, amplifying their industries, in 30 years they will have reached the same level of CO_2 emissions of developed countries.

According to the Kyoto Protocol, industrialized nations, historically responsible for the pollution and integrating what is known as Annex 1, shall reduce their emission levels by 5.2% of their total emissions, considering 1990 levels. The major difficulty is the enormous cost of this process and the need for radical changes in industry in order to adapt to these limits, which requires the development and adoption of clean energy technologies, in a short period of time. The international commercialization of emission credits or the reduction of the emission of greenhouse gasses was the solution proposed to reduce the global cost of this process. Countries or industries that manage to reduce emissions to levels below their limits will be able to sell their credit to other countries or industries that overpass their levels.

President Clinton proposed the voluntary agreement, or the voluntary adhesion of developing countries to the Protocol. These countries, however, do not want to assume responsibility for the reduction of emissions, which would be incompatible with their development needs.

The Brazilian proposal at Kyoto, which obtained the support of the G77, was the establishment of a penalty to the Annex 1 countries, charging each one according to its specific responsibility for global temperature increases, above the authorized limits. A Clean Development Fund would be created for peripheral countries; later, this fund evolved as the Clean Development Mechanism. Its role is to direct investments from developed countries to projects related to emission reduction in developing countries, in order to produce an additional reduction of present levels. This effort requires new enterprises, with implementation after 1990. A fundamental component of the Clean Development Mechanism is technology transfer.

For developing countries, and for Brazil, the use of clean energy sources such as hydroelectric, solar, aeolic, biofuels, and biomass constitute a considerable potential. To this may be added the possibility of Using the absorption of CO_2 by vegetation to compensate other countries' emissions, due to the conservation of carbon stocks in the soil, forests and other types of vegetation, or due to the establishment of new forests and agro-forestry systems, or even due to the recovery of degraded areas. By this means, instead of directly cutting its own emissions, a country such as the USA that, alone, is responsible for 25% of world carbon emission, could pay its quota of 7% simply by buying carbon credits from other countries. In addition to this, forestry investments in developing countries are much cheaper. For an enterprise such as BP-AMCO to reduce the carbon emission level at a sophisticated oil platform on the North Sea by one ton, the cost would be approximately US$ 150; the same level of carbon emission reduction can be achieved by investing only 15 cents in a reforestation project in Bolivia.

The conflicts involved in the construction of the new market, related to what can and what cannot be accepted, or in other words, in the process of institutionalization of this new market, includes the conflict between the industrialized nations. The American government defends the possibility of an unlimited purchase of carbon credits, while the European Union defends a 50% purchase limit, in order to assure that the rich countries will also assume their responsibility in the reduction of carbon emissions.

But the most serious conflict, which particularly interests Brazil – the Amazon region – is related to forest area. The United States wanted to consider the carbon derived from burned Amazonian forest over the last 30 years in emissions accounting. According to this proposal, Brazil would be considered the fifth carbon emitter in the world. But there are no means of adequately considering emissions related to deforestation since it is not known how much carbon is emitted, and, if this damage were considered, the benefits of the forest to global climate should also be taken into account.

The conflict involving the inclusion of standing forests in the Clean Development Mechanism exists within the country itself. Some understand that the country should accept the benefits of the Clean Development Mechanism by avoiding deforestation. For others, standing forests should not be considered as money, because they do not contribute to mitigate climate change and also because the rhythm of emissions generated by deforestation have been held at relatively stable levels over recent years. This would make their inclusion in calculations pointless, since the Kyoto Protocol establishes carbon reduction levels considering the year 1990.

Several projects already suggest the formation of this market in Brazil. To date, the best accepted options are projects related to the planting of forests, generally connected to interests of big oil corporations, mediated by the International Bank for Reconstruction and Development (IBRD), by the French government, and associated with national and/or international NGOs. Other alternatives are also being considered.

In this context, public and governmental opinions are polarized. One of the positions denies arguments about climate change. This was seen in the extensive advertising campaign in the United States in 1997, and might have been a decisive factor for this country's intransigence and recent decision at the Hague meeting in 2000, on not signing the Kyoto Protocol. Another group chooses to reformulate its activities, searching for alternatives and solutions for the global energy problem.

There is no doubt that useful transactions could take place with the marketing of air. But there is another order of questions that should be considered, such as:

- The social risk of transforming air and life into fictitious goods, whose destinies could be directed exclusively by market mechanisms;
- The lack of ethics in this market, that allows rich countries to continue polluting by means of the purchase of carbon credits, a distinct possibility since the reduction goals established were not achieved. Therefore, the position of NGOs that pressure industries to reduce pollution levels within their own countries is quite reasonable.
- The risk of privatization or internationalization of national territories by means of purchase and/or control of huge land areas and, mainly, by controlling the use of the territory in the case of including original forests in the Clean Development Mechanism;
- An ostensive pressure in international negotiations – particularly on Brazil, China and India – a result of the asymmetry in the number of members of the national representations, for adhesion to the "voluntary agreement;"
- The gaps in scientific knowledge on global heating, since the quantity of carbon effectively held by forests and vegetation in general is not known. Furthermore, the long-term climatic oscillations that influence the size of the ozone layer and that may cause global warming are also unknown. These gaps make the scientific explanations presented to date, nothing more than hypotheses, and the actions underway in precautions that, without doubt, must be considered.

The Life Market

Biodiversity emerged as an environmental question in the 1980s, and soon became the object of a Convention at the United Nations Conference on Environment and Development, held in Rio de Janeiro in 1992. It is an environmental problem under construction, which became a matter of public debate and action before science could provide knowledge capable of clarifying the actions and policies of international and national organizations.

Nevertheless, differently from air, diversity of life is not a purely physical and biological phenomenon, but also human, since its geographical location and particular appropriation mechanisms insert it, necessarily, in the context of social relations. This condition implies the recognition of the fact that there are different projects for biodiversity, corresponding to the diversity of meanings for societies, at different geographic scales, grounded on specific geographical and historical conditions, resulting in diverse ways of using and protecting diversity.

Today, not only the immensity of the Amazonian nature challenges contemporary science, but also the postmodern scenario generated by the conflicting but fascinating way by which modernity relates to social groups living in different times, generating a mosaic of alternative and diversified projects. It is a matter of new territorialities, forged through the resistance of social groups supported by international alliances, motivated by legitimate environmental interests and by geopolitical interests for controlling the natural stocks.

It is assumed that in the logic of contemporary science, these alternative projects cannot be dissociated from advances in biotechnology, one of the fields with significant scientific progress, where the biggest global scientific project is being developed, the Genome Project, that aims to map life; life that is codified in the Amazonian genetic bank, the planet's largest. In this context, the alternative projects being implemented in the region are also experiments, essential elements of the logic of biotechnology, a logic comprehended in this case, as a system of relations (Becker, 1999).

What are the nexus between clones and embryos, networks of which have permitted scientific reproduction of monkeys and sheep, with the experiments in the Amazon region? The progress of experimental research takes place *in vitro*, with sophisticated techniques in modern laboratories located in universities and industrial enterprises of developed nations – especially those in the United States and England – and also *in situ*, in the heart of the forest, where the genetic matrices are found, many of which are still not capable of being reproduced *in vitro.*

Social practices developed in the region, both traditional and new, constitute a crucial condition for the *in situ* research: they are a source of information due to their local knowledge, or in other words, due to the knowledge that they, throughout the generations, accumulated about the local ecosystems; they facilitate access to the genetic matrices and speed up the process of reviewing information; they protect biodiversity in both traditional and new ways.

Mediation between global and local laboratories is established, above all, by international NGOs, but also by scientists, churches and foreign government agencies, associated to international NGOs.

Between the two extremes there is a conceptual gap. Theory has not been able to follow the speed of experimental advances; consequently, the social practices lack conceptual and theoretical elaboration. The complexity of biodiversity includes ecological, economic, technological, sociopolitical, ethical and epistemological dimensions. It is at the center of the global debate involving the restructuring of the development standard, North-South relations, the commercialization of nature and human survival, and it expresses itself in the conflicts involving the uses and choices made by societies. Biodiversity is loaded with value norms. Protecting it may mean the elimination of human action, as proposed by the radical ecological proposal; it may mean protecting the populations whose production systems and cultures lie in specific ecosystems; and it may also mean defending the commercial interests of companies that use it as a source for the production of new merchandises. The concept therefore involves different functions that, by their turn, induce to different forms of use.

The Biological Diversity Convention made the problem more complex. It gave priority to the risks and to the need for protecting global biodiversity, subordinating the distribution of benefits to inhabitants of the ecosystems to secondary importance. But biological resources, for which the FAO defends the status of human patrimony, were declared national patrimonies at the Earth Summit, affirming the sovereignty of states in the exploitation of their own resources. Furthermore, the affirmation of this right lacked the necessary clarification of property rights. And, under this paradox, the North-South relations were tensioned.

To science, biodiversity presents a double challenge: to describe and quantify biological processes and states, and to attribute a value to nature that, until now, was considered out of the economic sphere (ORSTOM, 1996). Until recently, biological diversity had no value itself, but only use value to populations whose survival depended on the ecosystems where they lived. Today, natural capital has become a limiting production factor (Daly, 1991), and new technologies tend to change the notions of value associated to goods obtained by the use of labor. Furthermore, the threats facing species and the degradation of ecosystems associated the idea of scarcity and profit to biodiversity, inducing economists to attribute prices to the elements related to it.

For society, the valorization of genetic resources requires rules for controlling access to it, rules that are still under discussion in Brazil. To date, this access is still free, opening paths to "bio-piracy", that is, the illegal withdrawal of genetic material from the region through: a) foreign scientific expeditions or enterprises; b) eco-tourism; c) NGO activities; d) national research institutions located in the Amazon region, with scientific cooperation links with foreign countries (Albagli, 2001).

Three alternatives for the sustainable use of biodiversity are currently identified in the Amazon: vegetal extractivism and fishing, local processing of biological resources, and industrialization by means of advanced biotechnology and genetic engineering. An attempt was made for this last alternative, but the lack of regulation over access to genetic resources and political impasses frustrated this effort.

Attributing economic value to life and identifying the rights of property over living beings are obstacles to their transformation into fictitious merchandises, which obstruct the institutionalization of a real market for its control. To date, this is achieved by illegal methods.

NATIONAL INTEREST AND PUBLIC POLICIES

Behind the international conventions and multilateral organizations involved in the attempt to create global governance, lie powerful agents: corporations and companies associated to their home states, in a partnership that has been studied by Geo-economics. Economics and geopolitics are united, and national states develop strategies aiming at the conquest and/or defense of their markets, seeking the interests of their enterprises. These strategies demonstrate how the role of nation states remains unmodified in the process of globalization.

National interests are understood as the sum of interests shared by national society in its relations with the world, based on historically built, fundamental values and geo-economical and geopolitical determinants. In Brazil, the basic values are: the quest for economic development, autonomy, peace, and respect for cultural diversity. To these values, new demands related to citizenship, involving environmental conservation, human rights and democracy, are incorporated, and the importance of the peace and stability surrounding them are reinforced for economic complementarity and political alliance (Magnoli et alii). The concept of territoriality as a value of national interest is still worthy of attention (Becker, 2000).

Public policies related to the Amazon region reflect national interests, with historical values updated by the incorporation of citizenship demands. This transition is perceived in parallel public policies, unarticulated, that while implicit until the mid 1980's, became explicit in 1996. Both seek development in a selective territorial strategy, but the development foreseen by one or the other are not only diverse, but opposite and conflictive.

Environmental Policy

The Ministry of the Environment and of the Legal Amazon seeks sustainable development. Contrary to what the media says, there is a considerable national effort for environmental conservation. A comparison between Brazil's public policy in this area with those of other countries would certainly place the country in a favorable position.

It was after 1990, due to social struggles and environmental pressure, that the environmental variable was introduced in the territorial policy for the region. Since then, three types of action can be identified:

a. The importance of socio-environmental policies, emphasizing institutional changes for modernization of the codes regulating the use of natural resources, the expansion of the protected areas and alternative communitarian projects. The protected areas, on the one hand, are the indigenous delimited areas, a conquest of the collective rights of indigenous communities that have been recognized by the government since 1991. This has resulted in delimiting their territories and strengthened their roles as political actors on the regional scene. Between 1995 and 1998, the government recognized 58 indigenous reserves, corresponding to 26 million hectares, and delimited 115, in an area of 311.000 km2. According to these numbers, 63% of indigenous lands are now delimited, representing 78% of the total area of national reserves. On the other hand, protected areas include the Conservation Units (CUs), that have multiplied in the Amazon region since the mid 80's, but whose management constitutes a central problem, due to its large extension and to the lack of personnel. Today, indigenous lands occupy more then 20% of the Amazonian territory, while more then 6% of this is occupied by Conservation Units (Fig. 1). As for the alternative projects, these are localized and dispersd communitarian experiments that implement diverse forms of conservationist use of the forest.

b. Major environmental protection projects. The Pilot Program for the Protection of the Brazilian Tropical Forests (PP-G7), negotiated in Geneva in 1991 and formally launched in 1993, has been in operation since 1994.

It is financed by the European Union, Canada, France, Germany, Italy, Japan, Us, and Great Britain, and administrated by the World Bank, with investments estimated at US$ 250 million, deriving from the donor countries and the Brazilian counterpart. It is the largest environmental project implemented in a single country. The amount already invested – US$ 110.41 million, by 1999 – is much less than what was promised in Geneva, in 1991. The pilot projects that constitute the program had a late start, and currently find theirselves at various stages. A recent review of the institutional organization concluded that the Program reflects the lack of a joint strategy, weak management, complex design and financing plan, and responsibilities undefined and poorly assumed by participants. It is, however, worth registering the success of the Demonstration Projects in the Extractive Reserves, as well as the attempts to increase the scale of action and the partnerships among public, private and nongovernmental actors in the Program. The most recent and audacious proposal (1996) is the establishment of the Ecological or Conservation Corridors, that correspond to a revolution in the planning of conservation. Instead of conservation islands, networks are established, adding buffer zones, as well as other areas with various degrees of human use, and considering the inter-relations between the mosaics of protected areas.

The SIPAM/SIVAM Project – protection and vigilance over the Amazon – is a gigantic project of the Brazilian government to establish control over the Amazonian region, based on modern technology, based on a strategy against territorial intervention, in the name of drugs and environment. A national initiative, to be established over 5 years, with a total cost of US$ 1.4 billion and requiring advanced technology, the project promoted a selective process to choose the firm best able to satisfy this double necessity. A major handicap that favored Raytheone's victory was the financial facilities with the American proposal. For the first time in 15 years, the American Eximbank made a loan to Brazil, in advantageous conditions, providing 85% of the necessary financial support. It was due to the financing of the SIVAM Project that the US managed to participate in the War on Drugs in Brazil. With this program, the Amazonian region enters the 21st century with one of the most sophisticated information systems in the world.

Two other major projects focus directly on biodiversity and climate. The PROBEM – Brazilian Program on Molecular Ecology of Amazonian Biodi-

FIGURE 1: CONSERVATION UNITS AND INDEGENOUS LANDS

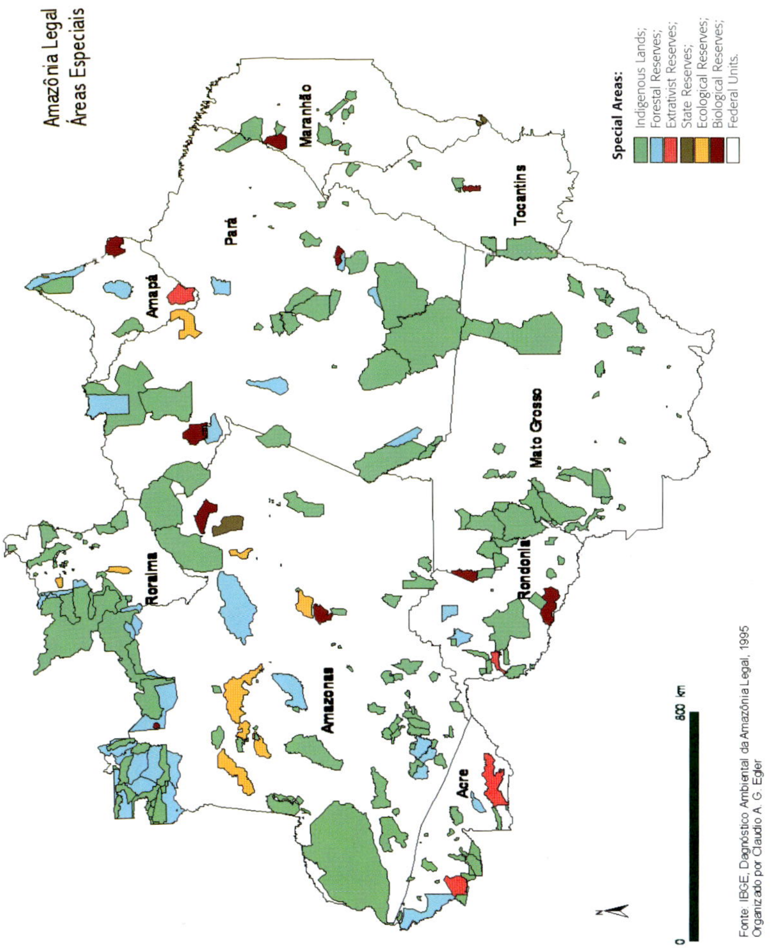

Source: IBGE, Environmental Diagnosis of the Legal Amazon, 1995.

FIGURE 2: NATIONAL INTEGRATION AND DEVELOPMENT AXES
"AVANÇA BRASIL" PROGRAM

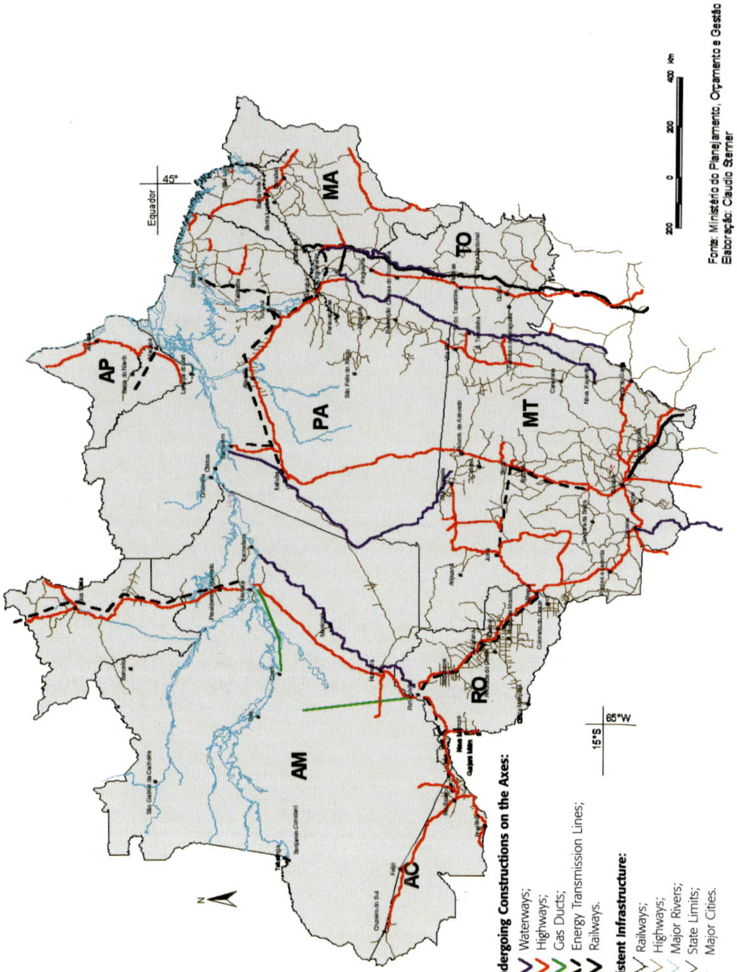

Source: Planning, Budget and Management Ministry.

FIGURE 3: URBAN DEMOGRAPHIC GROWTH 1996-2000

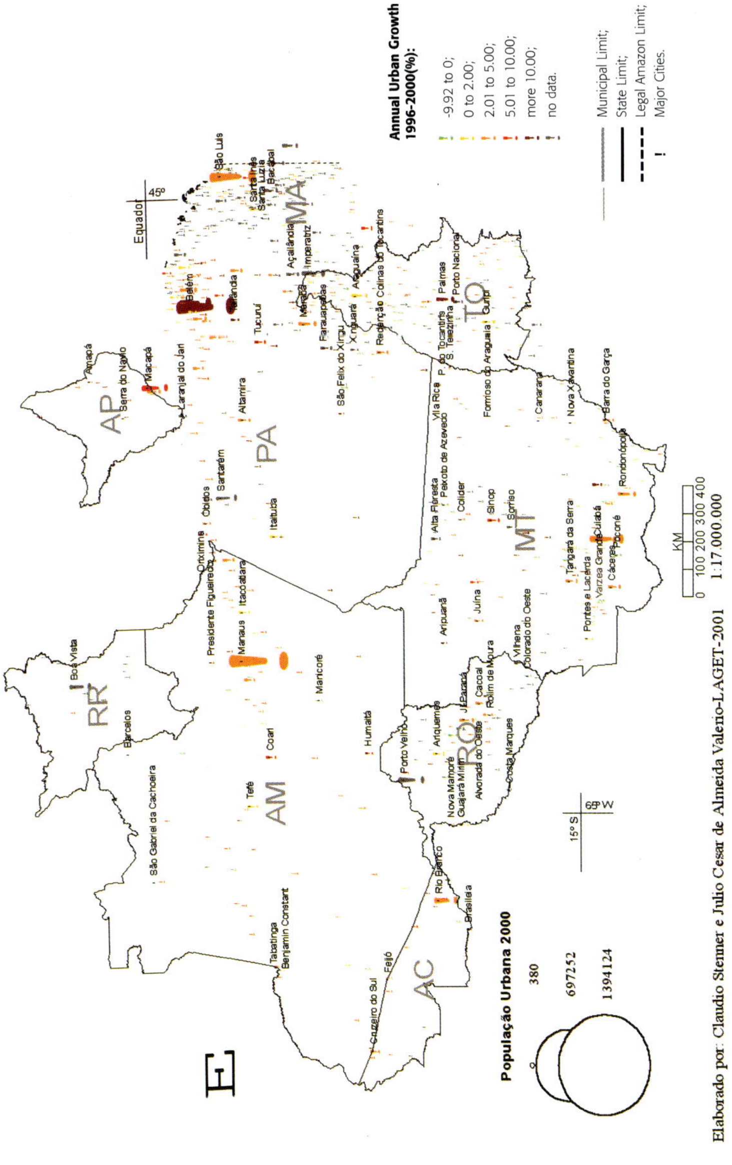

Source: IBGE

FIGURE 4: REGIONALIZATION AND MACRO-ZONING - 2003

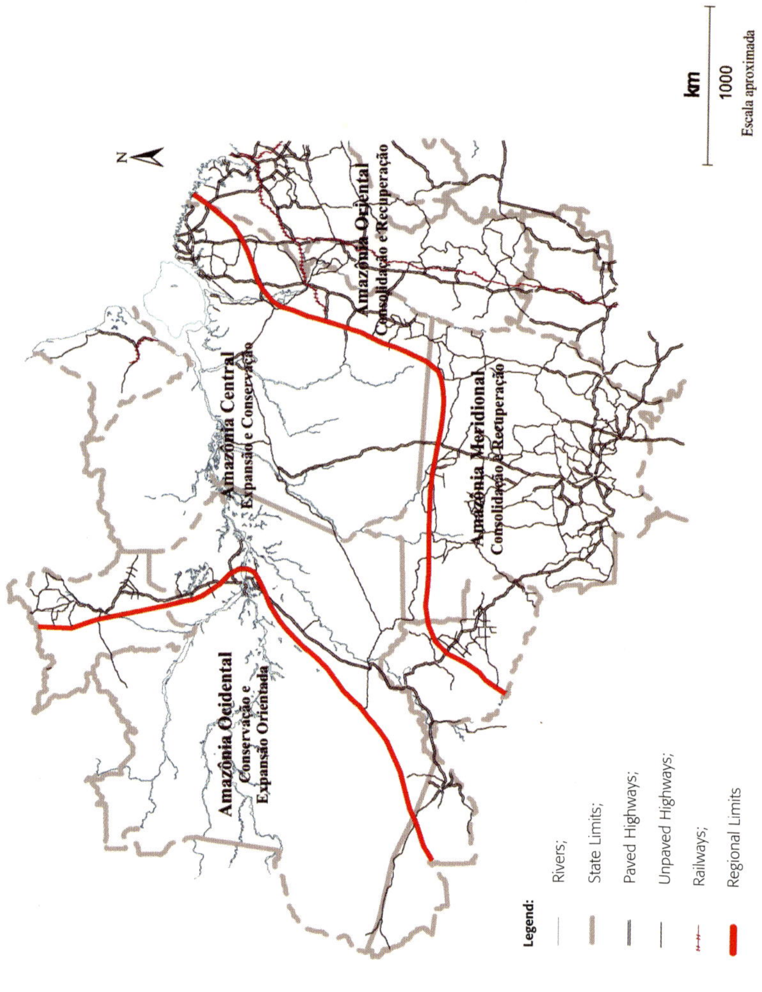

Org: Bertha K. Becker. Elab.: Claudio Stenner

versity – is a multi-institutional Brazilian program supported by the national and international scientific communities, by the private sector and by the Brazilian government, at various levels. Its main objective is to qualify the country in research and development (R&D) in the areas of Biotechnology and Chemistry of natural products, seeking especially the development of industrial products with higher aggregated value, besides contributing to sustainable development and the preservation of biodiversity. This program was supposed to be the baseline for the recovery of national decision-making capacity for transformation of natural capital into effective support for sustainable development, building up a strategic answer to the pressures that challenge the legitimacy of national authority over the Amazon. Nevertheless, the project was immobilized due to political impasses.
While the LBA – Large Scale Biosphere Atmosphere Experiment in the Amazon – is an international research initiative, headed by Brazil, that seeks the generation of new knowledge necessary for the comprehension of the climatological, ecological, biogeochemical, and hydrologic functioning of the Amazon; of the impact of the changes in soil use, in this functioning; and of the interactions between the Amazon and the global bio-geophysical system. The LBA made the Amazon a subject of the first project supported by the three most important research programs of the International Geosphere – Biosphere Programme.

c. International financial and/or technical cooperation, present in almost all projects in the Amazon region. The PP-G7 is one of the most significant expressions of the new form of international cooperation in partnerships and constitutes a changing process through which partners make concessions and adjustments between them. This process reflects a change of Brazil's position, that used to be reticent concerning external influence on the Amazon. It also reflects the modification of the initial preservationist goal of foreign partners. All are now committed to sustainable development, finally consolidated as a reference for the Program.
International cooperation therefore constitutes an instrument of pressure and deregulation, but also an instrument for the construction of a new development model, sustained by adequate negotiation and, above all, societal involvement and support. (Becker, 1997)

Infrastructure Policy

At the same time that the Ministry of Environment proposed Ecological Corridors, in 1996, the Ministry of Planning and Budget proposed the institution of Development Corridors.

After a decade of little planning activity, the Brazilian government, by means of the Ministry of Planning, is rehabilitating national economic and territorial planning, through the *Brazil in Action Program*, based on development corridors. The execution of the planning will take place through the *Advance Brazil Program* (PAB), which will be initiated between 2000 and 2003, and completed by 2007 (1999). This is the central strategy for regional development.

The government justifies the PAB asserting that it will propel economic development, by means of the integrated management of the Four-Year Investment Plan (PPA) and of the budget. The expectation is that R$1,113,000,000 will be available for an 8 year period. The government – national, state and municipal – will be responsible for 66,6% of this amount, the rest to be subscribed by the private sector, public companies and other funds. The growing participation of the private sector is expected.

The main instruments of the program are: the PPA, that expresses the federal strategy and establishes priority areas for allocation of monetary resources from the budget, for the next 4 years; and the *Integration and Development Axes* (ENID), that constitute the main agenda of the Program, and to which R$317 billion were planned for the next 8 years.

The PPA of the *Advance Brazil Program* (2000-2003) is quite ambitious, based on a complex logic that includes not only circulation and communication networks, but also energy, and their supporting points. The goal is to increase the speed and efficiency of the transportation and communication multimodal systems. The conception of the ENID considered fundamentally national logistic objectives, without considering regional strategies of insertion in the national productive structure and in the global economy. It also did not consider the environmental dimension. The national logistic objectives are mainly: a) stimulate and assure national exportation, increasing exports to the Northern Hemisphere; b) strengthen relations among South American countries, aiming for the consolidation of the "Mercosul" in South America as a whole.

Joining the North and Center-West regions, the Amazon region will have the major ENIDs, huge exportation corridors to support the new scale of production and circulation. These are the North and West axes (Fig 2).

A global overview of the ENIDs shows the formation of two immense multimodal corridors of circulation – the Center-North, in a N-S direction, involving the Eastern Amazon; and the Northwest, involving the Western Amazon axes, including the BR 174 highway, that opens a new regional circulation arc in the Western Amazon.

Considering the powerful logistics being installed in the Amazonian region, the priority established by the government for exportation, and soybeans' huge market potential, it is licit to deduce that these corridors will serve largely as a stimulus to its exportation. Departing from the production belt in the "cerrado" surrounding the forest, which it timidly penetrates, soybean culture represents a great risk to the forest, a risk also deriving from international conflicts. At the same time that the G7 and the European Union make donations through the PPG7 and other initiatives to protect the forest, at the World Trade Organization they stimulate the expansion of soybean culture in Brazil, in order to feed their bovine and swine flocks. In the same manner, Japan, though participating in PPG7, gives it little support; it is one the major investors in the expansion of soybeans in the Brazilian "cerrado".

It is important to understand the differentiated roles that might be played by soybean culture in the Legal Amazon region. Theoretically, it is a viable economic alternative for the "cerrado" region. It might also play an important role in the recovery of land degraded by pasture. What is unacceptable is expansion of soybean culture to the forest area.

REGIONAL TERRITORIAL DYNAMICS AND THE NEW AMAZONIAN GEOGRAPHY

Amazonian territorial dynamics expresses the impact of global, national, and local actions and demands.

In recent years, the increasing lack of integration between two public regional policies has accentuated. One is based on planning and facilitating new investments in infrastructure and the opening of markets. The other is directed towards sustainable development, local populations and environmental protection.

As a consequence, old standards coexist with new dynamics and can be aggregated in two major topics: land settlement and land use. Furthermore, a new Amazonian geography is delineated.

Settlement

On the aspects related to settlement, three elements must be considered:

a. Urbanization – accelerated urban growth has introduced significant changes in the structure of regional settlement. Between 1970 and 1996, urban population has grown more than the total population, at a rate twice the average national rate, from 35.5% in 1970 to 61% in 1996.
New nuclei have been formed, but the most important characteristic of this growth are the changes in urban hierarchy, breaking up the model prevalent until 1970. The Amazon is the only region in the country where there is population growth in cities with fewer than 100,000 inhabitants, especially expressive in cities from 20,000 to 50,000 inhabitants.
The cities currently represent 12 million consumers and may: constitute a market for natural products not related to lumber; contain emigration from the extractivist reserves; and extend services and governance to the Conservation Units of direct-use and their surrounding areas, articulating the two networks (GTA / Earth's Friends; Becker, 1998). But the urbanization rate tends to slow down, mostly because the multiplication of the nuclei throughout the decade was a result of an intense process of development of new municipalities, each of them having a town as a nucleus (no matter the size). Apparently this process has slowed down, mainly because of the decrease of the immigration of settlers from other regions to the Amazon region. (Becker, 1998)

b. Decentralization – the most recent urban dispersion has been a result of the emergence of new municipalities, stimulated by the 1988 Constitution that transferred assets and, in a smaller proportion, responsibility to the municipalities, in order to combat centralization. The basic point is that municipalities are financially highly dependent on the federal government (Fiqueiredo, 1998). Therefore, what is granted is not autonomy, but a "dependent emancipation".
Resulting from this process, municipalities do not constitute a financial and economic propelling force, although they represent a political force since they represent an expression of the pact between the new units of local power and the superior levels of the federative order. In other words, they must become political partners.

c. Migration – it is hard to precisely estimate migration to the region, since the most recent data are from 1996. Nevertheless, it is generally believed that immigration has declined in comparison to past decades and that, presently, it is mostly intra-regional. In any case, migration to the Amazon is still superior to that of other Brazilian regions.

Isn order to delineate scenarios, the major aspect to be considered is the direction of the migratory flows, an important framework for present and future settlement moves. Among these, stand out:

- Migration to the states of Mato Grosso and Pará, that have a more dynamic economy;
- The emergence of new areas attracting migrants in the extreme North: the frontier between the states of Amapá and Pará; Amapá and French Guiana; and Roraima's frontier with Venezuela, axis of the BR 174 highway;
- The delineation of new settlement corridors. Departing from Mato Grosso, one follows the Cuiabá-Santarém highway and the other penetrates the state of Amazonas, in the direction of Manaus and the BR 174 highway. In the state of Pará, immigration towards the state of Amazonas tends to develop to the North, with strong immigration to the state of Amapá;
- The significant reduction of immigration to the state of Rondônia, that currently is an area of out-migration, mainly to the state of Roraima.

The migratory process has a spontaneous and, also, a directed component, determined by the settlement policy associated to speeding up agrarian reform, a priority of the federal government which has been accelerated. While speeding up agrarian reform is an innovation, its implementation in the Amazon region preserves the old settlement and colonization pattern common to forest areas, where little support and orientation is given to settlers. This results in rapid and accentuated deforestation due to the high rate of land abandonment.

Land Use

As for land use, new and old patterns coexist.

a. The reproduction of the deforestation/exploitation cycle of lumber/cattle raising

The growth of the agrarian economy in forest areas, due to peasant families (posseiros) who cut forests, stimulated by lumber companies, and later occupation of the area by cattle farmers, was initiated in the states

of São Paulo and Minas Gerais in the 1950's, and expanded through the states of Goiás and Mato Grosso, characterizing Amazonian occupation beginning at the end of the 1960's, at a much more accelerated rate.
The analysis of the settlement process indicates the reproduction of this cycle throughout the areas with highest immigration levels: Mato Grosso, the Eastern Amazon and new settlement corridors.
The activity of lumber companies has reached a rate more significant than that of the peasants, and "wood mining" – the selective and predatory exploitation of the most valuable trees – occurs even in areas where the peasants still have not reached. While the certification process has been timidly adopted in the Western Amazon, in Itacoatiara (Mil Madeireiras and Gethal), eight exporting enterprises – four European and four Asian, among which there is a Chinese state company – advance over the Amazon Valley, reducing the primacy of exploitation in the Eastern Amazon, where the largest regional exploitation area, Paragominas, is losing importance toward the Lower Amazon region. Furthermore, lumber markets grow relentlessly, especially the local market. The major wood exploiters are still located in the states of Pará and Mato Grosso, followed by Rondônia. Most of this production, about 80%, is absorbed by Brazil's Southeastern region, especially by the state of São Paulo (20.1%), while foreign countries export only 14% of the total lumber production (Earth's Friends, Imaflora and Imazon, 1999).
Besides opening the way for settlement, this exploitation also represents an increasing exposure of the forest to the risk of fire, due to the renewal of neighboring pasture and agronomic practices. This new phenomenon is introducing fire as an actor in the evolution of the daily landscape in settled areas in the Amazon.

b. The Conservationist Pattern

The great novelty of the 1980s and 1990s were innovations introduced by the initiative of local communities and by the PPG7. If the 1980s were considered a lost decade, considering economic aspects, this is not true for social aspects. Civil society reached a level of organization never seen before. The conflicts in the 1970s and early 1980s became organized demands, with different conservationist development projects. These are related to experiences associated to Amazonian biodiversity, carried out in different ecosystems, by populations with different ethnic and/or geographical origins, with diverse techniques and partnerships, as well as distinct productive, social and political structures. The pioneering experience was, without doubt, the struggle for the creation of the Extrativist Reserves (Becker, 1999). The basic communitarian strategy is association with transnational social and political networks, forging local and global partnerships.

Presently, new agents and wider scale characterize conservationist projects. This is the case of state governments that manage the Integrated Environmental Management Projects (PGAI), executed in extensive areas chosen by them, as well as the Environmental Corridors.

Together with institutional areas – indigenous lands and Conservation Units – the communitarian projects form localized and/or extensive interventions, in a territorial techno-ecological web, forging a potential scenario for the Amazon, based on a sustainable development model.

c. *The New Model: Expansion of Capitalized Agriculture*

In the 1990s, a new scenario begins to be drawn, where small farmers find themselves threatened with losing their lands to soybean culture. Without an economic alternative, they may see deforestation of new areas as the only viable option for making a living.

The arrival of capitalized agriculture to the Amazon is, without doubt, a historical novelty to a region that has always been based on extractivism. Soybeans are emblematic of this new model. Until the mid 1990s, together with rice and corn, soybeans had a timid presence in the Amazon region. By 1999, its culture already occupied new and significant areas, reaching a production of 6.9 million tons, 2 million tons in the state of Mato Grosso.

Once again, the region finds itself threatened by outside actors and activities. A technified crop, with high financial costs, soybeans are cultivated by a new type of modern entrepreneur who has little connection with the traditional farmer and farming enterprises. A corollary of this situation is the new production scale that demands large and medium properties, of approximately 1.000 hectares. Most of these farmers come from the South of Brazil, many having passed through the Center-West region, are and now expanding their activities to the Amazon.

The rate with which this crop has increased is a matter of concern, since it may represent a significant impulse to a rapid and violent process of deforestation.

The New Amazonian Geography

The priority established for investments in infrastructure as a strategy for Amazonian development represents the reiteration of an old strategy and of a pattern that concentrates life in the region along the circulatory paths. The Integration and Development Axes will bring new migrants and investments, increasing the pressure on the forest. Networks will cover a larger portion of the forest and consequently contiguous forest area will be diminished.

But these axes also represent new economic opportunities for local populations. The success of this strategy will therefore depend on the application of policies and integrating efforts that take into consideration past experiences and the current complexity of the region. A determinant aspect will also be the way the project is carried out, which will determine whether the axes will act as ordering or predatory instruments.

The present challenge is how to make economic expansion compatible with principles of social and environmental sustainability or, in other words, how to conciliate two major concerns, in order to conduct the Amazon region towards a sustainable development model. (Becker, 1999)

A major reference for action is the recognition of spatial differences. This understanding will permit the redefinition of the Amazon region considering diverse spaces and times.

With different rates of transformation, the following are the space-time differentiations foreseen for the Amazon (Fig. 3):

a. Southern and Eastern Amazon Region – corresponds to a populated arc from the South to the East of the forest. It comprehends the extensive "cerrado" areas of the states of Mato Grosso, Tocantins and Maranhão, and the deforested areas in the states of Sergipe, Pará and Rondônia. This is a more accessible area, consequently with faster transformation capacity, from which the networks, interests and money spread out toward the Amazon. This area also corresponds to the deforestation arc and to the heat focuses. This area constitutes, in effect, a new region – the Center North – implying the dissolution of the Legal Amazon. The latter, an expression of a political regionalization, now disappears due to concrete techno-ecological regionalization, symbolized by agrarian technification in the savannahs, where grains are widely planted, and by planted pastures.

This large sub-region contemplates areas that are still to be consolidated.

In it, the urban-industrial economy is in an advanced stage. At the same time, there are areas degraded by deforestation and by the planting of pasture for extensive cattle raising, that must be recovered. In this sub-region, production predominates over conservation. The focus of policy must be the reduction of occupational instability, which may be achieved by guaranteeing the maintenance of the large number of small farmers that were introduced in the region either spontaneously or by the government.

b. Central Amazon Region – this is the area created by the new Integration and Development Axes, that extends from the middle of the state of Pará to the future Porto Velho-Manaus highway. It is an extensive region that, due to the axes, can expand the development of valuable activities at an enormous speed. This can be seen in the case of soybeans, which is already transported via the Madeira river and might well occupy the pasture area along the Cuiabá-Santarém highway. This also happens in the colonizing corridor along the Juma river (in the state of Amazonas), induced by an INCRA settlement program.
This area is covered by huge forest areas, indigenous lands and Conservation Units, which makes it an extremely vulnerable area. Here, therefore, the focus of policy, considering the need for conciliating both production and conservation, must be ordering the expansion along the axes and speeding up and making viable preservationist efforts.

c. Western Amazon Region – this is the area that, away from the Integration and Development Axes, still has its rhythm commanded by nature. It has huge potential, not only in forests but also in water resources, to which rich mineral resources and the social diversity of the Indian and "caboclo" populations are added.
The extreme concentration of the industrial economy in Manaus does not affect the dominance of extractivism and of fluvial circulation in the Western Amazon. Drug traffic and money laundering increasingly use this nature-dominated area. Manaus is today the capital of the Amazonian Frontier due to its location near the northeastern circulation corridor and large extensions of forest, not only of the Brazilian Amazonian region, but also of the South American Amazonian region as a whole.
Vigilance is a key task for policy in this area. For this, SIVAM is crucial and will play a major role.

Environmental concern does not mean that there will be no regional development, but it represents a need for considering it from a different perspective. Due to its still preserved natural potential, the Western Amazonian region is in a privileged situation to implement a sophisticated sustainable development, processing natural resources by the use of advanced technologies, that allows the production of certified products, highly valued in the international market. In the same manner, recognizing that the Amazon region must be considered at a South American scale does not mean that regional and national frontiers will be eliminated; in fact, it represents the possibility of developing new partnerships and synergies.

LESSONS FROM THE APPRENTICESHIP

The first lesson one may take from the present analysis, and the one here expected to be transmitted, is the complexity of the human dimensions of global environmental change, which requires the consideration of multiple interests and actions implemented at several scales, making them compatible to a logic that preserves the planet's health, respects national and regional interests, and promotes the development of local populations.

As a frontier of global processes, the Amazon reveals that this compatibility must be based on the advancement of science and on adequate actions and policies that involve society. Methods for protecting air and life from an unbridled mercantalization must be developed by societies in order to overcome the latent contradiction of degrading nature in the name of protecting it.

What limits can be established to protect the new fictitious goods?

First, demanding the reduction of environmental pollution in the countries responsible for most of this risk. Second, in the case of the trade in carbon-credits, as in Brazil, not including the original forests in the computation. Approving only projects with a determined period of implementation, without access to land ownership, with special emphasis on the conservation and/or planting of native species. By aggregating value to forest products and by incorporating the effective demands and projects of involved local populations, eliminating innocuous and paternalist actions. Concerning biodiversity, two main actions are urgent: regulate access to genetic resources and intellectual property, and promote further industrialization of the resources.

The Amazon region is also a frontier for the process of updating basic values of national interest, clearly expressed in the diversity of social demands and in the disarticulation of public policies. Making these demands and policies compatible will certainly contribute to consolidating the incorporation of the demands of citizenship into national interests, as well as approximating local and global interests and perceptions on the meaning and use of nature.

At the local and regional levels, the Amazon region teaches us that the climate-biosphere interaction and the protection of biodiversity cannot be dissociated from the social projects for their use. It would be more appropriate to refer to bio-sociodiversity, since this is the basis for practices that represent differentiated methodological paths that contribute locally to the solution of global problems, and represent diverse routes towards sustainable development.

Finally, technical problems tend to be solved. The great challenge to be met for the advancement of a new society-nature relation, able to use and protect nature simultaneously, is the sensibility of institutions and social groups to understand nature as a collective good. In other words, the greatest challenge is to advance in the relations among men.

BIBLIOGRAPHY

ALBAGLI, S. (2001). Articulation Biodiversité: Lámazonie Comme Frontière Géopolitique. *Courrier de la Planète*, Paris, n.59, 2000.

BECKER, B. K. (1990). *Amazônia*. Rio de Janeiro, Editora Ática, 1990. (Série Princípios)

------ (1996) *Construindo o Conceito e a Projeção da Biodiversidade na Prática*, 1996. (in print).

------ (1998) *A Especificidade do Urbano na Amazônia - Subsídios para Políticas Públicas Conseqüentes*, Brasília, SCA/MMA, 1998.

------ (1999a). Amazônia, Fronteira Experimental para o Século XXI, In: BECKER, B.(org.). *Bioética no Brasil*. Rio de Janeiro, Ed. Espaço e Tempo, 1999.

------ (1999b). *Cenários de Curto Prazo para o Desenvolvimento da Amazônia.*, Brasília, Secretaria da Amazônia/MMA, 1999.

------ (2000). *Reflexões sobre Políticas de Integração Nacional e Desenvolvimento Regional*, Mimeo., Brasília, Ministério da Integração Nacional, 2000. (mimeo)

MINISTÉRIO DO PLANEJAMENTO E ORÇAMENTO (1999). *Programa Avança Brasil.* Brasília, 1999.

DALY, H. (1991). From empty-world economics to full-world economics, In: *ENCIRONMENTALLY Sustainable Economic Development*: Washington, Building on Brundtland, Banco Mundial/Unesco, 1991.

FIGUEIREDO, A. H. (1998) *A Floresta Dividida*. 1998. Tese (Doutorado), Programa de Pós-Graduação em Geografia. UNIVERSIDADE FEDERAL DO RIO DE JANEIRO, Rio de Janeiro.

ORSTOM. (1996) *Coût Incremental et protection de la Biodiversité*, Paris, ORSTOM, 1996.

POLANYI, K. (1944) The Great Transformation, In: *A GRANDE Transformação*, (1980). Rio de Janeiro, Ed. Campus, 1980.

PORTER, G & BROWN, J. W. (1991) *Global Environmental Politics.*, Oxford, Westview Press, 1991.

Recent Changes in Land Use and Land Cover in Brazil

Claudio A. G. Egler
Laboratory for Territorial Management - LAGET
Federal University of Rio de Janeiro

INTRODUCTION

The objective of this paper is to describe and study recent changes in land use and land cover in Brazil, with the intention of evaluating their effect on global changes. By looking at the dynamic interface between human society and the environmental conditions that cause production and reproduction, through a range of differing geographical dimensions, this paper follows the line of thinking set out in the Land Use and Land Cover Research Plan (LUCC) , part of the International Geosphere-Biosphere Program (IGBP) and of the International Human Dimensions Program on Global Environmental Change (IHDP) (LUCC, 1999).

The importance of studies on land use and land cover is well known. At the theoretical-methodological level, due to its interdisciplinary character, establishing a bridge between natural and human sciences, LUCC is not only the first IHDP program, but also the one that is developed in conjunction with the IGBP. Secondly, in line with the best traditions of geographic studies, this paper is based on the study of real territorial conditions, that is the space occupied and worked by human groups, reflecting the development of their technical abilities and the conflicts and adjustments that are required among the various agents that interact within the social fabric.

TERRITORIAL PATTERNS OF BRAZILIAN DEVELOPMENT

From the viewpoint of relations between land use and global changes, Brazil has an important position on the world stage. First, its territorial and demographic dimensions place it among the 10 largest nations in the world. Second, the existence of its immense, and as yet largely untouched, rainforest, place Brazil in the front line in terms of the planet's remaining natural vegetation cover. Third, the extreme inequality of income distribution within the country, in both social and territorial terms, limits the options for social improvement and contributes towards the great mobility of Brazil's population, one of the main factors in determining the speed and size of land use changes.

It is exactly the speed and dimension of these changes that allow us to determine the study period, its starting point to be found in the accelerated industrialization and territorial expansion processes implemented by the Brazilian Government in the beginning of the 1970's, by means of the First and Second National Development Plans. These were decisive in redesigning the spatial distribution of the Brazilian productive base, and speeding up the use of resources in the Amazon frontier (Becker and Egler, 1992).

The effects of this process are still present today, although somewhat diminished due to the dimensions of the crisis that marked the following decades. The search for a greater role in the world economy, together with an expressive reduction in the fiscal and financial capacity of the State, resulted in the abandonment of a project for national development in favor of one for global macro-economic stability. Thus, planned interventions at national level encountered substantial difficulties, and were replaced with more selective and dependent actions by the private sector, and with the participation of international finance.

The reduced role of State planning in Brazil has, however, contradictory features. On one hand, far from reducing the pressure on the primary sources of wealth: land and labor, the latent forces within Brazilian society - used to the predatory exploration of natural resources while furnishing only deplorable living conditions for the vast majority of the population - were allowed to continue with their old practices, degrading the environment and concentrating income.

On the other hand, new players and forms of social organization appeared to fill the space left vacant by government entities, putting forward new development alternatives and solutions on a more sustainable basis than those in use. The land question - looked at from agrarian and environmental viewpoints - is the stage from which these new forces within Brazilian society prefer to manifest themselves, showing, in a most expressive form, the on-going conflicts within society, in a search for recognition on the national political scene.

This situation allows us to characterize the present moment as transitional in terms of the standards to be applied to territorial development, the long-term results of which are as yet undefined, depending, as they do, on the relative strength of the forces within society that do not yet have positions of hegemony over the national and international political scenarios. These circumstances show the risks and limitations for the formulation of long-term estimates on the future of Brazil, based on past trends.

This is especially important in terms of the changes in land use and land cover , as one of the main objectives of the LUCC is exactly to understand "how land cover has been altered by humans during the last 300 years?" Also to establish " how changes in land use will affect land cover during the next 50 - 100 years?" (IHDP, 1999: 12).

These questions, when related to the situation in Brazil, may receive quick, but not necessarily true, answers. It is well known that, during these last 300 years, Brazil took its place in the world economy, and vast areas of the Atlantic Forest were consumed, cut down and burned , as described by Dean (1998), while the remaining areas continue to be threatened by urban and industrial expansion.

Rapid urban growth is also one of the main features of Brazilian spatial dynamics. Based on a highly concentrated agrarian structure, where a rural property is a source of wealth and power, the exodus from the countryside is a direct consequence of the natural population growth, as the new contingents of the population, without access to rural land, are forced to seek shelter in the towns and cities, which grow very quickly without, however, minimum conditions for people to live in urban areas.

With this background, the same destination thus awaits the Amazon Forest, the largest rainforest in the world - with more than 3.5 million km^2 within Brazil's frontiers. Brazilian Government agencies started promoting the settlement of the region in the 1960's, building highways that acted as penetration routes for settlers and new economic activities.

The opening of highways through the forest intensified in the 70's, with the building of the Transamazônica, the Cuiabá-Santarém, and the Cuiabá-Porto Velho Highways, and in the 80's, with arrival of industrialized grain production to Brazil's Central Savannas. The crescent of fire spread into the Amazon forest, aided by the concentration of land and the construction of new towns in the south of the region, leading to the apparent conclusion that the time-honored traditions of forest devastation, and the concentration of wealth and land, are inexorable in Brazil.

There are, however, signs that the historical and geographical processes are not as linear and foreseeable as the mere mechanical extrapolation of trends would have it appear. This paper will examine precisely these signs, in order contribute to a better understanding of the relationship between the spatial dynamics and the changes in land use and land cover in Brazil.

Therefore, certain 'truths' need to be questioned. The first is that Brazilian farm and beef production should continue to grow based mainly on the incorporation of ever more land to the production process. The second, that, because 8 out of 10 Brazilians live in towns, the urbanization process is close to its stabilization in Brazil. By questioning these two apparent truths, this paper seeks to clarify some of the possible trends in the evolution of the use of land for society and its effects on land cover in Brazil

GROWTH AND EXPANSION OF URBAN AGGLOMERATIONS

The term urbanization, frequently used to describe the concentration of population in housing groups, is not very adequate to describe the central characteristics of the Brazilian spatial dynamics. Urbanization involves a radical alteration in the conditions of social reproduction, in relation to the standards that exist in rural areas, and that is not exactly what is happening in Brazil with the multiplication of small agglomerations dispersed throughout the country.

To affirm that more than 80 % of the Brazilian population live in urban areas is, at the minimum, fallacious. Most of the small agglomerations have no professionals, not a doctor, a dentist, or a lawyer. Most of them have only primary schools, and only a few junior high schools, which helps to explain why the Brazilian population has one of the lowest levels of schooling in the world.

This fallacious idea serves only to hide the true dimensions of the agrarian question in Brazil. It is placed at a secondary level, as it only affects 20% of Brazilians, when in fact, a significant number of them still depend on agriculture for part of their survival. A study carried out to try to understand the real dimensions of the Brazilian urban network showed that only about 65 % of Brazilians live in what could effectively be called towns (IPEA/IBGE/NESUR-UNICAMP, 1999). This is due to different reasons:

1. The fragility of the political-administrative concept of urban population adopted in Brazil for statistical purposes, means that settlements with little more than 100 inhabitants are considered to be towns, irrespective of the functions that they carry out or the services offered to the community. This simplification gives the distorted view that more than 80 % of the Brazilian population live in towns, hiding the reality that a considerable number of Brazilians live in rural areas, though they are not exactly peasants, at least not in the classic sense of the word.

2. The accelerated growth in the number of new towns, facilitated by the Constitution of 1988, occurs in frontier regions like the north of the States of Mato Grosso and Tocantins, and in more consolidated areas, such as the Colonial Zone of Rio Grande do Sul State or the west of Santa Catarina State. This disorganized multiplication of municipalities encounters reaction in the spatial dynamics of the population, as shown in the preliminary results of the 2000 census where many 'towns' throughout the country show considerable population losses.

In Brazil, the lack of advance in terms of the division of labor between town and country, and among the different towns, means that a large part of the population, residing in small settlements, still depend on rural activities for their livelihood, either in the form of subsistence farming on smallholdings, or in the form of work under various guises, normally unsalaried, on medium and large-scale agricultural undertakings.

This weakness in the territorial division of labor results from the high concentration of wealth and income. Its most immediate result is the lack of a market of a size compatible with Brazil's population. This has profound implications for urban development, in terms of the expansion of productive activities to supply towns and cities, where excess local production, resulting from natural resources or the low salaries paid to the work force, is appropriated by the many forms of agrarian-mercantile capital, that proliferate in the interior of Brazil. The limitations imposed by the reduced social and territorial division of labor also restrict market development for urban services. As a result, there are no doctors, dentists or lawyers in the great majority of Brazilian municipalities.

The expanding number of small settlements lacking in minimal urban services reinforces the concentration of population in cities with more than one million and less than five million inhabitants, that form the nuclei of the metropolitan areas located outside the Rio de Janeiro - São Paulo axis. Today, the population residing in those two megacities is less than the sum of the inhabitants that live in the so-called "million people cities" dispersed throughout the regions of the country. These are cities such as Manaus and Belém in the Amazon; Brasília and Goiânia in the Center-West; Salvador, Recife and Fortaleza in the Northeast; Curitiba and Porto Alegre in the South, and Belo Horizonte in the Southeast.

Is should be noted that, among the cities with more than one million inhabitants, both Belém and Manaus, the metropolises of the Amazon region, have the highest population growth rates. This shows that the attraction of the great cities has reached a national scale, regardless of the geographic region, and this has important implications for land use and land cover, as these cities are exactly the ones that use most land for building, exerting strong pressure on their surrounding areas.

TABLE 1: BRAZIL

CITIES WITH MORE THAN ONE MILLION INHABITANTS IN THE YEAR 2000

City	Urban Population (Inhabitants)		Growth Rate
	1991	2000	(% p/a)
Guarulhos	777.921	1.048.280	3,37
Goiânia	913.485	1.083.396	1,91
Belém	849.187	1.271.615	4,59
Porto Alegre	1.247.529	1.320.069	0,63
Manaus	1.006.585	1.394.724	3,69
Recife	1.298.229	1.421.947	1,02
Curitiba	1.315.035	1.586.898	2,11
Brasília	1.515.889	1.954.442	2,86
Fortaleza	1.768.637	2.138.234	2,13
Belo Horizonte	2.013.257	2.229.697	1,14
Salvador	2.073.510	2.439.881	1,82
Rio de Janeiro	5.480.768	5.850.544	0,73
São Paulo	9.412.894	9.785.640	0,43

Source: IBGE, Preliminary Census Results for 2000

To evaluate the extent of the pressure exerted by urban growth on land cover and land use, 100 inhabitants per square kilometer was established as the municipal population density benchmark at or above which land use would be considered predominantly urban. This benchmark is less than the one defined by industrialized countries (150 persons/km^2) and tries to arrive at a better approximation within the very large land areas of Brazilian municipalities.

The result, which can be seen in Map 1, shows some important aspects to be considered in evaluating the role of urban agglomerations with regard to changes in land use and land cover in Brazil.

First, it should be pointed out that, even taking into account the continental dimensions of Brazil, the built-up area for urban use already covers significant areas of land originally used for farming and cattle raising. This is already flagrant in the main industrial state of Brazil, São Paulo, where the large built-up area made up of three metropolitan areas (São Paulo, Campinas and Baixada Santista) denounces what is already practically a conurbation.

This area extends to the north in the direction of Araraquara, to the west along the old Sorocabana rail track, and to the east following the valley of the Paraíba do Sul River, with Sorocaba and São José dos Campos as nuclei for pre-metropolitan growth. The role of the highways as axes for the urbanization process is clear, and the fact that the land over which this process will occur is one of the most fertile in Brazil, the red earth along the Paulista crests, is unfortunate.

In Rio de Janeiro, as well as with the expansion of the built-up area along the Paraíba do Sul Valey, in the direction of São Paulo, and constituting, according to some writers, the embryo of the Brazilian world-city, attention should also be paid to the rapid expansion along the coast. An almost continuous urban area as far as Macaé, where the support for the off-shore oil rigs at Bacia de Campos is based, and which, in the short period of less than 20 years, has become Brazil's main production facility for natural gas and petroleum processing.

The construction of the bridge linking Rio de Janeiro to Niterói, over the Guanabara Bay, opened the Atlantic side of the Fluminense Lagos Region to land speculation, where first and second home buyers, working in the Rio de Janeiro metropolitan area, could have their houses and apartments. The pipelines and the intense movement of heavy vehicles associated with the oil industry must also be added, together with the construction of nuclear power stations in the south of the state, and the construction of Sepetiba Port, in the bay of the same name. All these transformed this urban agglomeration into one of the areas at greatest environmental risk in Brazil, and exert strong pressure on the remaining areas of Atlantic forest (Mata Atlântica) in this zone.

The land cover and land-use in this coastal zone are also presently under pressure from the urbanization process in the southern region. Urban advance is easily verified between Curitiba (PR) and Porto Alegre (RS), along the Itajaí Valley and Florianópolis (SC). The Curitiba metropolitan area, an alternative to the now less economically attractive metropolitan area of São Paulo, has been receiving industrial investments, which have transformed the smaller towns around it, e.g. São José dos Pinhais, in new industrial centers for metal working and tool making.

The indirect effects of Mercosul integration can also be felt, for example at

Paranaguá Port, which has become the main seaport for soybean exports from Paraguay. The increase in value of the coastal region of Santa Catarina, today one of the most sought after tourism and leisure time destinations for Argentines, who buy houses and seek out the region during the summer months, intensifies the building process along the sea coast.

The metropolitan area of Porto Alegre, which, due to the emigration of local farmers to plant soybeans in the savannas of central Brazil, has seen a reduction in its growth rate. The State of Rio Grande do Sul has also been affected by the building of metal and tool working facilities for the General Motors assembly plant at Gravataí, and by Mercosul, with the significant increase in the volume of road freight crossing the state along its main highway. Within the state, other important agglomerations are also growing - Santa Maria and Pelotas/Rio Grande, the latter due to investments made in the port terminal, where they hope to have a role as a marine transportation hub for containers and bulk cargo in the southern cone of the Americas.

Still in the south of Brazil, the urban growth of agglomerations, such as Londrina, Maringá, Ponta Grossa and Cascavel, in the interior of Paraná, should be noted. These cities have always had, and continue to have, an important role in supporting farming and cattle raising, but today their dynamic economies are essentially urban based and advance over fertile land, originally used by agriculture.

A similar process can be seen in Minas Gerais, especially in the so-called Mineiro Triangle, where cities such as Uberlândia and Uberaba have increased their urban areas to receive industrial and service activities transferring from São Paulo. Belo Horizonte is heading in the same direction, aggravated by the impact of iron mining, and has been transformed, in a short period of time, into a metal-working and tool-making industrial complex, due mainly to the installation of the FIAT assembly plant in Betim in the middle of the 1970's.

Goiânia, Anápolis and Brasília form a new and dynamic urban complex in the heart of the Brazilian Central Plains. Goiânia, founded in 1942, and Brasília, in 1960, are recent events in the history and geography of Brazil. However, together, they add up to more than three million inhabitants and show a strong trend towards maintaining this growth. Located on flat land, without obstacles of any note to impede urban growth, the spatial pattern tends towards the creation of new nuclei around their peripheries, generally housing lower-income workers, and that expand very quickly over the savannas.

In the Northeast the trend has always been to occupy the coastal regions, although there are some signs of growth towards the interior of the region, in the direction of Feira de Santana in Bahia, towards Caruaru in Pernambuco, and Campina Grande in Paraíba. The three metropolitan areas of the Northeast, Salvador, Recife and Fortaleza, are the main centers for population growth and urban expansion in the region, although, as can be seen on the map, they have distinctly different layouts.

Salvador expands along the coast, following the curves of Todos os Santos Bay and branches out in the direction of Feira de Santana in the interior. There are no other important urban centers in its immediate area of influence, the nearest state capital, sparingly populated Aracaju, being situated some distance away.

Recife, on the other hand, is located inside an area of much greater urban density, in the center of the forest region of the Northeast (Zona da Mata), part of the remaining Atlantic forest (Mata Atlântica). Centers such as Natal, João Pessoa, Campina Grande and Maceió are also located in this region, together with various other agglomerations of lesser importance in the immediate surroundings.

Fortaleza is a metropolitan area that has shown rapid urban growth over the last decades, but its development has not gone beyond neighboring municipalities, due to, in great part, the semi-arid nature of this stretch of the Brazilian coastline. The limitations imposed by the lack of water restrict the growth of the urban area and today act as a brake on the continuance of a high urban growth rate. In the Northeast, the availability of water for human consumption is a decisive factor for urban growth. Metropolitan areas such as Fortaleza and Recife are suffering directly from water shortages, the first depending on water from the Orós Reservoir, located some hundreds of kilometers away, in the middle of the Northeast's semi-arid zone, where water is a scarce and fought-over resource.

The expansion of the urbanized area in the Amazon is as yet mainly restricted to the regional capitals. However, Belém and Manaus were the metropolitan areas that grew most between 1991 and 2000, showing that the process that quickly transforms cities into metropolitan areas is already advanced in this resource frontier. As well as these metropolitan areas, special attention should be given to the expansion of Imperatriz and Marabá, which act as centers for growth in the eastern Amazon region. This area, where logistical networks are being implemented, is speeding up the land use and land cover changes on the eastern edge of the Amazon Forest.

Thus, even though the dimension of the urbanization phenomenon is not exactly that described in government statistics, the expansion of urbanized areas in Brazil is a real problem. It exerts pressure on natural resources, principally land and water, both of which Brazilians devour voraciously, and which, due to the lack of efficient liquid and solid waste treatment systems, are subject to pollution and contamination on a scale that has reached national proportions.

THE EXPANSION OF FARMING AND CATTLE RAISING

In historic terms, the expansion of farming and cattle raising has been responsible for the main alterations in land use and cover in Brazil. The new frontiers, already well described in Brazilian scientific publications, constitute the main cause for the increase of cleared areas for agricultural and cattle raising. The advance of agriculture over areas of natural forest, together with extensive cattle raising on open lands and savannas, lead to the cutting down and burning of natural forests. These have been the principal causes for the loss of natural vegetation cover and for the large-scale changes in land use in Brazil over the last 50 years (Map 2).

The agricultural system based on the practice of slash-and-burn, together with large-scale cattle raising, continue to be very large consumers of land and of its natural fertility. More than this, the inclusion of new land had an important role during the industrialization process, guaranteeing the supply of wage goods, mainly food, required to maintain the low real-wage levels of city workers, and avoid social pressure after land reform and land distribution.

This movement continues today. However, there are signs that a profound transformation of the Brazilian social structure, in relation to the dynamics of land, is underway. This transformation requires more study if the real meaning of its dimension and how deeply it will impact on the way in which Brazilians use land, are to be better understood.

TABLE 2: BRAZIL - MAIN USES OF LAND IN FARMING AND CATTLE RAISING - 1975-95
(IN THOUSANDS OF HECTARES)

Categories	1975	1980	1985	1995
No. of farms/ranches	4,993,247	5,159,851	5,801,809	4,859,865
Total area	323,896,378	364,852,419	374,924,929	353,611,246
Open area (1)	115,411,007	148,128,023	167,390,594	171,512,594
Continuously under crop	8,385,435	10,470,167	9,903,487	7,541,626
Not continuously under crop	31,615,982	38,630,167	42,244,221	34,252,829
Fallow land	2,206,316	8,617,901	10,662,715	8,310,029
Productive land not in use	30,637,470	24,794,718	24,519,143	16,360,085
Man-made pastures	39,701,465	60,600,459	74,094,402	99,652,009
Man-made forests and woodlands	2,864,339	5,014,611	5,966,626	5,396,016
Native grasslands	125,950,905	113,895,395	105,094,029	78,048,463
Native forests and woodlands	67,857,692	83,150,078	83,016,973	88,897,582
Unusable land	14,676,774	19,678,923	19,423,333	15,152,607

(1) The sum of permanent and temporary crop lands, fallow land, planted grasslands, and unused productive land.

Source: IBGE, Agricultural Census -1975, 1980, 1985 and 1995/96

Today, however, there are indications that the largest part of the products that make up the basic food needs of the Brazilian population are supplied by industrialized food companies, and that the process for acquiring unoccupied government land, has led to the relative sealing off of the resource frontier. This movement has led to an inversion of the historic trend with ever growing numbers of farms/ranches and the areas occupied by them (Table 2).

At state level, in terms of the total area occupied by farms and cattle ranches, the trend is towards stabilization, and even decreasing in most Brazilian states, with the exception of Mato Grosso, Rondônia and Pará, where the area occupied advances towards the Amazon.

Graph 1 shows that, while the total area occupied by farming and cattle ranching has fallen, newly cleared areas, that is, land that has suffered changes to its original cover, continue to grow, although at a slower speed than during the previous decade.

GRAPH 1: BRAZIL - OPEN AREA FOR AGRICULTURAL USE - 1975/1995
(IN THOUSAND OF HA)

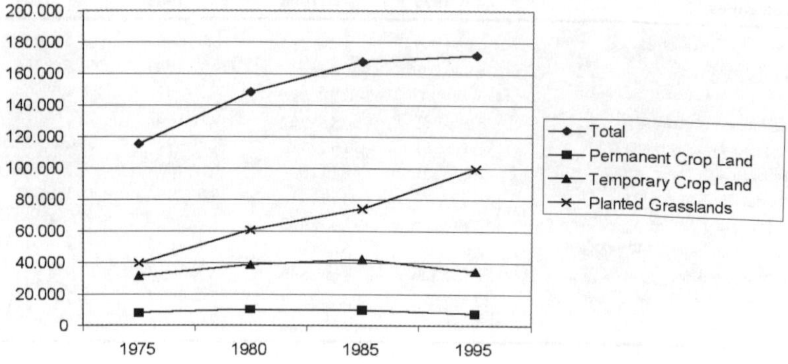

Source: IBGE, Agricultural Census of 1975, 1980, 1985 and 1995/96

The increase in the area of planted grasslands was the principal reason for this expansion, as shown in the graph. However, it is also clear that starting in the early 80's, there has been a separation of trends between farmland and pastures. While the area of pastures has continued to expand, especially after 1985, farmland has done the opposite, with a decline in land given over to this use starting at about the same time.

Two processes help to explain this inversion in the spatial dynamics of agriculture. The first is associated with the expansion of urban areas, including urban perimeter zones and urban activities in rural areas, destined not only for building, but also for uses associated with the expansion of towns and cities, such as leisure and recreation areas. This process, as described in the previous item, has been intense in the Southeastern and Southern regions and on a lesser scale in both the Northeast and Center-West.

The second refers to the increased returns obtained from the main crops planted in Brazil, resulting from new farming techniques. Looking at only the five main crops produced in Brazil: corn, soybeans, rice, sugar cane, and coffee; there has been a reduction in the area under cultivation from 33,083 to 29,100 thousand hectares between 1985 and 1995/96. Without taking into account any peculiarities of the crops, and the lack of credit for agriculture in 1995, all of them show increased productivity during the period in question, especially the grain crops, i. e., corn, soybeans and rice. In the case of corn, production increased 43.5% between the two censuses, at the same time that the area under cultivation decreased by 11.9%. Production of soybeans increased by 29,4%, with practically no increase in the area under cultivation from 1985 to 1995/96. With regard to rice, which is not grown for export, while the total of rice produced decreased by about 10%, the area under rice cultivation shrank by two-thirds from what it was in 1985 (Table 3).

TABLE 3: BRAZIL - PRODUCTION, AREA UNDER CULTIVATION AND TON/HA FOR THE FIVE MAIN CROPS: 1985 - 1995/96

Crops	Production (tons)		Area cultivated (ha)		Ton/ha	
	1985	1995/96	1985	1995/96	1985	1995/96
Corn	17,774,404	25,511,987	12,040,441	10,603,646	1,476	2,406
Soybeans	16,730,087	21,650,696	9,434,686	9,478,823	1,773	2,284
Rice	8,986,289	8,076,751	5,173,330	2,989,209	1,737	2,702
Sugar cane	229,882,037	259,806,703	3,798,117	4,216,427	60,525	61,618
Raw coffee	3,700,004	2,838,195	2,636,704	1,812,250	1,403	1,566

Source: IBGE, Farming and cattle raising censuses from 1985 to 1995/96

With regard to cattle raising, even though the number of cattle in Brazil is second only to India, the rate at which they are slaughtered is still relatively low, showing it to be semi-extensive, i.e., this activity requires ever more land for it to develop.

The growth of cattle raising in the North and Center-West regions has transformed land for farming and cattle raising. At the same time that Brazilians consume a lot of beef, the country has increased its participation in the world beef market. As shown by Michels (2000), Brazil is the only country in the world that has three different control zones for foot-and-mouth disease. The North, the Northeast and part of the Southeast are in areas contaminated by the disease, the Center-West and São Paulo are considered free of foot-and-mouth disease with vaccination, and the South is defined as free without vaccination.

The expansion of cattle raising was given only a paragraph in the text on Sustainable Agriculture for Brazil's Agenda 21, while grain production was amply covered. It is interesting to note that, in a country with a pastoral culture, as described originally by Sodré (1941), even though farming has advanced, cattle raising for beef and milk production continues to occupy more land throughout the country. In the Southeast, of the 841,000 rural establishments identified in 1995-96, 541,000 raised cattle for beef or milk on native grasslands or on man-made pastures, covering 37.7 million of the 73.2 million hectares belonging to agricultural establishments in the region. In Minas Gerais, for example, native grasslands and man-made pastures cover almost 62% of the area of the state (see: IBGE, 1998).

Graph 2 shows that the expansion of cattle livestock has been vertiginous in the Center-West and in the North Regions. This expansion came about as a result of the availability of land suitable for both native grassland and man-made pasture, the incentive policies of the 1970's (rural credit, new highways); and by the symbiosis with lumbering. A consequence of this was the clearing and burning of dense forests to form new pastures, the main ecological problem of the Amazon. Cattle raising in the area concentrates on beef production and the raising of buffalo, introduced in 1982 to the island of Marajó. In 1994, 11% of all the cattle in Brazil were in the North; and when Mato Grosso State is included, this number rises to 20%. Mato Grosso has the fourth largest number of cattle of any Brazilian state.

GRAPH 2: BRAZIL - CATTLE LIVESTOCK BY REGION - 1975/1995 (IN MILLION)

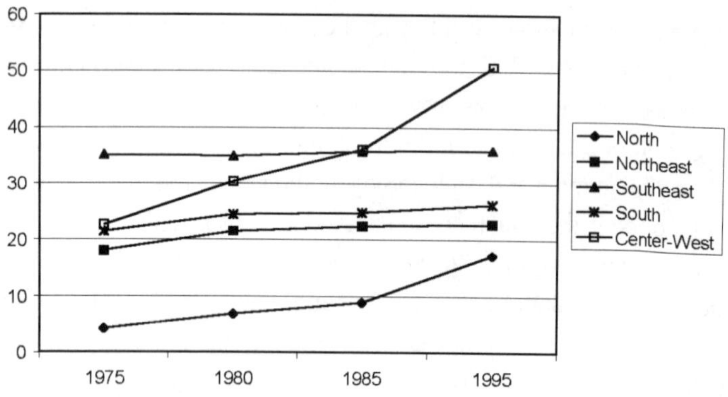

Source: IBGE, Agricultural Census of 1975, 1980, 1985 and 1995/96

Cattle raising is considered by land owners to be a low risk, high liquidity business, which requires little manpower and investment, other than those related to the creation of pastures. In time-honored Brazilian tradition, this is normally done by smallholders through the use of slash-and-burn techniques to clear land for subsistence farming.

While agriculture has gone through a process that changed its technical base, mainly in order to compete in the world market, cattle raising is still in a transitional phase, with very different conditions from region to region. The possibility of increasing their competitive position in the international market is an element that could accelerate the modernization process, and is in fact being done to control foot-and-mouth disease, as the present three-zone situation within the same country is not sustainable over the long term.

LAND USE AND THE VULNERABILITY OF NATURAL SYSTEMS

For the purpose of carrying out this impact study of recent changes in land cover and land use in the main Brazilian ecosystems, the Physiographic Zones, groups of municipalities with similar environmental characteristics, were chosen as territorial units. This is justified as follows: 1) it presents a clear environmental component for spatial classification, allowing an approximation to larger Brazilian ecosystems; 2) their mesoregional dimensions, which allow you to estimate data between censuses, overcoming the chronic Brazilian problem of successive municipal divisions, which impede the building-up of historical series at this level of spatial desegregation.

The main advantage of this spatial classification is the possibility of quickly establishing relationships among the social processes and their affects on land use and land cover in the last two decades (Map 3a, 3b and 3c). A brief resume of these relationships on the most important Brazilian ecosystems are described below:

The Amazon Forest

Ecosystems in the Amazon take up a surface area of 4,005,082 km^2. Water and native vegetation still cover approximately 92 % of the area. Land use and occupation in the Amazon concentrate on plant and animal extraction, including and dominated by the lumber industry, cattle raising, and subsistence farming, as well as the cultivation of some tree and bush species with medium-to-long life cycles.

Continued use of traditional exploitation methods will lead to more deforestation, leaving the region's main environmental problem still unresolved. The large ranch owners consider deforestation and cattle raising to be activities that confirm their legitimate right over the properties; for the smallholders, cattle raising is a fast way to increase the value of the land, as it is unable to recycle nutrients after the first few years of cultivation.

The substitution of woodland by pasture leads to a build-up of phosphorous in the soil, with more erosion - as water runs off pastures 10 times faster than in the forest - leading to more intense flooding during the rainy season and less water for the rivers during the dry season.

The practice of lumbering opens up routes for land settlement and represents an increased risk from fires set to renew pastures, or for other agricultural practices. This new phenomenon is introducing fire as part of the evolution of the landscape in populated areas of the Amazon. The dry, recently cleared areas around the forest are burned every year for planting and most of the heat sources are found there. However, this dryness, resulting from the clearing of the native forest, brings the possibility for disastrous fires such as the one in Roraima in 1998.

The introduction of grain crops to the Amazon is a new thing for a region that has always lived from the extraction of the forest's wealth, and more recently from large-scale cattle raising. The leading culture for this new model is the soybean, which, together with rice and corn, have conquered the savannas of the Amazon region and now advance to the north along roads cut through the forest. The largest grain plantations are in the State of Mato Grosso, mainly on the Parecis Plain, where the weather and topography are especially favorable for mechanized farming. The high productivity obtained in the Pre-Amazon region is unlikely to be repeated at competitive costs in the forested areas, due to the high rainfall, which facilitates the proliferation of insects and fungus, and to the high acidity of the leached forest soil.

The Savannas *(Cerrados)*

The savannas cover some 2.04 million square kilometers (about 23% of the area of Brazil), occupying a large part of the country's central region, bordering on all the other main ecosystems in Brazil. The savannas covers most of the territory of the states of Goiás, Tocantins, Mato Grosso and Mato Grosso do Sul, and also the Federal District.

The savannas are among the largest land areas still available in the world and could in the short term increase grain production and supply grazing land to meet present demand. The region has thus become a new frontier in agricultural terms, with high settlement growth rates.

This intensive exploitation of the savannas has led to various types of environmental problems. The rapid expansion of farming and cattle raising that is happening in the region is causing large-scale environmental damage: erosion and compacting of soil, chemical contamination of waterways and the 'biota' by the use of pest controls, unsuitable irrigation techniques, and slash-and-burn clearing.

The use of technologies designed for different environments, such as the irrigation of the so-called highlands using intensive, pivoted irrigation systems, require water volumes which the savannas are, at the moment at least, unable to supply. Questions are being asked about the use of this type of equipment on the savannas, as when used improperly, they can cause the water table to go down significantly and generate serious future water supply problems, including for human consumption. The pivot method is also not very efficient as it causes high water losses due to dispersion.

The slash-and-burn clearing method is used mainly to create large areas of pasture for cattle. During the preparation of the pastures, which are a single crop, various instruments are used intensively, such as soil correction, fertilizers, herbicides, pesticides and a lot of heavy machinery. The immediate effect is to make the ecosystem poorer through the loss of native species of vegetation, creating the conditions for the proliferation of pests, insects and weeds.

Technological innovations incorporated into the production systems for soybeans, corn and rice have led to new forms of production in the savannas. The value of the crops has been enhanced by the industrialized production of chaff, feed-cake and vegetable oil, which enter the economy, leading cattlemen to modernize their practices through the use of balanced feeds, made from chaff and grain residuals, which also affect pig and chicken farming.

Production of soybeans in the savannas rose, by the end of the 90's, to about 4% of total production in Brazil, with higher than national average productivity rates, especially in Mato Grosso. Corn production is also important in the region, with productivity rates 47% above national average (3,476 kg/ha *versus* 2,362 kg/ha), as seen in Goiás State. Cotton and rice are also important crops in the region, being responsible for up to 23% of the total produced in the country, with productivity higher than the national average.

Cattle raising in the region has also experienced improvements due to new technology, increasing its participation in national production. In 1970, the Center-West had17.2 million head of cattle, growing to 39.6 million in head in 1985, and to 52.2 million in 1993, or 34% of all the cattle in Brazil in that year. Cattle raising is to be found all over the region, although increased numbers and more intensive methods are more frequently seen in the eastern part of the region, especially in Mato Grosso do Sul, along the border with São Paulo, and in the central part of Goiás.

The Caatingas

The Caatingas or the deciduous forests of the semi-arid Northeast cover an area of 939,391 km2. Approximately 60% of that area is covered by native vegetation that has been altered to some degree. This degradation - which must affect more than 50% of the 'natural' areas - continues to be caused by the production of firewood; by overgrazing; and sometimes by fire. It should be noted that, as the herb cover disappears during the dry season - different from the savannas - fire is not used to improve natural grassland. Man's use and occupation of land is mainly for farming, using short-cycle methods, and cattle raising. Medium and long-term methods are only used in a few locations and are not representative. Fruit and grain is produced in certain irrigated areas. Mining and the areas destined for regional infrastructure represent only a very small part of the region.

The agriculture practiced in this part of the semi-arid was organized around the cattle-cotton-food crop complex in effect up to the middle of the 1970's. Labor productivity has always been low, although the population that took their living from the activities of this complex, such as smallholders, residents and workers with no land title, managed to survive, making their living from cotton and food crops. Any disorganization of these activities, such as the drought of 1979-83 or the *bicudo* pest that has attacked cotton plantations since 1980, and agriculture in the Northeast loses one of its main sources of income and employment, and on which those categories of rural workers depend.

Modern agriculture, while expanding in the semi-arid region, has not yet attained sufficient volume or area to be an option for the workers that are losing their source of income because of the gradual extinction of the activities of the cattle-cotton-food crop complex. The attempts to introduce higher production varieties of herb cotton, a single crop culture, have not been successful. The cultivation of fruit, which has started to appear in the interior, around the few irrigation projects - public and private - are an alternative of great importance. It is, however, still restricted to small extensions of the irrigated lands along the humid valleys of the semi-arid region, and requires capital and a qualified workforce, both of which are in short supply among the workers from the areas of the cattle-cotton-food crop complex. Cattle raising, restricted for technical reasons to farms of at least 200 ha, but preferably with more than 500 ha, continues to be carried out using traditional methods.

Environmental degradation of the semi-arid region has been happening for

MAP 1: BRAZIL - URBAN LAND USE - 2000

Source: FIBGE: Demographic Census of 2000 - Preliminary Results
Organized by Claudio A. G. Egler

MAP 2: BRAZIL - AGRICULTURAL LAND USE - 1995

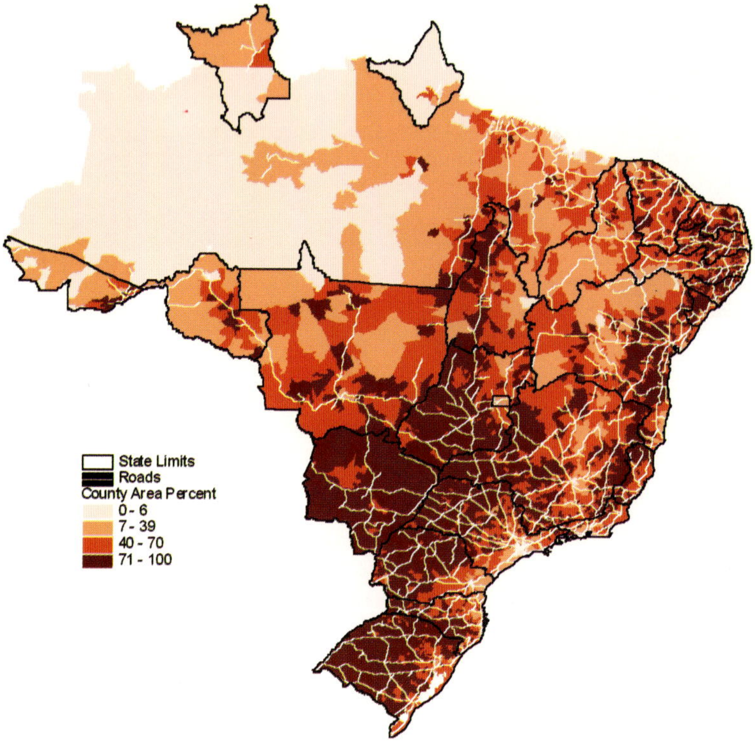

Source: FIBGE: Agricultural Census of 1995/96
Organized by Claudio A. G. Egler

MAP 3A: BRAZIL - ECOSYSTEMS BY GROUPING PHYSIOGRAPHIC ZONES

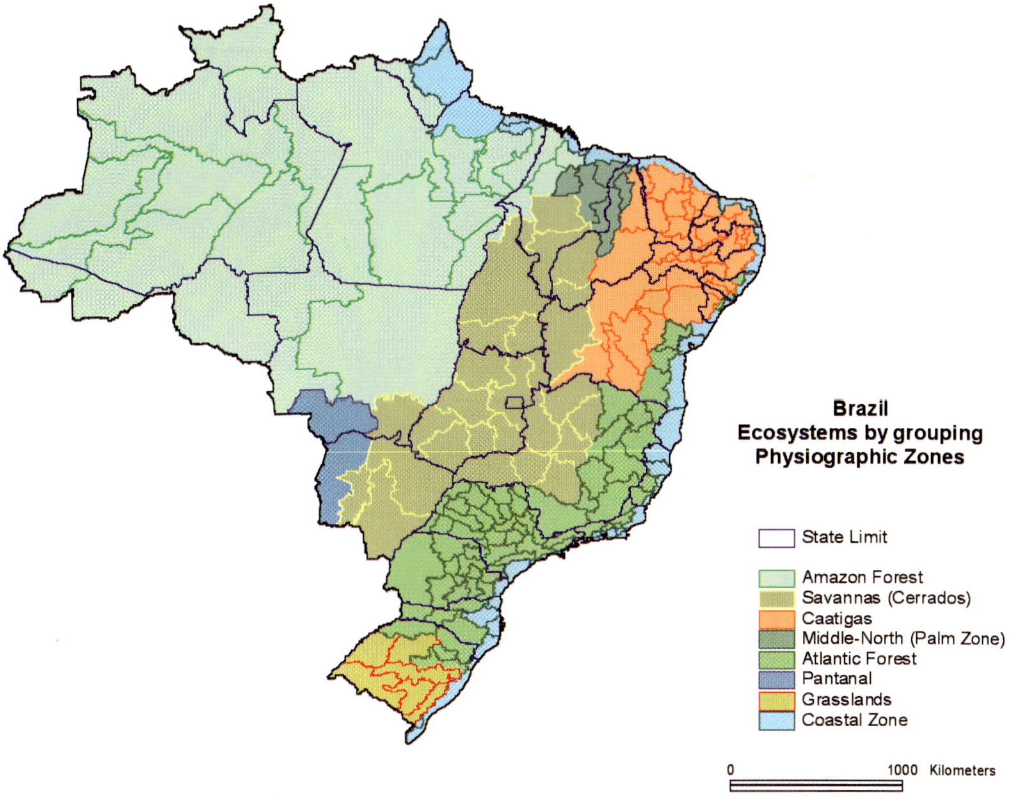

MAP 3B: BRAZIL - DOMINANT AGRICULTURAL LAND USE BY PHYSIOGRAPHIC ZONES - 1975

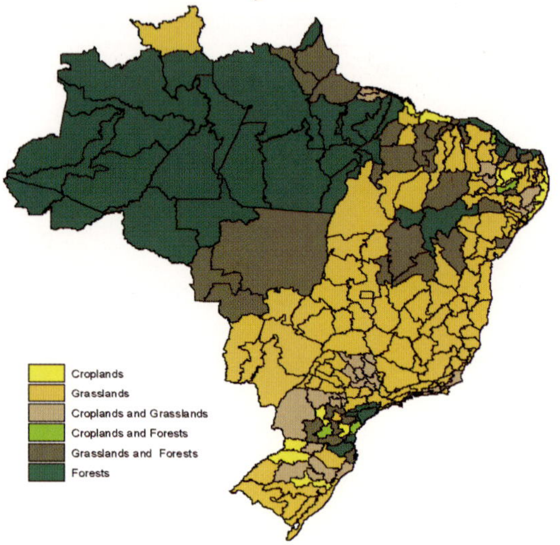

Source: FIBGE: Agricultural Census of 1975, 1980, 1985 and 1995/96
Organized by Claudio A. G. Egler

MAP 3C: BRAZIL - DOMINANT AGRICULTURAL LAND USE BY PHYSIOGRAPHIC ZONES - 1995

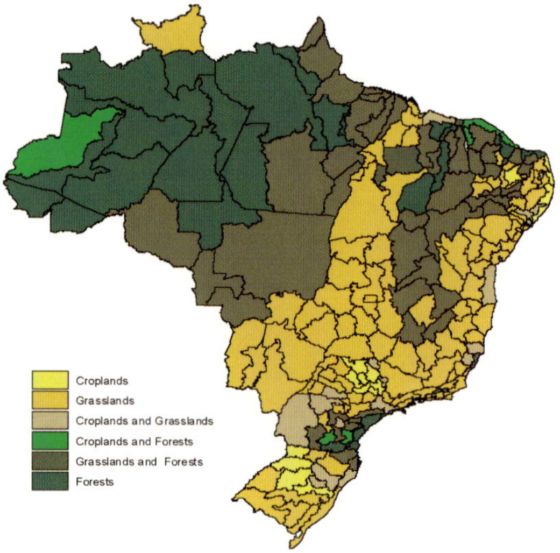

Source: FIBGE: Agricultural Census of 1975, 1980, 1985 and 1995/96
Organized by Claudio A. G. Egler

a long time, due mainly to the continued extensive cutting down of tree cover and the resulting soil erosion. The semi-arid region has lost its capability to retain water in the soil or sub-soil. Consequently, the runoff of rainwater, year after year, feeds the process whereby the remaining fine layers of soil are swept away. The small streams and rivers used to run, during a few months of the year, with water from the water tables that were replenished during the rainy season. They no longer exist.

A number of factors contribute to this degradation. Among them, the high population density of the semi-arid Northeast, resulting in the occupation of large areas by subsistence farming, with even greater areas being used to raise cattle, sheep and goats, with no preoccupation with the need to conserve soil and water. This shows clearly the lack of sustainability. The situation tends to get even worse as the local populations continue to ignore the need to conserve soil and water in the Drought Polygon (Polígono das Secas).

The Pantanal

The Pantanal region takes up an area of 165,000 km^2. Native vegetation covers more than 97% of the area, already partly altered by cattle raising and improvements made to the native pastures. Land use and occupation is almost exclusively related to cattle raising; at a secondary level, fishing and the (not always legal) capture of wild animals, tourism and leisure activities, and mining, which is found only at Corumbá. The productive system in the Pantanal works with those in adjoining areas. As a counterpart, the use and occupation of these adjoining lands have a negative effect on the ecosystems of the Pantanal. (MMA, 1995.)

The Pantanal has certain peculiarities that characterize it as a specific sub-region, and is considered to be the largest wetland on the planet. The sub-region extrapolates Brazil's frontiers, and occupies an area equivalent to 165,000 km^2, of which 130,000 km^2 are in Mato Grosso do Sul and 35,000 km^2 in Mato Grosso. It is one of the largest natural fish-breeding areas in the world, with its own geomorphological formation, and has the largest fauna of the Americas. Its large expanses of water are a traditional fishing area, attracting sportsmen from around the world.

Economically, the region is a cattle area, using intensive and predatory methods. Although this activity has, in a way, become adapted to the environment, the indiscriminate increase in the number of cattle and their constant movement have unbalanced the regional ecosystem, based on alternating floods and dry periods.

The Atlantic Forest *(Mata Atântica)*

The agricultural exploitation of the group of ecosystems that make up the Atlantic Forest intensified at the start of the last decades of the 17th.C and beginning of the 20th.C. During this period, seeds were planted on the ashes of the recently burned forest, dispensing with the need to plough the soil, to weed, or to use organic or chemical fertilizers. This soil gave a very good return for about two or three years, after which time the area was left fallow. A secondary natural growth would then cover this land and some years later, it too would be burned to give way to crops for two or three more years. However, the increasing population and the resulting demand for crops forced a reduction in the 10 year period land was left fallow, with the result that the forest didn't have time to grow back.

The limits of this type of productive system, where the recovery of the soil depended, fundamentally, on land rotation, became more and more evident. At the end of the 1960's, the use of chemical fertilizers rejuvenated the tired soil, and later, with the arrival of the Green Revolution, a homogeneous group of new technological procedures were put in use. This substitution at the technical base of farming and cattle raising, permitted the implementation of large-scale, single-crop systems, starting a period of radical change in the South and Southeast of Brazil. This modernization process was also helped by the existence of subsidized credit for agriculture, and by very large investments by the public sector in research and teaching in the field of agronomy. Improvements in international agricultural processes also helped, as did the so-called 'miracle' period for the Brazilian economy. During the initial years, the new technologies resulted in surprising increases in yields for almost all crop types. However, the euphoria of the large yields was undermined by the social and environmental problems that, even today, are evident in the agriculture practiced in these ecosystems.

Another characteristic of the end of the 20th century is the advance of agribusiness in almost all the South-Southeast regional complex. The production of soybeans, sugar cane, oranges, coffee, corn and fowls and pigs are among the most important examples of this process - not only because of changes they brought to the farming community, but also because of their very strong dependence on chemical, mechanical and genetic industrialized inputs, and the fact that most of the production is now destined for industrialization. These same 'giants' of agribusiness are also major 'consumers' of the natural resources of the ecosystems of these regions and in some cases are responsible for environmental depletion on a similar scale. (Brasil, MMA, 2000)

The Coastal Zone

The original vegetation cover, mainly mangroves and the primitive vegetation found on sand shelves, was seriously affected by human occupation, especially considering that 45% of the Brazilian population lives, works and plays in the Coastal Zone.

There is a considerable concentration of production and power equipment in industrial zones and centers. The association of power generating plants with specialized terminals and industrial plants increases significantly the risk of accidents, as well as long-term exposure of the population to toxic substances in the water and air.

There is a major concentration of industries along the Coastal Zone, with oil and natural gas fields, terminals and pipelines, thermoelectric and nuclear power generators, and large concentrations of chemical and other plant. Accidents involving oil spills and leakage of gas and toxic effluents are recurrent events along various stretches of the Brazilian coastline, with serious results, affecting people's lives at every level.

FINAL CONSIDERATIONS

Data regarding Brazil's role in the generation of so-called greenhouse gases shows that more than 70% come from the burning of biological material, with a significant part originating from changes in land use and land cover.

This paper has sought to show that the territorial standards behind these changes are in transition, and that its long-term results are as yet not very clear.

This transition can be described as a double movement. On one hand, there is the spread of urbanized areas - the most densely built-up occupy vast areas of arable land, a situation typical of mature economies. On the other hand, there is a cooling-off in the movement to incorporate new land for farming and cattle raising, due, in large part, to the gains in productivity made in both agriculture and cattle raising.

It is, however, premature to state that these trends will come to full consolidation in the future. In like manner, it is not useful to build scenarios for future changes to land use and land cover in Brazil, based only on past estimates. The lack of definition, typical of a transition period, makes the building of reliable estimates very difficult, but, in compensation, it affords numerous opportunities for conscious action.

BIBLIOGRAPHY

AB'SABER, A. N. (1992). Domínios da Natureza no Brasil: ordens de criticidade. O caso da Amazônia. In J. P. d. R. VELLOSO (Ed.) *A ecologia e o novo padrão de desenvolvimento no Brasil.* São Paulo: Nobel. p. 167-184.

ALVES, D. S. (1999) An Analysis of the Geographical Patterns of Deforestation in Brazilian Amazonia in the 1991-96 Period. INPE, mimeo.

AMIGOS DA TERRA, IMAFLORA E IMAZON (1999) *Acertando o Alvo.* Consumo de Madeira no Mercado Interno Brasileiro e Promoção da Certificação Florestal. São Paulo. mimeo.

ANDERSON, A. (1990) Deflorestación de la Amazonia: Dinámica, Causas y Alternativas. In A. ANDERSON (Ed.) *Alternativas a la Deflorestación*. Quito: Abya-Yala. p. 13-43.

ANDERSON, A. (Ed.). (1990). *Alternativas a la deforestacion.* Quito: Abya-Yala.

ARNT, R. A. (Ed.). (1994). *O destino da floresta. Reservas extrativistas e desenvolvimento sustentável na Amazônia.* Rio de Janeiro: Relume-Dumará.

BECKER, B. K. (1995) A Amazônia nos Cenários para a Gestão Ambiental. In: MINISTÉRIO DO MEIO AMBIENTE DOS RECURSOS HÍDRICOS E DA AMAZÔNIA LEGAL. *Os Ecossistemas Brasileiros e os Principais Macrovetores de Desenvolvimento.* Subsídios ao Planejamento da Gestão Ambiental. Brasília: MMA.

BECKER, B. K. (1999) Amazônia, Fronteira Experimental para o Século XXI. In: BECKER, P. (org) *Bioética no Brasil.* Rio de Janeiro: Espaço e Tempo.

BECKER, B. K.; EGLER, C. A. G. (1992). *Brazil; a new regional power in the world-economy.* Cambridge: Cambridge University Press.

BECKER, B. K.; EGLER, C. A. G. (2000) Sustentabilidade e Desenvolvimento Regional da Amazônia. Opções Estratégicas para a Dimensão Ambiental do Plano de Desenvolvimento da Amazônia (PDA- 2000-03)., Belém: SUDAM. mimeo.

BELTRAO, V. d. A. , & LAMOUR, C. (1984). *Uso atual e potencial dos solos do Nordeste.* Recife, PE: SUDENE.

BRASIL, MMA, Ministério do Meio Ambiente, dos Recursos Hídricos e da Amazônia Legal (1995). *Os Ecossistemas Brasileiros e os Principais Macrovetores do Desenvolvimento.* Subsídios ao Planejamento da Gestão Ambiental. Brasília: MMA.

BRASIL, MMA, Ministério do Meio Ambiente, dos Recursos Hídricos e da Amazônia Legal (1996). *Macrodiagnóstico da Zona Costeira do Brasil na Escala da União.* Brasília: MMA.

BRASIL, MMA, Ministério do Meio Ambiente, dos Recursos Hídricos e da Amazônia Legal (2000). Agenda 21: *Agricultura Sustentável.* Brasília: MMA.

CARVALHO, R. (1999) A Amazônia Rumo ao "Ciclo da Soja". *Amazônia Papers,* v.1, n.2, Amigos da Terra Programa Amazônia.

DEAN, W. (1996) *A ferro e fogo. A história e a devastação da Mata Atlântica brasileira.* São Paulo: Companhia das Letras.

DOUGLAS, I. (1994) Changes in Land-Use and Land-Cover: Human Settlements. In: W. B. MEYER; B. L. TURNER (Eds.) *Changes in Land Use an Land Cover: a Global Perspective.* Cambridge, UK: Cambridge University Press. p. 149-169.

EGLER, C. A. G. (1995) *Os impactos da política industrial sobre a zona costeira.* Brasília: MMA.

EGLER, W. A. (1951) Problemas Agrários do Brasil. *Boletim Carioca de Geografia,* v.4, n.2-4, p.39-61.

FEARNSIDE, P. M. (1990). Usos Predominantes de la Tierra en la Amazonia Brasileña. In: A. ANDERSON (Ed.) *Alternativas a la Deflorestación* p. 363-393). Quito: Abya-Yala.

FIBGE, Fundação Instituto Brasileiro de Geografia e Estatística (1988). *Brasil: uma visão geográfica dos anos 80.* Rio de Janeiro: FIBGE.

FIBGE. Fundação Instituto Brasileiro de Geografia e Estatística. *Censos Agropecuários* (Vários anos)

FIBGE. Fundação Instituto Brasileiro de Geografia e Estatística. *Censos Demográficos* (Vários anos)

FUNATURA. Fundação Pró-Natureza (1996). *Alternativas de Desenvolvimento dos Cerrados: manejo e conservação de recursos naturais renováveis.* Brasília: FUNATURA.

GRAETZ, D. (1994). Changes in Land-Use and Land-Cover: Grasslands. In: W. B. MEYER; B. L. TURNER II (Eds.) *Changes in Land Use an Land Cover: a Global Perspective.* Cambridge, UK: Cambridge University Press. p. 125-148.

GUSMÃO, R. P. d. (Ed.). (1990). *Diagnóstico Brasil: a ocupação do território e o meio ambiente.* Rio de Janeiro: FIBGE.

IPEA/IBGE/NESUR-UNICAMP (1999). *Caracterização e Tendências da Rede Urbana no Brasil.* Campinas, SP: IE-UNICAMP.

LUCC, Land-Use and Land-Cover Change (1999). *Implementation Strategy.* Stockholm: IGBP, International Geosphere-Biosphere Program.

MAGALHAES, A. R. (Ed.). (1994)..*Projeto ARIDAS: políticas de desenvolvimento sustentável no Nordeste Semi-árido: documento final de trabalho.* Brasília: IICA-OEA.

MCNEILL, J. e. a. (1994). Toward a Typology and Regionalization of Land-Cover and Land-Use Change. In: W. B. MEYER; B. L. TURNER II (Eds.) *Changes in Land Use an Land Cover: a Global Perspective*, Cambridge, UK: Cambridge University Press. p. 55-72.

MEYER, W. B.; TURNER II, B. L. (1994). Global Land-Use and Land-Cover Change: an Overview. In: W. B. MEYER; B. L. TURNER II (Eds.) *Changes in Land Use an Land Cover: a Global Perspective* p. 3-10). Cambridge, UK: Cambridge University Press.

MEYER, W. B.; TURNER II, B. L. (Eds.) (1994) *Changes in Land Use and Land Cover: a global perspective.* Cambridge, UK: Cambridge University Press.

MICHELS, I. L. (2000) *A bovinocultura de corte brasileira e o mercado externo: regiões sanitárias e a cadeia produtiva da carne bovina de Mato Grosso do Sul.* Tese (Doutorado) Universidade de São Paulo, São Paulo.

MORAES, A. C. R. d. (1999). *Contribuições para a Gestão da Zona Costeira do Brasil.* São Paulo: HUCITEC/EDUSP.

MUELLER, C. C. (1992). Centro-Oeste: evolução, situação atual e perspectivas de desenvolvimento sustentável. In: VELLOSO, J. P. d. R. (Ed.) *A ecologia e o novo padrão de desenvolvimento no Brasil* p. 89-128). São Paulo: Nobel.

NEVES, C. F., MUEHE, D.; FIALHO, G. O. M. (1991) *Coastal Management and Sea Level Rise in Recife*, Brasil. Long Beach, CA: ASCE.

ROBINSON, J. M. E. A. (1994) Land-Use and Land-Cover Projections. In: MEYER, W. B.; TURNER, B. L. (Eds.) *Changes in Land Use an Land Cover: a Global Perspective.* Cambridge, UK: Cambridge University Press. p. 73-92.

ROCKWELL, R. C. (1994). Human Driving Forces: Culture and Cultural Change. In: MEYER, W. B.; TURNER, B. L. (Eds.) *Changes in Land Use an Land Cover: a Global Perspective.* Cambridge, UK: Cambridge University Press. p. 357-382.

ROSS, J. L. S. E. A. (1995). *Plano de Conservação da Bacia do Alto Paraguai.* Brasília: PNMA.

SERRÃO, E. A., & TOLEDO, J. M. (1990). La Busqueda de Sustentabilidad en Pasturas Amazonicas. In: A. ANDERSON (Ed.) *Alternativas a la Deflorestación.* Quito: Abya-Yala. p. 303-336.

SODRE, N. W. (1941) *Oeste : ensaio sobre a grande propriedade pastoril.* Rio de Janeiro: José Olympio.

UHL, C., & NEPSTAD, D. (1990). Perturbaciones Naturales y Antropogénicas en la Amazonia. In A. ANDERSON (Ed.), *Alternativas a la Deflorestación.* Quito: Abya-Yala. p. 45-76.

VALVERDE, O. (1979). A organização do espaço na faixa da Transamazônica: 1. Rio de Janeiro: FIBGE.

VALVERDE, O. (1989). *Grande Carajás : planejamento e destruição.* Rio de Janeiro: Forense Universitária.

VELLOSO (Ed.), *A ecologia e o novo padrão de desenvolvimento no Brasil*. São Paulo: Nobel. p. 79-88.

VELLOSO, J. P. d. R. (Ed.). (1992). *A ecologia e o novo padrão de desenvolvimento no Brasil.* São Paulo: Nobel.

WILLIANS, M. (1994). Changes in Land-Use and Land-Cover: Forest and Tree Cover. In: MEYER, W. B.; TURNER, B. L. (Eds.) *Changes in Land Use an Land Cover:* a Global Perspective. Cambridge, UK: Cambridge University Press. p. 97-124.

Urbanization and Global Environmental Changes

HENRI ACSELRAD
IPPUR/Federal University of Rio de Janeiro (UFRJ)

INTRODUCTION

Urbanisation processes affect ecological dynamics as a result of the town spatial distribution patterns, distribution patterns of the activities inside the towns and the expansion of the urban areas. The spread of these areas causes changes to the vegetation and encroaches on mangroves, makes soils impermeable, shifts land along the banks of water courses and may occupy valley floors. In its relation with the so-called global environmental processes,[1] urbanisation may be a factor that changes the biodiversity by interrupting ecological corridors that link ecosystems and permit genetic flows. It also may change the climate through the effect of greenhouse gases on the sun's rays, temperature, wind velocity and direction, rainfall and humidity.[2] The impacts of urbanisation may, then, be characterised through the material configuration of towns in their relation with immediate ecological processes as described above. Yet the nature of such impacts may also be discussed on a much broader and more relational basis when considering the urbanisation process increasingly as a part of an archipelago-network of large cities, knots in a chain of multiple production flows and exchanges. These then are the urbanisation trends in the context of productive restructuring and a more financing form of capitalism, as adopted in the last decades of the 20th century: city networks moving increasingly to more centralised logistic architectures, through which all currency and information, matter and energy flows to fewer central platforms, although endowed with greater economic and territorial scope.[3] From this viewpoint, it is hard to clearly distinguish the boundaries between what would be the environmental impacts associated with the urban fact itself from those originating from the more general dynamics of actual development models from which the cities are poles radiating economic processes and environmental impacts. It may, in fact, be considered that processes are released in the main knots of the city archipelago-network, which cause evidential ecological alterations far beyond the immediate boundary of the cities, including those that could be called "biosphere impacts". For in the same way as the so-called "ecosystem peoples" and "biosphere peoples" as defined by M. Gadgil and R. Guha, it may be considered that the social forms of appropriation of the material basis for development may also be divided in "ecosystem" and "biosphere" – that is, founded respectively on the exploration of local resources or by economic agents whose power allows them to drain resources from a distance.[4] Cities, in particular, play a significant role in the "biospheric" social practices of appropriating the environment. The discussion on the environmental impact of cities necessarily raises the question of the intensity of the flows from the rural hinterland responsible for providing commodities to meet urban demand. As the flow of those resources has to be constant in order to keep the cities functioning, the "biospheric effects" may often cause a break in material flows as well as in the extra-urban cultures whose existence is linked to ecosystem reproduction where such resources are found.[5] The extraction of inputs from increasingly remote places may then cause the spatial displacement of "ecosystem" social groups to the poor peripheries of large towns, causing them to be dependent, albeit on a quantitatively less significant scale, on the "biosphere economic systems".

Similar processes may arise from the imposition of values and symbolic practices of city dwellers on extra-urban social spaces that are intended, also for environmental reasons, to be preserved. Levy et alii and Adams, for example, mention that urban-originated aesthetic concepts, when applied to landscapes that are to be transformed into parks and reserves, for environmental reasons, can have dramatic and disruptive effects on the people affected by their implementation.[6] "The dangers in keeping a continuing colonial comprehension of 'resources', 'environment' and 'Nature' – adds Willems-Braun – are directly linked to the mistaken idea of an environment separated from culture, of the countryside separated from the urban and of 'civilised' urban settlements separate from Nature".[7]

Thus, the cities, through either their material or symbolic practices, have been participating in what MacNeill, Winsemius and Yakushiji call a "shadow ecology"[8], in which the main urban centres in the economically more powerful countries are central points to "enable the western nations to straddle the ecological capital of all other nations to provide food for their populations, energy and materials for their economies and, moreover, land, water and air to absorb their sub-products and waste".[9]

Before the 17th century, urbanisation was restricted by a very specific metabolic relationship between the towns and their productive hinterlands, as well as by the possibilities of surplus extraction that would sustain the urban masses. Harvey claims that medieval towns were similar to what contemporary environmentalists would consider "virtuous bio-regionalist forms"[10]. The stability of the economies of towns depended crucially then on local metabolic systems, with their environment-sustaining quality and potential surplus production, apt to be transferred to the towns. By early 19th century, what today is called the "ecological footprint" of urbanisation on the surface of the Earth was relatively smaller. What increased this "footprint" was the wave of new technologies that considerably helped overcome spatial and temporal barriers which, until then, had set constraints against the capitalist accumulation in the cities.[11]

This was how the development of a global network of towns connected to a number of hinterlands permitted, since the mid-20th century, a process of a radically greater aggregate urban growth than achieved by each town on its own .[12] For this reason, scholars of global ecology suggest that it is important to understand the contemporary environmental and urban processes, in terms of the biosphere, without restricting analyses to individual ecosystems of nation states, but rather to endeavour to learn the links between the different types of processes, regardless of geographic conventions or mere physical distances. In the context of these eco-geographic dynamics causing consequences at multiple distances, the environmental practices of the agents may be understood from the viewpoint of extraction of remote resources. But they may also be understood as the results of decisions to locate hazardous toxic waste in a context of "global neighbourhood" relations, including practices of the NIMBY ("Not in my Backyard") attitude, by which urban elites endeavour to move away from the damaging ecological consequences of their own actions.[13] It is common knowledge that exporting undesirable environmental impacts does not eliminate them from the face of the Earth. On the contrary, Harvey recalls that the growing and ruinous inter-territorial competition (between nations, regions and cities) associated with the supposed need for them (in this case, cities) "to sell themselves" at the lowest cost for the good of the maximum mobility of capital on Earth, has only succeeded in aggravating the lack of co-ordination between the decision making processes of individual capitals and, consequently, the risks of developing uncontrollable global impacts.[14]

CITIES AND GLOBAL ENVIRONMENTAL CHANGE

The discussion on global environmental changes raises significant drawbacks relative to the separation between natural and social dimensions of environmental change. The research agenda on global change has been dominated by efforts to set up predictive-type models. For the global model that uses the theories of Natural Sciences, climate, for example, should be translated in a large number of mathematical equations to predict the complex interactions between the atmosphere, oceans and Earth's surface, based on physical principles validated by a vast and still imperfect range of observations.[15] Calbo et alii, for example, attempt to establish interactions of the global and urban aspects of the anthropogenic emissions in the troposphere chemistry. This study attempts to confront the problems of parameterised dispute between different spatial and temporal scales: while the greenhouse effect phenomenon requires working with the Earth's diameter and a 30-year period, the studies of atmospheric pollution in the cities are on much smaller scales.[16] M. L. Bell, in turn, attempts to explore the relations in the opposite direction, through the hypothesis that climate change may alter the concentrations of atmospheric pollution, considerably affecting human health. He, therefore, sought to integrate data from remote sensing, mesoscale modelling and epidemiological data.[17]

From the dialectic angle of authors such as Harvey, however, environmental change is ongoing and the question, consequently, would not be actually that of characterising a concurrent urban event for a change understood as abandoning stability, but rather in search of a correlation between the sense of the urbanisation processes and that of "environmental change". So it is that by considering the risk, including the risk of energy consumption for global environmental changes, the environmental question has become a means of expression of a widespread social malaise, consisting of a central dimension of new institutionalised spatial practices. As the concerns with the environment come from the multiple facts of social experience and are linked to each other in the formation of urban environmental agendas, a new set of rules is emerging to outline the changes towards city planning.[18] Research and debate, then, on global environmental change may be viewed as capable of providing a certain number of justifications for the options relating to alterations in the urban development model, either with regard to its energy matrix, transportation, locational pattern of activities, but also concerning the social formations underlying the configuration of the urban model. Sonja Boehmer-Christiansen demonstrated, for example, how in England, Mrs. Thatcher was converted to the environmental cause, and particularly to the global warming in 1984, when she suspected that the environmentalists were using the environmental problem to attack capitalism: "in the belief that a free scientific debate alone would offer means to overcome environmental threats" – states Boehmer-Christiansen, "Mrs. Thatcher recognised in the investment in science part of the battle against the socialist environmental policy that she understood to be repressive"[19]. Moreover, as a follower of nuclear energy and deadly enemy of the coal miners union organisations, Mrs. Thatcher could not but be attracted by moves to put an end to coal burning[20].

Some signs of effects of global climate change on cities may be found from studies on the repercussions of the climate events El Nino and La Niña on Brazil[21]. Between 1982 and 1983, El Niño caused much heavier rainfall than considered normal, rivers overflowing and flooding plains in South and Southeast Brazil. Between 1997 and 1998 higher temperatures were recorded at the beginning of summer in the South, Southeast, Mid-West and southern part of the Northeast. With regard to its repercussions on the urban environment, El Niño may be blamed for below average rainfall and, consequently, a crisis in the power supply in the North and East Amazon; for the lack of water and pollution of springs, with the consequent migration of the population to the large towns along the coast in Northeast Brazil and the over-demand for local urban services; for the deluge and floods in cities in the South. La Niña, in turn, was blamed for the abnormally abundant rainfall in the North and East Amazon and for the exceptional drop in rainfall in Rio Grande do Sul between September and February[22]. When it significantly altered the climate and meteorological patterns of large urban areas, causing alternatively drought, torrential rain, rising temperatures, then climatic phenomena on the El Niño and La Niña scale add new and inevitable conditioning factors to public policies on urban infrastructures of water supply, sanitation and energy[23]. Brandão, for example, in the case of Rio de Janeiro, claims that "the disordered growth of the city, which resulted in the occupation of areas that should have been preserved, such as the mountain slopes and various hills and riverside areas, makes the city increasingly vulnerable to anomalous climatic events, and thereby increases the risks of floods and flooding"[24].

Based on SEHAB and FIPE/USP data, Marcondes mentioned that in 1991, 24 551 homes had been built in 298 slums on water source areas in the city of São Paulo[25]. According to Figueiredo[26], in 1994, in the city of São Paulo, 808 slums were built on the banks of water courses, almost one third of them liable to be flooded. Sharply sloping land with a high erosion rate was at that time the site of 466 slums. Between 1987 and 2000, the capital of São Paulo lost 30% of its vegetation as a result of urban growth, namely in clandestine allotments of the poor population displaced from central areas because they were unable to pay the high rents[27]. Erminia Maricato states that "while the properties have negligible or no value as merchandise, illegal occupation spreads without state interference"[28]. Occupations tend then to occur in areas of no interest for nor possible economic exploitation by the real estate capital, such as hillsides, water sources, areas of environmental protection in general. As a result of a segregating urbanisation, Maricato continues, it was estimated that in 1999 20% of the population in Rio de Janeiro, Belo Horizonte and Porto Alegre, and 28% of the population of Fortaleza lived in areas of irregular occupation. In the early 1990s, FIPE/USP estimated that almost 20% of the São Paulo population lived in slums. IBGE data showed that Recife had 40% of its population living in slums[29].

By the fact that victims of climate catastrophes are, as a rule, the social groups living on the outskirts, hillsides and river banks due to the excluding urbanisation, there is little propensity for a social consensus to be made in favour of reformulating the urban planning pattern. Mike Davis already mentioned, in the case of Los Angeles, how the fact that all major floods and earthquakes in the 20th century having coincided with recessive conjunctures had contributed to concealing the socio-political extent of urban disasters: in addition to carrying on the business growth cycle without interruption, the Keynesian measures of an anti-cyclical cut, convenient in a period of recession, were widely justified on the basis of the need to recover from the damages[30].

URBAN FORM AND GLOBAL ENVIRONMENT

Urban expansion and the successive landfills that usually accompany it reduce the riverbeds and, together with the erosion of the unprotected slopes from land clearance, transform streams into raging torrents in the seasons of heavy concentrated rainfall. Many river bends disappeasr because of the straightening and channelling works that reduce water seepage, thereby contributing to increasing the flow velocity[31]. The concentration of pollutants in the atmosphere, in turn, together with the built-up density of the towns, leads to the formation of so-called "heat islands", responsible for the urban thermal discomfort[32] ; these heat islands associated with the surface being made impermeable by asphalt and concrete encourage the flooding phenomenon; heat discomfort caused by the heat island encourages, in turn, further use of air conditioners, leading to a contrast between the microclimate inside and high temperatures outside the buildings. Environmental pollution and heat discomfort tend to affect the lower income residential areas to a much greater extent, and which at the same time have few trees, are close to pollution sources and use building materials that emit more heat.[33]

Cities are, on the other hand, recognised to be major energy consumers. In OECD countries, cities consume 60% to 80% of the energy balance. Current research finds that gasoline consumption would be inversely proportional to urban density[34]. In Canada, USA and England, studies have shown that doubling the population density or homes would cause a drop of 20% to 30% in the annual number of kilometres travelled by car per person[35]. The outcome would be the predominant propensity to reduce the energy consumption by designing more compact cities. Moreover, despite the common sense suggesting that improving traffic efficiency, by creating expressways and technical enhancement of vehicles, is an effective way to reduce urban pollution levels, research performed in New York and Perth, Australia, shows that there is an inverse correlation between efficient traffic in fuel (fast track) and efficient cities in fuel (requiring less distance travelled)[36]. Although vehicles circulating in denser downtown areas use more fuel per kilometre than the urban average, they would still use relatively less fuel since they circulate less. Inversely, vehicles circulating in low density areas, although they travel less kilometres per litre of fuel than the urban average, would circulate much more, and possibly use much more fuel.

Other approaches, however, although they imply a global drop in consumption, find that the spatial concentration of residents and activities does not do away with a spatial concentration in fuel use nor, therefore, a concentration in pollution from mobile sources. In densely used areas, this would tend to cause an increase in the exposure of the populations to the hazards of atmospheric pollution. This is because energy consumption linked to the alternating migrations drops only slightly in an urban concentration, given unvarying transportation structure. The most powerful source of a fuel economy would then be the change in the method of transportation in a densely populated zone, that is, actually preferring a modal transfer towards less polluting public transport.[37]

CO_2 emissions in Brazil are basically due to deforestation, and the volume is seven times more than the emissions from an energy origin. CO_2 energy emissions in Brazil are much lower than those found in other countries due to the level of industrial production in the country and to the fact that the energy matrix is hydroelectricity-based.[38] The role that urbanisation has played in the clearing tropical rain forests in general is considered to be very small in scientific literature. In a survey carried out by researchers from the Human Ecology Department of Rutgers University on 825 articles discussing the causes of deforestation, urbanisation was the 19th of the 20 main causes.[39]

Motor vehicles are the main sources of atmospheric pollutants in Brazil. According to a study on Greater São Paulo in 1990, they contributed with more than 90% in cases of CO, HC and NOx, 64% in SOx and 40% in particulates. The quality of air in urban centres is closely associated with the public transport system, which causes a large part of all emissions and encourages individual transportation to cause environmental losses and energetic inefficiency.[40] In large towns in industrialised countries, there is a close correlation between the population density and proportion of public transport routes made by public transportation.[41] This correlation may only be found, however, in cities where the public transport system is the subject of appropriate policies and investments.

Indicators of CO_2 emission per inhabitant and energy consumption unit in Brazil are quite few when compared to more developed countries, the first indicator expressing the low level of income per capita in the country and the second demonstrating efficient use of endogenous renewable resources.[42] A study by COPPE/UFRJ shows that between 1990 and 1998 Greater Rio de Janeiro increased its methane (CH_4) emissions by 57.7%, probably as a result of the improved garbage collection and more accumulated residues on waste landfills. A slight increase was also found in CO_2 emissions as a result of the increased number of vehicles in circulation and the drop in use of alcohol as fuel. Despite the trends towards the increase in greenhouse gas emissions found in Rio de Janeiro, the city of Berlin in 1990 had four times more greenhouse gas emissions than in the city of Rio de Janeiro in 1998.[43]

A glance at the data on greenhouse gas emission by sector of end energy use shows that the transportation sector is responsible for a substantial share (estimates varying between 32.5% and 42%) of all CO_2 emissions in Brazil.[44] Nevertheless, despite its car-oriented urban model, the inhabitant/car ratio is, in Brazil, extremely high when compared with those of other producing countries. Although Brazil's automobile industry is over 40 years old and it is the eleventh vehicle producing country in the world, its inhabitant/vehicle ratio (10.3 in

1996) places it eighteenth among the producing countries, behind Mexico, South Korea and Argentina.

Heavy investments in the Brazilian automobile industry in recent years has especially increased its installed capacity, estimated at between approximately 2.5 million and 3.5 million/units a year.[45] Data for the 1990s shows that the growth rates for new car licensing in Brazil is very high compared to more developed countries.[46] Nevertheless, in 1998, only 1 535 000 vehicles were registered in Brazil, which shows that there is quite a large idle capacity in this industry.[47] The Brazilian automobile market, although small, is greatly affected by the international crises and economic conjunctures. In 1998, for example, after the Russian crisis, 400 000 vehicles less were registered than the year before.

The car industry data shows a high spatial concentration in demand for cars, which points to the possibility of constant impacts on the Brazilian cities. São Paulo city, for example, today has the largest number of vehicles in Brazil, and it is growing at an average of 6% p.a. Each day, 1 500 vehicles are added to it. The average velocity on the main highways has systematically decrease to 20 km/h in 1994. On the other hand, environmentalist NGOs estimate that around 300 children die each year as a result of the pollution from car exhaust.[48] These figures express the environmental implications from the car option in Brazilian cities.

The mass influx of cars to urban city centres has not been accompanied by alterations in the historic pattern of land occupation in the city. Urban investments, both in infrastructure and aesthetics, during the increased supply of cars in the country, have concentrated on urban nuclei. The urban concept that, in the USA, permitted the absorption of a growing number of cars in the cities, resulting in low population density suburbs and wide highways for vehicle circulation, was not repeated in Brazilian cities. On the contrary, the urban nuclei population density increased, principally through accelerated verticalisation. In some points of Rio de Janeiro, namely Copacabana, houses built less than 30 years ago have been demolished to give way to high-rise buildings. In Copacabana, with an area equal to 0.4% of the city's territory, today live 2.9% of its population.[49]

In the early 1950s, constructing buildings without a garage for the middle class demonstrated that the car had not yet become an asset in the consumer goals of this class and that it was only to become part of its way of life and social climbing project as a status symbol after the car industry started up in the country. The adaptation of the middle class dwelling space to the car was only to occur in Rio de Janeiro, for example, in the late 1950s, when the law providing for new buildings to have space reserved for car parking was approved at the end of the 1950s.[50]

The 1960s were characterised by the effort to adapt urban space of the major Brazilian cities to car requirements, therefore benefiting the classes with a higher purchasing power. The start of a highway craze in those cities meant, on the one hand, the destruction or degradation of public or residential areas, mainly in neighbourhoods that were in the way of the expressways opening up at that time. On the other, the focus on roads was also reflected in the public transport system being based almost solely on the bus in the urban outskirts, with the suppression of trams and limited use of the underground railway.[51]

In 1994, Rio de Janeiro buses carried 1.3 billion passengers[52], that is, close to the number of passengers travelling in the Paris metro. Although the major Brazilian cities had not adopted the layout of American cities to absorb cars, ownership and use of this means of transport became part of the social status mechanisms of the Brazilian middle class. In the 1970s, a study by Fiat demonstrated that a Brazilian driver drove on average almost twice as much as an Italian in a year. A car consumption pattern was thus adopted in Brazil like that of the North American modern Ford attitude, without the corresponding change in land use methods that could contribute towards preventing traffic jams and reducing the high levels of pollutant emission levels.

The urban land use method and mobility pattern characterising the Brazilian cities since the 1950s was, as shown below, closely associated with the more general pattern of spatial distribution of the activities conditioned by the specific dynamics of the Brazilian style of the so-called "peripheral Fordism".

SPATIAL DISTRIBUTION OF BRAZILIAN DEVELOPMENT AND URBANIZATION

The pattern of spatial distribution of the economic activities and cities is historically related to the different ways of the Brazilian economy's inclusion in the international trade and investment flows. By the first decades of the 20th century, the different regions in the country were not closely linked to each other, but rather connected to the overseas market. Railways and ports ensured the integration of the producing regions to the world trade flow. "Drove roads" connected the capitals to the interior: the old roads opened by the cattle herds in the 17th century were used to carry slaves and manufactured goods to the gold mining regions in Minas Gerais in the 18th century. The inland railroads could not raise capital, interested mainly in the banking sector and textile manufacture after the stop of the slave trade. Landowners also offered resistance to railroads crossing over their land.[53] Despite the signs that some authors consider a movement of endogenous accumulation, led by a mercantile elite operating on the internal market, the international demand continued in control of the productive dynamics of the country, establishing the main spatial contours of the production flow.

With the crises affecting the exporting regions and the rise in sustainable conditions of an industrial growth in the 1930s, a growing commercial connection, however, arose between the regions. Inter-regional complementary facts were developing from the industrialisation process. A relative specialisation of the regions kept pace with the increasing national integration of the economy. By the end of the 1940s, international trade was still well ahead of interregional trade, mainly through coastal trade. Investments in infrastructure built bridges in the old "archipelago" to intensify the accessibility between the inland parts of a genuine transportation system designed effectively in the 1950s. With the huge expansion of interregional trade in the fifties, coastal trade diminished, and the truck became the prevailing means of haulage in the internal circulation of commodities.[54]

As an articulating centre for the demand, the Southeast began to lead the way to spatial concentration of production, as well as the concomitant movement of oligopoly concentration of the economy. São Paulo industrial development began to establish regional labour division in Brazil.

Since the 1960s, however, the integration of the national market then attracted regional development policies, by which the State would encourage peripheral industrialisation through tax incentives and support for installation of state-owned and private industrial plants. The State-encouraged growth of the economic infrastructure also contributed to the industrial deconcentration process: the federal and state paved highway grid increased fourfold between 1960 and 1970. Since the 1970s, there was, therefore, a trend to industrial deconcentration in Brazilian territory, with a simultaneous increase of the intraregional heterogeneity, given the different types of integration established with the national market. The link, so far of a commercial nature only, was complemented by a genuinely productive link between the sub-national areas. Capital was transferred to the less industrialised regions in search for new investment fronts and occupation of economic areas. The availability of natural resources and the encouraging role of the State were the determining factors of mobility of capital.[55] The circulation of commodities was then accompanied by even further dispute for capital in the interregional sphere, establishing an increasingly integrated accumulation logic, which made the regional economies more complex and internally more heterogeneous. Between 1970 and 1980, the number of industrial areas with more than 10 000 employees increased from 33 to 76 for the whole country. There was also a reversal process of the polarising movement in the São Paulo metropolitan area, and its share in employment and industrial production dropped from 34% and 44% to 28% and 29%, respectively, between 1970 and 1985.[56]

The deconcentration process did not, in fact, significantly alter the spatial competitive pattern of productive activities in Brazil, nor the prevailing interregional division of labour by decades, and this is why it is often described as a "centralised deconcentration" process. A selective deconcentration was then found, by which the less developed regions again joined the more industrialised regions by restating the specialised pattern of their economies. The, to a certain extent, limited nature of the deconcentration process undoubtedly presented a close relationship with the bankrupt developmentalist State and the policies that had previously supported the construction of the domestic market.[57]

The spatial concentration process of production, and the so-called "centralised deconcentration" process tended to accentuate in the space the environmental impacts of the activities, whether arising from their own technical profiles or from the effects of agglomeration that they could cause. The impacts of the technological options – such as, for example, those of the devastation of areas in the Atlantic rain forest due to the spatial concentration of the charcoal-fuelled steel industry – and the impacts of agglomeration arising from site options – such as the intensive pollution in centres like Cubatão – were also magnified when spatial concentration was associated with the high rates of economic growth.

Since the 1970s, on the other hand, there has been an increase in the influence of international trade on the dynamism of producing regions. In the early seventies, the export performance was linked to tax and government financial incentives, while in the 1980s, in turn, exports were associated with the shrinking home market and economic instability. At the same time, the relative intensity of the relations of the São Paulo economy with the overall national economy changed, and between 1975 and 1985, the former became progressively "extrovert" towards the international market and "closed" in relation to the other Brazilian regions.[58] But even without yet meaning a breakdown of the national economy and its home market, the drop in commercial competition of São Paulo with the home market indicated that, since the eighties, new forms of spatial controversy of the economy began to appear.

In the 1990s, the policy of opening up trade and giving priority to the so-called "competitive insertion" of the country split away from the process of forming a nationally integrated economic base which would decentralise activities along the margins. Globalisation seems to have been thus reinforcing the trends to regional specialisation: companies began adopting site strategies directed by the supply of skilled human resources, the closeness to technology-producing centres, the largest and most efficient offer of economic infrastructure and the closeness to higher income consumer markets. Throughout the 1990s, evidence was found of dynamic foci in different points situated inside the Brazilian macro-regions, reflecting the spatially selective character of investments that outlined the division of labour between the regions and setting a trend to the geographic re-concentration of the production in the region stretching from the middle of Minas Gerais to north-eastern Rio Grande do Sul. In this subspace is where new industrial areas began to emerge in the nineties in medium-size towns, characterised by a close productive and commercial integration in an intra and interregional network.[59]

TABLE 1 – SPATIAL DISTRIBUTION OF DEVELOPMENT,
URBANISATION AND CIRCULATION OF MERCHANDISE – SUMMARY TABLE

Period	Spatial Distribution of Development	Merchandise Circulation Network
Before the 1930s	Economic "archipelago"	Railroads and ports concentrating on exports. "Drove roads" and coastal trade attend the home market
From 1930s to 1950s	Commercial dispute between the regions. Relative regional specialisation	Coastal trade carries raw materials and food. Roads carry manufactured goods
1950s	Integrated and regionally located national economy. Industrial concentration in the Southeast	Trucks now dominate the internal circulation of merchandise. Highway construction into the North, Northeast and Mid-West
1960s	Regional development policies and peripheral industrialisation	Spread of paved highway grid. Peak of road transportation
1970s	Productive dispute between regions. Increase in intra-regional heterogeneity. Spread of the farming and mining boundaries. Concentrated industrial deconcentration. Greater trade dispute of the SE economy with the international market	Highways are part of the geopolitical project of the dictatorship. Transport corridors concentrating on agricultural exports and effort to upgrade railroad transportation.
1980s	Extroversion of the economy through shrinking home market and foreign debt crisis	Public financing crisis and drop in investments in expansion and maintenance of the transport system
1990s	More open trade strengthens regional specialisation. Drop in rate of economic deconcentration from the Mid-South. Disconnection of certain areas in relation to national economic drive.	Highways are presented as a component of "Brazil Cost" and restraining factor in the competitiveness of the economy. Road concessions to the private sector. Privatisation of the railroad grid. Leasing of port terminals and facilities

Period	Urbanisation **URBANISATION (cont. from Table 1)**
Before the 1930s	agribusiness and seafaring cities are the marketing locus for wealth produced in the countryside;
From 1930s to 1950s	cities are created based on the start of restricted industrialisation; rural-urban migration from Northeast and Minas Gerais to the Southeast; creation of small towns associated with the spread of the agricultural frontier to Paraná and Mato Grosso.
1950s	start of the creation of an urban network as evidence of the growing home market; population growth by drop in death rate and fertility rate in the cities; modernisation of transport and communication infrastructure.
1960s	confirmed population and industrial concentration in São Paulo.
1970s	growth of medium-size towns in São Paulo state, associated with the centralised deconcentration from São Paulo. Faster urbanisation in North, Northeast and Mid-West since 1975, with diversification of services and creation of locally-based light industries.

1980s	drop in the growth rate of cities due to the tax and financial crisis of the State; new dynamic centres appear as the result of industrial investments in primary transformation of natural resources; medium-size towns appear between mid-Minas Gerais and north-eastern Rio Grande do Sul, centralising new industrial areas with close productive and intra and interregional commercial and productive integration. Economic crisis reduces migration rate to Southeast and keeps populations in the small towns and family farming business.
1990s	de-industrialisation of large towns by social and environmental pressures, and mass diseconomies; development of service sector in towns increasingly involved in inter-town competition for private funds; economic deconcentration process exhausted. Important non-metropolitan urban growth

Throughout the 20th century, spatial distribution of activities was associated with the very striking process of urbanisation of Brazilian society. The proportion of the population living in the cities almost doubled between 1960 and 2000: representing 44.67% of the total Brazilian population in 1960, the urban population rose to 82% in the year 2000. From the point of view of its spatial implications, the urbanisation process underwent two simultaneous movements: the creation of new cities and the continuing growth of the existing cities.[60] An expression of these movements was the increase in the number of towns with more than 500 000 inhabitants from three in 1940, only in the Southeast and Northeast, to 17 in 1996, in all the major regions of the country.[61]

In the 1940-70 period, there was an ongoing and substantial growth in the participation of the larger cities in the urban and total population. In the 1970s, there was a break in this trend and in the 1980s its inversion.[62] In the eighties a change was found in the Brazilian urbanisation pattern, given the diminishing rural-urban and inter-regional migration, especially in the Southeast. The growth of the rural areas was remarkable, from the small towns and north-eastern urban network. It was important in that period to reduce the demographic "*metropolisation*" process; that is, the participation of the growth of the metropolitan regions dropped in the total population growth of the country. Alongside the decline of the process of concentration in large towns, the population growth was found to be more "peripheral" – the growth rate of the border counties was significantly higher than that of their respective nucleus counties. Such processes were explained by the industrial deconcentration, the so-called "counter-urbanisation" (a trend towards seeking homes and work outside the large towns), a drop in the fertility rate and the cumulative effect of occupation of the borders with the founding of cities distributed in the interior of Brazilian territory.[63]

It is since the 1980s that a well structured urban network is set up in Brazil, in which the great predominance of one city alone over others at an intra-state level becomes much less important. In the last thirty years of the 20th century, the medium-size towns in Brazil – those with a population between 100 000 and 500 000 inhabitants – absorbed the migratory flows that previously would aim towards the large centres. They were, in this way, a major contribution to the configuration of a less polarised urban network. The creation of small urban nuclei also contributed to this in areas on the edge of occupation and modernisation after dynamics developed by firms dedicated to valorising land through urban allotments in areas of agricultural spread – this is the case of several nuclei linked to adding value to the grasslands ("Cerrados") of Bahia. Other small towns specialised in certain productive activities, reasserting themselves uniquely in the globalised urban network through activities concentrating on modern agriculture in its hinterland, which gave these centres a character of "country towns".[64]

The role of "buffer" of the medium cities was mitigated, but during the 1990s. In contrast to the 1980s, the metropolitan regions grow in the 1990s at a much faster rate than the whole country. Although the medium-size towns – with a population of between 100 000 and 500 000 inhabitants – have grown at a faster rate than that of the Metropolitan Regions, this growth differential has dropped in the 1990s, compared to 1980/1991. The government preference for the support to exportable product flow, the loss of public investment capacity, the search for effects of closeness and outside economies involved in the flexible production are factors that seem to have contributed towards strengthening the role of the large centres in the 1990s.[65]

During the 1980s and 90s, the further opening of the economy competed for an increase in the economic and social heterogeneity of the country and for a change in the mobility pattern of the population, consequently altering the configuration of the Brazilian system of cities. Some of the main transformations found in the Brazilian urban network during the 21st century are said to be the creation of new urban centres in association with the industrialisation of the countryside – the creation of centres for marketing and processing farm produce, retail distribution and provision of services or as a reservoir of temporary labour. The number of cities grew that act as centres of drainage for agrarian income bound to the activities associated with the modernisation of the countryside. The urban and interactions between different sizes of centres in different regions, on the other hand, became more complex. To such multiple interactions between different urban circuits is added the increasing emergence of empty or under-occupied spaces that demonstrate the relative separation between urban dynamics and the surrounding countryside.[66]

The new industrial activities tended to be located outside the metropolitan areas, bearing in mind the growing business readiness to avoid mass diseconomies. Many small and medium-size towns have grown up in the period due to the development of agribusiness, irrigated agriculture and urbanisation of the frontier, which, together with the destabilisation of family farming, has encouraged short distance migration as an alternative to the displacement current until then to the large cities, which since the 1980s have concentrated the damaging effects of productive restructuring on the labour market. The trends towards deconcentration of the economy confirmed in the 1980s, namely to the interiorisation of São Paulo industry, alongside a slower growth rate of the metropolitan urban agglomerations, have contributed to the setting up of a more scattered and relatively deconcentrated city network.[67] Except for the Northeast, where the growth of some metropolitan areas has shown to be still high in the 1980s and 1990s, the medium-size towns had higher average growth rates than the cities. The trends to outsourcing and uncertain labour in the large towns not only blocked the mobility to the large centres but also encouraged return migrations, contributing to the consolidation of a subsystem of non-metropolitan urban agglomerations.

Regarding the fact of the cities continuing to concentrate substantial volumes of population, they have lost ground in the past twenty years to other urban agglomerations such as Campinas and Vitoria, for example, and to a certain number of urban centres with more than 100 000 inhabitants.[68] Hence a more significant growth in population in the former economic peripheries and medium-size urban centres, and urbanisation stretching westwards. New urban agglomerations were also added to the national urban network that was becoming more complex.[69] This complexity lies not only in the structural formation of the grid's mesh but also in the very socio-spatial configuration inside the towns. On one hand, a certain number of relatively small urban nuclei are characterised by housing population groups fully integrated in the national market, either due to their consumer pattern or connections to the productive and cultural circuits. On the other hand, the large and medium-size urban centres are increasingly crossed by processes of social differentiation and socio-spatial segmentation, sheltering significant social contingents of those excluded from the modernisation processes. These dichotomies in the towns certainly turn back on the urban grid itself, which then expresses the prevailing division of the modern and monetarily integrated parts of the cities and their social more skilled and higher income segments, more susceptible to join the extrovert and globalised circuits of accumulation. Significant parts of the populations in the cities and large towns will, in turn, be linked to the circuits of informal labour, dwellings of risk and precarious access to the urban infrastructures.

The spatial dynamics of the urban systems, as shown by the results of the study on Characterisation and Trends of the Urban Network in Brazil, consists of the following linked processes:[70]

a. "higher density around metropolitan nuclei or large urban centres in the forefront of the systems";

b. "spread of their areas of influence beyond the political-administrative boundaries of the original Brazilian states"

c. "shaping of new systems based on the consolidation of emergent polarisation centres, principally in the areas of recent settlement, as in the case of Cuiabá";

d. a differentiated metropolising process, ranging from macrocephalic systems, such as Rio de Janeiro, to an articulated system of regional centres, such as that nucleated by São Paulo;

e. spatial dispersion of small urban centres, responsible for the organisation of the space in the areas recently opened on the frontier of resources, namely in the Mid-North region.

The spatial dynamics described above competes for the configuration of quite a differentiated spatial urban population density distribution in the interior of Brazil. It is possible to identify the distribution of these densities from the form of metropolitan and non-metropolitan urban agglomerations portrayed by the aforementioned study. Within these agglomerations different distribution patterns of the densities are found, as follows:

a. High population density nuclei with surrounding municipalities that have even higher density populations. This is the case of São Paulo, which has around it municipalities with an even higher population density, such as Diadema, Osasco, São Caetano do Sul and Taboão da Serra, for example, and Rio de Janeiro, whose higher density municipalities around it are São João de Meriti, Nilópolis and Belford Roxo. This configuration also applies to Santos and Campinas (with lower densities and only two comparatively higher density municipalities around them), Belém, João Pessoa and Maringá (with lower densities and only one nearby municipality with a comparatively higher density than the agglomeration nucleus).

b. Agglomerations formed by a group of municipalities with moderate density and similar levels between each other, such as the agglomeration nucleated by Porto Alegre, surrounded by the municipalities of Alvorada, Esteio, Canoas and Cachoeirinha. This would also be the case of the agglomerations of Londrina, Caxias do Sul and São José dos Campos.

c. Agglomerations formed by a dense nucleus and sub-centre with a lower demographic density. These are the urban agglomerations of Salvador, Belo Horizonte, Fortaleza, Curitiba, Goiânia, Vitoria and Jundiaí.

d. Agglomerations where the nucleus has no significant sub-centre nearby, such as São Luis, Maceió, Natal, Aracaju and Ribeirão Preto.

If the first two distribution patterns of demographic densities in urban agglomerations are considered, then it is shown that they presume different dynamics of intra-agglomeration displacement. Should a dense agglomeration suggest shorter distances/travel times between home and work, the very high density levels can, in turn, suggest, on the contrary, greater possibilities of traffic congestion, longer travel times and more spent on energy with transportation. The Study on Standards of Living [*Pesquisa sobre Padrões de Vida*], for 1996-1997, carried out by IBGE in six metropolitan regions, three of which were in the Southeast and three in the Northeast, has revealed that 36.4% of the people walk to work while 24.8% use public transport and 14.3% use a car or motorbike. The survey also revealed that in the São Paulo metropolitan region, those who use public transport to go to work spent an average of 63 minutes to do so; in the metropolitan region of Rio de Janeiro, this time is 54 minutes and in Belo Horizonte, 47 minutes. In the towns in the Northeast, with lower population densities, this average time is 52 minutes in Fortaleza, 42 in Recife and 39 minutes in Salvador.[71] In the cases of São Paulo and Rio de Janeiro, and also Santos and Campinas, the existence of a certain number of higher density sub-centres than the actual nucleus of urban agglomeration, rather than the juxtaposition of a number of compact towns with prevailing intra-urban mobility of short duration, suggests the prevalence of stabilised structures, where real estate dynamics in the agglomeration nucleus force back the lower income populations to sub-centres around them. This presumes that the workers are constantly travelling between the surrounding sub-centres and agglomeration nucleus. This would not be the case, of course, for the second population density distribution pattern described above, shaped by the urban agglomerations of Porto Alegre, Londrina, Caxias do Sul and São José dos Campos, where greater moderation and spatial homogeneity of the urban population densities between the different municipalities would suggest the prevalence of a greater degree of compaction in the home-work dynamics, basically intra-municipal and, consequently, the possibility that expenses with energy are lower in the transport systems

The setting up of higher density urban sub-networks concentrating a considerable part of the populations and economic activities in Brazilian towns seems, therefore, to combine the processes of population growth towards the periphery and metropolisation of poverty. This contributes to the increase in number and time of travel between the workers' homes and their work places, constantly providing, presumably, further contribution by transportation to CO_2 emissions.

FINAL COMMENTS

The study on climate changes undertaken in the European Community only began to address the relevant technological, social and economic aspects as a basis for formulating policies, including the definition of the focus on energy efficiency, as a result of politicising the greenhouse effect in 1986. To formulate policies against climate change, this environmental effect had to be translated in the terms of a "tractable" and politically "manageable" problem.[72] In this way the procedure was set up which Hajer called "discursive closure of the problem", through which the speeches consists of the global environmental change as a subject for policies, in a sens that it can be presented as able for solution.[73] The transformation of climatological evidence in terms of a political plot thus underwent the selection of actions relating to the search for energy efficiency, which allowed the expected environmental benefits to be associated with the simultaneous enjoyment of economic benefits.

Buttel & Taylor claim that after an initial period of general acceptance in the late 1980s, the global climate model, estimates of biodiversity losses and other studies on the implications of environmental changes have become the subject of scientific and, consequently, political debates. They say, however, that for a long time a "moral construction of the global environmental problems emphasising the common interest in the efforts to face up to them, diverting the attention from political setbacks resulting from the diversity of social interests and nations involved in this confront" has prevailed.[74] Yet as early as in 1988, the Swedish Perspective on Human Dimensions of Global Change report called attention to the scientifically known social construction processes on global change, underscoring the role of history and culture in the definition of scientific and political issues. It is in this context of social construction of the problem that, in 1992, the U.S. National Research Council report on global environmental changes pointed out "the importance of Geography – for example, the distances between human settlements – and Demography – for example, the dispersion of populations in suburbs – in establishing the standard for energy consumption".[75]

Compared to the findings in the European experience, would it be worth asking on what would the construction of this "manageability" of global environmental changes in Brazil depend? The justifications relating to climate change seem to take relatively small role in the Brazilian discussion on urban policies. These policies do not seem to be substantially integrating the political topics in which the questions of global climate changes have been approached in Brazil.

In the study quoted by *The Economist*, the quality of the air only becomes a concern of public policy when the GDP per inhabitant reached 5 000 dollars. From this it could be inferred that the low development rates inhibit the fight against pollution.[76] In the hypothesis that such a correlation may be extended to the less immediately visible problems, such as that of global environmental change, it could be expected that the socio-political mobilisation on such a topic may increase alongside the increase in per capita income. This mobilisation may be associated, as shown, with possible particular involvement of urban elites who will eventually distinguish in the impacts of global change a problem for them, that seems to substantially affect their projects, that is sufficient reason to engage their capacity in being heard in the public sphere. Nothing however prevents representations from popular sectors to also distinguish and make evident the global dimensions of local struggles – namely for safe homes, urban sanitation and appropriate public transport – fought by them in the urban environment. The previous experience of Chico Mendes and global environmental disputes of the rubber tappers in the Amazon seem to be its witness.

NOTES

1. *According to Vitousek, global environmental changes "are those that alter the wrappers of the onshore system and are therefore felt globally, and those that occur in more restricted areas but acquire a more global character by being widely disseminated", cf. Vitousek, 1992, apud U. Confalonieri et alii, Mudanças Globais e Grandes Empreendimentos, in I Brazilian Seminar on Health and Environment in the Development Process, Fiocruz series, Eventos Científicos, RJ, 2000, p.35. A possible objective characterization of the environmental processes considered global is questioned, in turn, in V. Shiva. The Greening of Global Reach, in W. Sachs (ed.), Global Ecology a new arena of political conflict, ZED Books, London, 1993, p149 – 156 and L. Lohman, Local versus Global Ecology: two ways of looking at the conflict, mimeo, s.d., 18 p.*

2. *cf. S. Mota, Urbanização e Meio Ambiente, ABES, RJ, 1999, 352p.*

3. *cf. P. Veltz, Tempos da Economia, Tempos da Cidade: as Dinâmicas, in H. Acselrad (org.), A Duração das Cidades – Sustentabilidade e Risco nas Políticas Urbanas, DP&A Ed., RJ, 2001, p.149.*

4. *cf. M. Gadgil – R. Guha, Ecology and Equity: the use and abuse of nature in contemporary India, London, Routledge, 1995*

5. *John Vidal quotes the episode of mass suicide of the U'Wa tribe faced with the oil field development of Shell in Colombia, as a consequence of the growth of "urban footprint" associated with the car civilization, cf. J. Vidal, "A Tribe's suicide pact", in Manchester Guardian Weekly, October 12, 1997, p.8-9, apud S. Dalby, Ecological Metaphors of Security – World Politics in the Biosphere, in Alternatives, 23 (1998), p. 310.*

6. *cf. A.Levy – C. Scott-Clark- D.Harrison, "Save the Rhino, but kill the people" in Manchester Guardian Weekly, March 30, 1997, p.5 and J.S.Adams-T. O. McShane, The Myth of Wild Africa: conservation without illusion, Berkeley:University of California Press, 1996, apud S. Dalby, op.cit., p.308.*

7. *cf. B. Willems-Braun, "Buried Epistemologies: the politics of nature in (post)colonial British Columbia", in Annals of the Association of American Geographers 87, n.1 (1997): pp.3-31, apud S. Dalby, op.cit., p.308.*

8. *"Essentially, the ecological shadow of a country consists of environmental resources extracted by it from other countries and global common wealth", cf. J. MacNeill et alii, Para além da Interdependência – A relação entre a Economia Global e a Ecologia da Terra, Jorge Zahar ed., RJ, 1991, p. 71. Similarly, M. Wackernagel and W. Rees call "ecological footprint: capacity the area of land required to provide resources and assimilate the waste of a certain group of people –, housing, city or country", cf . M. Wackernagel, La Huella Ecológica de las Ciudades – Como Asegurar el Bienestar Humano dentro de los Límites Ecológicos?, mimeo, Mexico,11p.*

9. *cf. J. MacNeill et alii, ibidem.*

10. *cf. D. Harvey, Justice, Nature & the Geography of Difference, Routledge, New York, 1996, p.411.*

11. *cf. D. Harvey, ibid.*

12. *cf. D. Harvey, op.cit., p.413.*

13. *cf. S. Dalby, op.cit., p.306.*

14. *cf. D. Harvey, op.cit., p.413.*

15. *cf. S. Boehmer-Christiansen, Global Climate Protection Policy: the limits of scientific advice", in Global Environmental Change, 1994, 4 (2), p.141.*

16. *cf. J. Calbo – W. Pan, M. Webster – R.G.Prinn – G. McRae, A Parametrization of Urban-sacle Photochemical Smog for its Use in Global Atmospheric Chemistry Models, MIT, Joint Program on the Science and Policy of Global Change, Report 20, July 1997. The seismologist Thomas Heaton asks himself, for example, about "how can we intend to prevent future events from happening in the next 1000 years, when our ideas are normally not sustained even for 20 years", cf. M. Davis, Ecology of Fear – Los Angeles and the Imagination of Disaster, Vintage Books, New York, 1998, p.35.*

17. *Cf. M.L.Bell, The Impact of Global Climate Change on Urban Air Pollution and Human Health, John Hopkins University, STAR Pprogram, in http://es.epa.gov/ncerqa_abstracts/fellow/99/belmi.html*

18. *cf. P. Brand, The Environment and Postmodern Spatial Consciousness: a Sociology of Urban Environmental Agendas, in Journal of Environmental Planning and Management, 42 (5), p. 646.*

19. *cf. S. Boehmer-Christiansen, op. cit., p.177.*

20. *cf. S. Boehmer-Christiansen, ibid.*

21. *The El Niño phenomenon is characterised by anomalous heating and the subsequent cooling of the surface water of the Eastern Equatorial Pacific Ocean for a period from 12 to 18 months, causing changes in the atmospheric circulation and rainfall on regional and global scales. Cf. M. A. V. Freitas – M. Santos – A. Hamada, Vulnerabilidade Climática e Consumo de Eletricidade em Grandes Áreas Urbanas, 2000, mimeo.*

22. *Cf. M. A. V. Freitas – M. Santos – A. Hamada, , op. cit., pp.4-5.*

23. *INPE researchers are now also studying the hypothesis that the increase in atmospheric pollution in cities such as Rio de Janeiro, São Paulo and Belo Horizonte would be attracting more frequent and stronger lightning, with the consequent falling of trees, traffic congestion and other accidents. Cf. A. L. Azevedo, Poluição Faria Raio Afetar mais o Rio, in O GLOBO, 20/11/2000, p.29.*

24. *Cf. A.M.P. M. Brandão, As Alterações Climáticas na Área Metropolitana do Rio de Janeiro: uma provável influência do crescimento urbano, in M. de A. Abreu (org.), Natureza e Sociedade no Rio de Janeiro, Secretaria Municipal de Cultura, Turismo e Esporte, RJ, 1992, p.183.*

25. *cf. M. J. A. Marcondes, Cidade e Natureza – Proteção dos Mananciais e Exclusão Social, FAPESP/EDUSP/Nobel, 1999, p. 165.*

26. *cf. R.B. Figueiredo, Engenharia Social, soluções para áreas de risco, SP, Makron Books, 1994, 251p.*

27. *cf. Atlas Ambiental de São Paulo apud Radiografia Completa da Paulicéia, in Pesquisa FAPESP, n. 60, dez. 2000, São Paulo, p. 44.*

28. *Cf. E. Maricato, O Urbanismo na Periferia do Capitalismo: desenvolvimento da desigualdade e contravenção sistemática., in M. F. Gonçalves (org.) O Novo Brasil Urbano – impasses, dilemas, perspectivas, ed. Mercado Aberto, Porto Alegre, 1995, p.269.*

29. *Cf. E. Maricato, As Idéias Fora do Lugar e o Lugar Fora das Idéias – Planejamento Urbano no Brasil, in O. Arantes – C. Vainer – E. Maricato, A Cidade do Pensamento Único – Desmanchando Consensos, Ed. Vozes, Petrópolis, 2000, pp. 153-154.*

30. *Cf. M. Davis, Ecology of Fear – Los Angeles and the Imagination of Disaster, Vintage Books, New York, 1998, p.38.*

31. *Cf. A.M.P. M. Brandão, As Alterações Climáticas na Área Metropolitana do Rio de Janeiro: uma provável influência do crescimento urbano, in M. de A. Abreu (org.), op. cit., p.188.*

32. *"The larger the city, the greater the heat contrast effect between the city and countryside", cf. M. Lombardo, Ilha de Calor nas Metrópoles – o exemplo de São Paulo, Hucitec, SP, 1985, p.39, apud C. Cabral, Clima e Morfologia Urbana em Belém, UFPA, Belém, 1995, p.1.*

33. *cf. Marcelo Lopes de Souza, O Desafio Metropolitano – um estudo sobre a problemática socio-espacial nas metrópoles brasileiras; ed. Bertrand Brasil, 1999, RJ, p.126.*

34. *Cf. UITP, Des Villes à Vivre, Bruxelles, 1996, apud J. Bindé,Villes et Environnemnt au XXI Siècle, in Environnement au XXI Siècle, vol I, les Enjeux, GERMES, Paris, 1998, p. 98.*

35. *cf. J. Bindé, op. cit., p. 98.*

36. *Cf. M.D.Lowe, Shaping Cities: the Environmental and Human Dimensions, Worldwatch Paper 105, Washington, October 1991, p. 16.*

37. *Cf. P. Mathis. Consommation d'Énergie et Pollutions Liées à l'Étalement des Densités, in J. P. Gaudemar (org.), Environnement et Aménagement du Territoire, DATAR/La Documentation Française, 1996, Paris, p.106.*

38. *Cf. R.S. da Motta, Política de Controle Ambiental e Competitividade, MCT/FINEP/PADCT, Estudo de Competitividade da Indústria Brasileira, Campinas, 1993, p. 65.*

39. *Cf. T. K. Rude et alii, Tropical Deforestation Literature: Geographical and Historical Patterns, Forest Resources Assessment Program, Working Paper n. 27, FAO apud C.S.Teixeira, Florestas Sociais – uma Resposta à Destruição das Florestas Tropicais?, Diss. Mestrado CPDA/UFRRJ, RJ, 2001, p. 39.*

40. *Cf. R.S. da Motta, Indicadores Ambientais no Brasil – Aspectos Ecológicos, de Eficiência e Distributivos, IPEA, Textos para a Discussão n. 403, RJ, 1996, p. 46.*

41. *Cf. E.A. Vasconcellos – I. M. O. Lima, Quantificação das Deseconomias do Transporte Urbano: uma resenha da experiência internacional, IPEA, Texto para a Discussão n. 586, RJ, 1998, p. 13.*

42. *cf. L. P. Rosa- M. R. Tolmasquim, An Analytical Model to Compare Energy-efficiency Indices and CO2 Emissions in Developed and Developing Countries, in L.P.Rosa et alii, Carbon Dioxide and Methane Emissions: a Developing Country Perspective, RJ, COPPE/UFRJ, 1996, pp. 30-32.*

43. *Cf. D. Nogueira, Rio Avalia os Gases do Efeito Estufa, in Jornal do Brasil, 13/8/2000, p. 21.*

44. *Cf. E. L. La Rovere, "A Sustentabilidade da Produção de Energia no Brasil, PNUD, mimeo, Oct. 1995, p. 17 and I. Klabin, O Mecanismo de Desenvolvimento Limpo e as Oportunidades Brasileiras, in Parcerias Estratégicas, October 2000, n.9, Brasília, p.47. Also see L. A.B. Uria, Emissão de Gases de Efeito Estufa no Setor de Transportes e seu Potencial de Aquecimento Indireto: o caso dos automóveis e veículos comerciais leves no Brasil, Masters thesis COPPE/UFRJ, 1996, and A G. Monteiro, Estratégia de Redução de emissões de Poluentes no Setor de transportes por Meio de Substituição Modal na Região Metropolitana de São Paulo", Masters thesis, COPPE/UFRJ, 1998.*

45. *Cf. Vigevani & Veiga, p. 348; BNDES. AO2, GGSCA, April 2000, Indústria automobilística no Mercosul, p. 5, apud F. Limoncic, A Insustentável Civilização do Automóvel – A Indústria Automotiva Brasileira em Tempos de Reestruturação Produtiva; Projeto Brasil Sustentável e Democrático, Cadernos Temáticos, RJ, 2001, p.31.*

46. *cf. W. Meiners, Novo Ciclo de Investimentos da Indústria Automobilística no Brasil e seus Desdobramentos regionais, in Cadernos IPPUR/UFRJ, vol. XIII, n.1, jan. jul. 1999, pp.199.*

47. *As Jack Smith, chairman da General Motors claimed: "We have many factories, we need to sell cars" in Valor, 18/7/2000, apud F.Limoncic, op.cit., p.31.*

48. *Cf. Greenpeace, Automóveis: saúde agredida e alterações no clima do planeta; 7.7.2000 apud F. Limoncic, op. cit.p.35.*

49. *cf. Prefeitura da Cidade do Rio de Janeiro. IPLAN-RIO. Op. cit. apud F. Limoncic, op. cit.*

50. *cf. M. Malin e I. Junqueira, Ivan. "Negrão de Lima" in Beloch, Israel & Abreu, Alzira Alves de. Dicionário Histórico Biográfico Brasileiro, 1930-1984. Rio de Janeiro: Forense Universitária FGV/CPDOC/Finep, 1984, p. 1851-1854, apud F. Limoncic, op. cit.*

51. *cf. F. Limoncic, op. cit.*

52. *Rio de Janeiro City Hall. IPLAN-RIO. Anuário Estatístico da cidade do Rio de Janeiro 1993/1994, apud F. Limoncic, op. cit..*

53. *cf. K. M. Mattoso, Bahia – Século XIX – uma província do Império, Ed. Nova Fronteira, RJ, 1992.*

54. *cf. O. J.A. Galvão, Comércio Interestadual por Vias Internas e Integração Regional no Brasil, in Anais do XXI Encontro Nacional da de Economia, ANPEC, Belo Horizonte, 1993, p.278.*

55. *cf. – Guimarães Neto, L., Desigualdades Regionais e Federalismo, in R. B. Alvares Affonso – P.L.B. Silva (eds.), Federalismo no Brasil, Fundap/UNESP, 1995, SP, p.14.*

56. *cf. – Campolina Diniz, C., "Reestruturação Econômica e Impacto Regional: o novo mapa da indústria brasileira", in Nova Econômica, BH, vol. 6, n.1, July 1996, p.84*

57. *cf. C.A. Pacheco, – Pacheco, C.A., Fragmentação da Nação, Instituto de Economia, UNICAMP. Campinas, 1998, p. 229*

58. *cf. Diniz e Santos, 1993, apud Guimarães, op. cit., p. 24-26*

59. *cf- C.Campolina Diniz, "Reestruturação Econômica e Impacto Regional: o novo mapa da indústria brasileira", in Nova Economia, BH, vol. 6, n.1, July 1996, p.100.*

60. *cf. Ministério do Meio Ambiente, Os Ecossistemas Brasileiros e os Principais Macrovetores do Desenvolvimento – Subsídios ao Planejamento da Gestão Ambiental; PNMA, Brasília, 1995, p.30.*

61. *cf. UNICAMP/IPEA/IBGE, Caracterização e Tendências da Rede Urbana do Brasil, vol. I, Campinas, 1999, p.323.*

62. *Cf. G. Martine, A Evolução Espacial da População Brasileira, in R.B.A .Affonso – P.L.B. Silva (orgs.), Federalismo no Brasil – Desigualdades Regionais e Desenvolvimento, UNESP/FUNDAP, 1995, SP, p. 78.*

63. *Cf. G. Martine, op. cit., p. 83.*

64. *Cf. R. L. Correa, Globalização e Reestruturação da Rede Urbana – uma nota sobre as pequenas cidades, in Território, n. 6, jan.jun. 1999, p. 47.*

65. *Cf. T.A. Andrade – R.V. Serra Fluxos Migratórios nas Cidades Médias e Regiões Metropolitanas Brasileiras – a experiência do período 1980/96, IPEA, Texto para a Discussão n. 747, RJ, 2000.*

66. *Cf. R.L.Correa, Reflexões sobre a Dinâmica Recente da Rede Urbana Brasileira, in IX Encontro Nacional da ANPUR, RJ, 2001, pp. 10-12.*

67. *cf. UNICAMP/IPEA/IBGE, Caracterização e Tendências da Rede Urbana do Brasil, vol. I, Campinas, 1999, p.11.*

68. *UNICAMP/IPEA/IBGE, op. cit., p.23.*

69. *cf. UNICAMP/IPEA/IBGE, Caracterização e Tendências da Rede Urbana do Brasil, vol. I, Campinas, 1999, p292.*

70. *cf. UNICAMP/IPEA/IBGE, op.cit., p.380.*

71. *cf. IBGE, Pesquisa sobre Padrões de Vida, 1996-1997, RJ, 1999, pp.140-141.*

72. *cf. Liberatore, A., Facing Global Warming: the interaction between science and policy-making in the European Community, in M. Redclift – T, Benton (eds.), Social Theory and the Global Environment, Routledge, London, 1994, p. 192.*

73. *Cf. M. Hajer, Politics of Environmental Discourse: Ecological Modernization and the Policy Process, Oxford, 1995.*

74. *cf. F. Buttel – P. Taylor, How Do We Know We Have Global Environmental Problems? Science and the Globalization of Environmental Discourse, in Geoforum v. 23, n.3, 1992, p. 406.*

75. *cf. U.S. National Research Council, Committee on the Human Dimensions of Global Change, Global Environmental Change: Understanding the Human Dimension, Washington D.C., National Academy, 1992.*

76. *cf. J. Bindé, Ville et Environnement au XXI Siècle, vol. 1, Les Enjeux, GERMES, Paris, 1998, p. 105.*

BIBLIOGRAPHY

ARAUJO, T.B., "*Dinâmica Regional Brasileira nos anos noventa: rumo à desintegração competitiva*", In: CASTRO, I.E. de, MIRANDA, M., EGLER, C. A. G. (eds.), Redescobrindo o Brasil - 500 anos depois, Bertrand Brasil Editora, RJ, 1999, p. 73-91.

ARAUJO, T. B., "Brasil nos Anos 90: opções estratégicas e dinâmica regional". *Revista Brasileira de Estudos Urbanos e Regionais* n.2, March 2000, pp. 9-24.

BECKER, B., Cenários de Curto Prazo para o Desenvolvimento da Amazônia, *Cadernos IPPUR/UFRJ*, vol. XIV, n.1, Jan. Jul. 2000, pp. 53-85.

BOEHMER-CHRISTIANSEN, S., Global Climate Protection Policy: the limits of scientific advice". *Global Environmental Change*, 1994, 4 (2), p.141-.

BRAND, P, The Environment and Postmodern Spatial Consciousness: a Sociology of Urban Environmental Agendas. *Journal of Environmental Planning and Management*, 42 (5), p. 646-658.

BUCH, M.N., The Environmental Impact of Cities, in Asia-Pacific Peoples Environment Network, *Global Development and Environment Crisis - Has Humankind a Future?*, Penang, 1998, pp. 628-641.

BUTTEL, F. , TAYLORr, P., How Do We Know We Have Global Environmental Problems? Science and the Globalization of Environmental Discourse. *Geoforum* v. 23, n.3, 1992, p. 406.

CAMPOLINA DINIZ, C., "Reestruturação Econômica e Impacto Regional: o novo mapa da indústria brasileira". *Nova Economia*, BH, vol. 6, n.1, July 1996, p.77-103.

CANO, W., "Concentração e Desconcentração Econômica Regional no Brasil". *Economia e Sociedade*, n.8, June 1997, p. 101- 142.

CORREA, R.L., Globalização e Reestruturação da Rede Urbana - uma nota sobre as pequenas cidades,. *Território*, n. 6, Jan.Jun. 1999.

COSTA, S., ALONSO, A., TOMIOKA, S., *Itinerários do Conflito: do rodoviarismo à negociação dos riscos ambientais*, São Paulo, CEBRAP, 1999, 173 p. (mimeo)

DAVIS, M. *Ecology of Fear:* los Angeles and the Imagination of Disaster, Vintage Books, New York, 1998.

GALVÃO, O. J. A., Desenvolvimento dos Transportes, Integração Econômica e Crescimento Regional. In: ENCONTRO NACIONAL DA ANPUR, 6, Brasilia, 1995. **Anais**. p.360-374.

GUIMARÃES NETO, L. Desigualdades e Políticas Regionais no Brasil: caminhos e descaminhos. *Planejamento e Políticas Públicas*, June 1997, p. 41-95.

GUIMARÃES NETO, L., Desigualdades Regionais e Federalismo. In: AFFONSO, R. B. Alvares, SILVA, P.L.B. (eds.). *Federalismo no Brasil*, Fundap/UNESP, 1995, SP, p.14-59.

HARVEY, D. *Justice, Nature & the Geography of Difference*, Routledge, New York, 1996.

HOGAN, D.J., Migração, ambiente e saúde nas cidades brasileiras. In: HOGAN, D.J., VIEIRA, P.F. (orgs.). *Dilemas Socioambientais e Desenvolvimento Sustentável*, Ed. Unicamp, Campinas, 1992, p. 149 - 170.

LIBERATORE, A. Facing Global Warming: the interaction between science and policy-making in the European Community, In: REDCLIFT, M.; BENTON, T. (eds.) *Social Theory and the Global Environment.* London: Routledge, 1994.

LIMONCIC, F. A insustentável civilização do automóvel em tempos de reestruturação produtiva. Projeto Brasil Sustentável e Democrático. *Cadernos Temáticos*, Rio de Janeiro, 2001.

LOWE, M.D. Shaping Cities: the Environmental and Human Dimensions, *Worldwatch Paper*, Washington, v.105, Oct. 1991.

MARICATO, E. As Idéias Fora do Lugar e o Lugar Fora das Idéias: Planejamento Urbano no Brasil, In: ARANTES, O.; VAINER, C.; MARICATO, E. *A Cidade do Pensamento Único:* Desmanchando Consensos. Petrópolis: Ed. Vozes, 2000.

MARTINE, G., A Evolução Espacial da População Brasileira. In: AFFONSO, R.B.A.; SILVA, P.L.B. (orgs.) *Federalismo no Brasil:* Desigualdades Regionais e Desenvolvimento, UNESP/FUNDAP, São Paulo. 1995.

MATHIS, P. Consommation d'Énergie et Pollutions Liées à l'Étalement des Densités, In: GAUDEMAR, J. P. (org.) *Environnement et Aménagement du Territoire.* Paris: DATAR/La Documentation Française, 1996.

MINISTÉRIO DO MEIO AMBIENTE. *Os Ecossistemas Brasileiros e os Principais Macrovetores do Desenvolvimento* - Subsídios ao Planejamento da Gestão Ambiental; PNMA, Brasilia, 1995,

MONTEIRO, A.G. *Estratégia de Redução de emissões de Poluentes no Setor de transportes por Meio de Substituição Modal na Região Metropolitana de São Paulo*, Dissertação (Mestrado) - COPPE/UFRJ, 1998.

OSÓRIO, L. M. Urbanização e Mercado de Trabalho na Amazônia. *Cadernos IPPUR*, Rio de Janeiro, 1999/1, p. 109-138.

PACHECO, C. A., *Fragmentação da Nação.* Campinas, Instituto de Economia/ UNICAMP. Campinas, 1998, 291 p.

UNICAMP/IPEA/IBGE, *Caracterização e Tendências da Rede Urbana do Brasil*, Campinas, 1999, v.1 p.11

TOLMASQUIM, M.R., ROSA, L. P. An Analytical Model to Compare Energy-efficiency Indices and CO_2 Emissions in Developed and Developing Countries. In: ROSA, L.P. et al. *Carbon Dioxide and Methane Emissions: a Developing Country Perspective*, Rio de Janeiro, COPPE/UFRJ, 1996

UNICAMP/IPEA/IBGE. Caracterização e Tendências da Rede Urbana do Brasil, 2 vols., Campinas, 1999. 2v.

URIA, L.A.B. *Emissão de Gases de Efeito Estufa no Setor de Transportes e seu Potencial de Aquecimento Indireto: o caso dos automóveis e veículos comerciais leves no Brasil*, Master's thesis, COPPE/UFRJ, 1996.

VELTZ, P.. Tempos da Economia, Tempos da Cidade: as Dinâmicas. In: ACSEL-RAD, H. (org.). *A Duração das Cidades:* Sustentabilidade e Risco nas Políticas Urbanas, DP&A Ed., Rio de Janeiro, 2001 (in press).

WACKERNAGEL, M.. *La Huella Ecológica de las Ciudades:* como Asegurar el Bienestar Humano dentro de los Límites Ecológicos?, Mexico, 11p. (mimeo.)

AUTHORS

Human Dimensions of Global Environmental Change
Brazilian Perspectives

BERTHA K. BECKER is Full Professor at the Department of Geography and Director of the Laboratory for Territorial Management, of the Federal University of Rio de Janeiro. She is Vice President of the International Advisory Group of the Pilot Program for the Protection of Brazilian Tropical Forests and member of the Scientific Council of the Large Scale Atmosphere Biosphere Experiment in the Amazon. Her research areas are political geography and the Brazilian Amazon.

CLAUDE COHEN has a M. Sc. degree in economic development from the Université de Paris X – Nanterre, France, and is presently concluding her doctoral dissertation on "Patterns of Consumption, Development and Environment" at the Energy Planning Program of the Post-Graduate School of Engineering of the Federal University of Rio de Janeiro (PPE/COPPE/UFRJ). Her main subjects of research are energy consumption and CO2 emissions of the residential sector and the macroeconomic impacts of consumption on development and the environment.

CLAUDIO ANTONIO G. EGLER is full professor of Economic Geography and Regional Development in the Institute of Geo-sciences of the Federal University of Rio de Janeiro. He has his master's degree in Urban and Regional Planning from the Federal University of Rio de Janeiro and his doctorate in Economics from the State University of Campinas. His research interests include the relationships between regional development and environmental change.

DANIEL JOSEPH HOGAN, Ph.D. in development sociology and demography from Cornell University, is full professor at the Institute of Philosophy and Human Sciences of the State University of Campinas, where he has been Director of the Environmental Studies Center and is currently Director of the Population Studies Center – Nepo. He has been President of the Brazilian Association for Population Studies and a member of the National Commission on Population and Development. His research interests include the relationships between demographic dynamics and environmental change.

DONALD ROLFE SAWYER, president of the Institute for Society, Population and Nature (ISPN), an independent, non-profit research center, founded in 1990 and located in Brasília. He coordinates projects and carries out research and consulting in the socio-environmental field, especially in the Amazon and in the Cerrado. He has been professor in the graduate program in Demography, of the Federal University of Minas Gerais (Cedeplar). He holds a Ph.D. in Sociology from Harvard University. Born in the United States, he has resided in Brazil since 1974.

EDUARDO VIOLA is Full Professor of International Relations, University of Brasilia. He has published extensively on international and Brazilian environmental politics, globalization and governance. He has been visiting professor at Stanford, Colorado, Notre Dame (USA), Amsterdam (The Netherlands) and San Martin (Argentina) universities, among others. His current research focus is on the international climate change regime.

HÉCTOR RICARDO LEIS, has his Doctorate in philosophy from the Catholic Pontifical University of Rio de Janeiro. He is adjunct professor at the Sociology and Political Science Department of the Federal University of Santa Catarina, where he is currently responsible for Interdisciplinary Human Sciences Studies. His research interests include globalization, environmental change and social and political theory.

HENRI ACSELRAD has his doctoral degree in Economics from the University of Paris I. He is a professor at the Institute for Urban and Regional Planning of the Federal University of Rio de Janeiro; assistant fellow of the Lincoln Institute of Land Policy; editor of the Brazilian scientific journal Cadernos IPPUR; and a member of the International Rio + 10 Memorandum Group, established by the Heinrich Boll Foundation (Germany), 2001-2002. His recent research interests focus on environmental conflicts and planning.

HENRIQUE RATTNER is Full Professor of Economics at the School of Economics and Management of the University of São Paulo. Currently he is International Project Coordinator at the Brazilian Association of Leadership Development. rattner@nhi.lead.org.br

JULIA SILVIA GUIVANT has her doctorate in Sociology from the State University of Campinas. She is Senior Lecturer in the Department of Social Sciences and in the Graduate Program on Political Sociology at the Federal University of Santa Catarina. She is currently Visiting Professor at the Graduate Program in Development, Agriculture and Society (CPDA), Federal Rural University of Rio de Janeiro. Her research interests include environmental sociology, with a particular focus on food risks.

LEILA DA COSTA FERREIRA, Professor of Sociology at the State University of Campinas, has her doctorate in Social Sciences from the same university. She has been a post-doctoral fellow at the University of Texas at Austin and the Director of Graduate Studies at the Institute for Philosophy and Human Sciences at Unicamp. Her research interests include environmental sociology and environmental policies.

MARIA MANUELA CARNEIRO DA CUNHA, professor of Anthropology at the University of Chicago, holds her doctorate in Anthropology from the State University of Campinas. Her research concentrates on indigenous populations in Brazil.

MAURO WILLIAM BARBOSA, Ph.D. in Anthropology from Cambridge University, is professor of Anthropology at the State University of Campinas. His research concentrates on populations of rubber-tappers in Amazonian regions.

MAURICIO TIOMNO TOLMASQUIM has a Doctor of Science degree in Economics from the École des Hautes Études en Sciences Sociales, in Paris. He is presently Chair of the Energy Planning Program (PPE) of the Institute for Research and Graduate Studies (COPPE) at the Federal University of Rio de Janeiro (UFRJ). He is also President of the Brazilian Society for Energy Planning (SBPE) and the Chairman of the Brazilian Academy of Science Committee on the Human Dimensions of Global Environmental Change. He has published widely in both books and academic journals. His research interests include the reform of the electricity and petroleum sectors, energy efficiency, energy demand and supply simulation models, options to mitigate climate change and environmental economics.

SOLANGE SIMÕES, Ph.D. in Sociology from the London School of Economics is currently Adjunct Faculty Associate at the Institute for Social Research/ University of Michigan and professor of Sociology at the Doctoral Program in Sociology and Political Science at the Federal University of Minas Gerais. She is co-coordinator of the Global Environmental Survey and her research interests include attitudes and behavior concerning environmental issues of mass public and decision makers.

ULISSES EUGENIO CAVALCANTI CONFALONIERI, M.D., DVM, D.Sc is Full Professor at the Department of Biological Sciences of the National School of Public Health, Oswaldo Cruz Foundation, Rio de Janeiro. His research interests include the relationships among environmental factors – especially climate variability, climate change and deforestation – and disease profiles in the tropics.

This book was composed in Formata Light 9/14,
printed in off-set paper 75g with cover in Supremo
paper 250g, in September of 2001.